Manufacturing technology

Manufacturing technology volume 2

R L Timings

Addison Wesley Longman Limited
Edinburgh Gate, Harlow
Essex CM20 2JE, England
and Associated Companies throughout the world

First published 1993
Second impression 1995
Third impression 1997
Fourth impression 1998

ISBN 0 582 41183 1

British Library Cataloguing in Publication Data
A CIP record for this book is available from the British Library

Set by 4 in Compugraphic Times 10/11 pt
Produce through Longman Malaysia, PP

Contents

Foreword

Manufacturing Technology: volume 2 has been written not only to satisfy the requirements of the Business and Technician Education Council standard unit for Manufacturing Technology at levels three and four, but also to satisfy the main requirements of those students following 'college devised' syllabuses. At the design stage of this text, most of the colleges and polytechnics offering HNC/HND in manufacturing engineering were sent a questionnaire requesting information concerning the topic areas and treatment they would like to see incorporated. The response must be a record for questionnaires and the author was swamped with information and spoilt for choice. However after analysing the returns and asking supplementary questions of the respondents, a specification covering the main requests was drawn up and became the synopsis of this book. Unfortunately commercial pressures prevented the inclusion of many interesting minority processes where colleges were catering for highly specialised local requirements.

The text follows on naturally from *Engineering Fundamentals* and *Manufacturing Technology: volume 1*. Students who have made a 'direct entry' onto their HNC/HND course will need to read volume 1 in conjunction with volume 2 for full understanding of the processes described.

The broad coverage of *Manufacturing Technology: volume 2* not only satisfies the requirements of Higher Technician Engineering Students, but also provides essential background reading for undergraduates studying for a degree in Production Engineering, Mechanical Engineering, or Combined Engineering.

The author wishes to thank Mr. Richard Duffill of Coventry University for his contributions to those chapters concerned with metal cutting, advanced manufacturing technology, and automated assembly, and also Mr. Frank Race for his major contributions to the Management of Manufacture.

The Author also wishes to thank Mr. Tom Watters of GKN Birfield Extrusions for his time, advice and help concerning cold and warm forging techniques, Mr. Chris Kenward of the GKN Powder Metallurgy Division for his time, advice and help concerning sintered powder metal manufacturing techniques, and Maurice Bonney, Professor of Production Management, University of Nottingham.

Roger L. Timings
1993

Acknowledgements

We are grateful to the following for permission to reproduce copyright material:

Alfred Herbert Ltd. for our Fig. 5.14; B & S Massey Ltd. for our Fig. 2.3(a, b & c); Bridgeport Machines Ltd. for the cover photograph; British Standards Institution for our Fig. 12.6; Bulpitt Engineering Ltd. for our Figs 2.34 and 2.35; Butterworth-Heinemann Ltd. for our Figs 6.23, 6.24 and 8.34 from *Mechanical Engineer's Pocket Book*; Cassell PLC for our Fig. 6.15 from *An Introduction to CNC Machining* by David Gibbs; Colchester Lathe Co. Ltd. for our Fig. 5.5; Davy Loewy Ltd. for our Fig. 2.3(d); Fordath Ltd. for our Figs 1.5 and 1.7; GKN Powder Metallurgy Division for our Figs 3.1, 3.2 and 3.11; Hayes Shell-Cast Ltd. for our Fig. 1.3; Hilger Watts Ltd. for our Figs 8.17, 8.18 and 8.19; H W Ward Ltd. for our Fig. 5.13; J E Nanson Gauges Coventry Ltd. for our Figs 8.22 and 8.27; National Machine Tool Corp. (USA) for our Fig. 2.7; P I Castings Ltd. for our Fig. 1.11; Rank Taylor Hobson Ltd. for our Fig. 8.30; Sweeney and Blockridge (Power Presses) Ltd. for our Fig. 2.18; Taylor & Challen Ltd. for our Fig. 2.17; Visual Component Engineering Systems for our Fig. 11.1; John Wiley & Sons Inc. for our Fig. 6.25 from *Handbook of Industrial Robotics (1985)* edited by Shimon Y. Nof, reprinted by permission of John Wiley & Sons Inc.

Part A Technology of manufacture

1 Casting processes

1.1 The casting process

The process of casting metal to shape is the oldest and still one of the most widely used metal-forming processes. A casting is formed by pouring molten metal into a mould whose cavity is the shape of the required component. For ease of casting, the metal should have a high fluidity (ease of flow) and a high fusibility (low melting point). For example cast iron, with its high carbon content, has a high fluidity and high fusibility compared with low-carbon steel. Therefore cast iron can be readily cast into complex shapes, whereas steel can only be cast into simple shapes as it has a low fluidity ('treacly' consistency) and does not flow easily. Again, the high melting temperature of the steel means that the moulding sand has to be carefully selected or it will itself be melted on coming into contact with the molten steel.

Unlike other primary forming processes, the crystal structure of a casting is not homogeneous but varies from the core of a casting to its surface, depending upon the rate of cooling. Such a lack of uniformity and refinement in the grain structure can lead to planes of weakness in the casting, as shown in Fig. 1.1(*a*), if the design is not matched to the casting process. A generous radius at the corner of the casting, as shown in Fig. 1.1(*b*), helps to remove the plane of weakness.

In all casting processes the pattern has to be made oversize to allow for any contraction of the metal that occurs during the change of state from liquid to solid and also to allow for the continued contraction as the solid

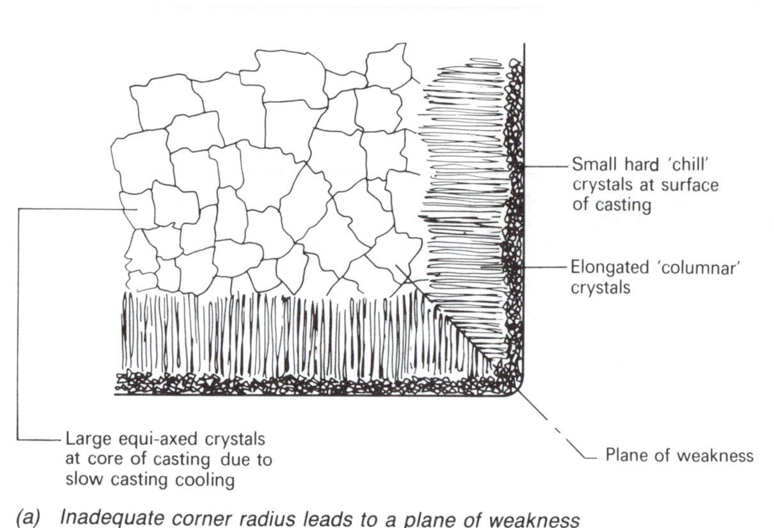

(a) Inadequate corner radius leads to a plane of weakness

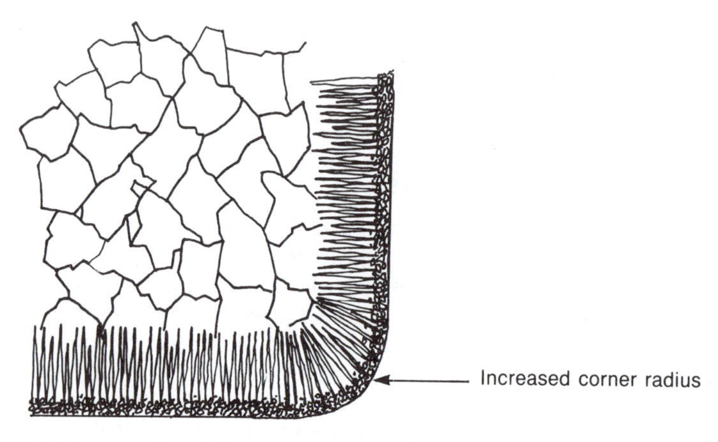

(b) Increased corner radius helps to remove any plane of weakness

Fig. 1.1 Crystal structure of a casting

metal cools to room temperature. Note that cast iron is exceptional in that it expands as it solidifies and this helps it to take a sharp impression from the mould. However, after solidification, it contracts like any other metal. In addition to shrinkage allowance, a *machining allowance* also has to be provided wherever a surface is to be machined. Not only must sufficient additional metal be provided to ensure that the surface 'cleans up' during machining but also that there is sufficient depth of metal for the nose of the cutting tool to operate below the hard and abrasive skin of the casting, as shown in Fig. 1.2.

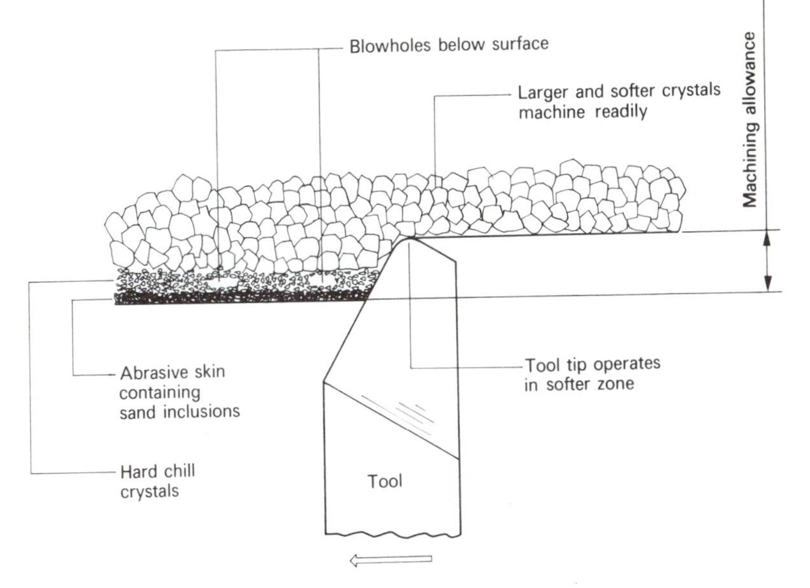

Fig. 1.2 Machining allowance

1.2 Green sand moulding

Green sand moulding was introduced in *Manufacturing Technology: volume 1*. The main advantages of the process can be summarised as follows:

- Most metals and alloys can be cast in green sand moulds.
- The size of the castings that can be produced in green sand moulds ranges from small components for model engineers to those used in the manufacture of power station turbines.
- Castings produced by this process can be very simple or extremely complex, requiring multiple internal cores, external cores, draw backs, loose pieces and other sophisticated foundry techniques to produce re-entrant surfaces.
- Quantities can range from single components to batch production.

The main disadvantage of the process is that the mould, which can only be used once, has to be made by highly skilled and highly paid moulders. This results in a process with relatively high unit costs. However, there are many large and complex components that cannot be made by any other casting process.

1.3 Moulding sands

Moulding sands used in green sand moulds must possess the following properties:

- *Cohesiveness*: the ability to retain its shape after the pattern has been removed and during the pouring and solidification of the metal.
- *Refractoriness*: the ability to withstand the high temperatures associated with molten metals without itself melting or burning.
- *Permeability*: the ability to remain sufficiently porous after ramming so that the steam and other gases evolved during pouring can escape. The permeability of the mould as a whole can be modified by the degree of ramming; so also can the rigidity of the mould. A mould that is rammed too hard will lack permeability and gasses will be trapped in the casting, causing porosity. Also, a mould that is too rigid and has insufficient 'give' will cause the casting to crack as it shrinks on cooling. On the other hand, a mould that is too loosely rammed may collapse as the metal is poured.

Moulding sands usually contain from 5 per cent to 20 per cent of colloidal clay to give cohesiveness, and will usually contain up to 8 per cent moisture. Because of the moisture content, moulding sands are referred to as 'green' sands despite the fact that their colour may range from red through dark brown to black.

1.4 Core sands

Cores are made from moulding sands to which an additional binding material has been added, or from a sand that is free from clay and that is bonded entirely by the addition of synthetic binders. Typical binders are:

- *Water-soluble binders* such as the by-products of flour milling, called glutin and dextrin. These are added to ordinary moulding sands.
- *Oil binders* such as linseed oil, soya-bean oil and fish oils added to clay-free sands. These oils undergo a chemical change during the baking of the core and set-off hard.
- *Resin binders* such as phenolic syrups which are added to clay-free sands and are *cured* when the core is baked. Such resins are also the basis of thermosetting plastics and produce very strong cores.
- *Carbon dioxide gas*: the core sand is mixed with sodium silicate which, after ramming into the core box, is exposed to carbon dioxide gas. This causes a chemical reaction in the sodium silicate which then provides a very strong bond. Less shrinkage occurs than for oven drying and more accurate cores can be produced. Greater productivity is also achieved and this offsets the expense of the materials and equipment used.

1.5 Casting defects

Blowholes

Blowholes are smooth, round or oval holes with a shiny surface usually occurring just under the surface of a casting. Because they are not normally visible until machining is underway, their presence can mean the scrapping of a casting on which costly machining has already been carried out. They are caused by gasses being trapped in the mould. This may result from inadequate venting, insufficient provision of risers, excessive moisture in the sand, excessive ramming reducing the permeability of the sand, inadequate degassing of the molten metal immediately prior to pouring, or a combination of these causes.

Porosity

Porosity is also due to inadequate venting and inadequate degassing of the molten metal immediately prior to pouring. However, in this instance, the trapped gases do not form large bubbles just below the surface of the casting. Instead, a mass of pinpoint bubbles are spread throughout the casting rendering it porous and 'spongy'. Apart from being a source of weakness, porosity renders castings useless where pressure tightness is required, as in fluid valve bodies and pipe fittings.

Scabs

Scabs are blemishes on the surface of the casting resulting from sand breaking away from the surface of the mould. This may be caused by lack of cohesiveness due to insufficient clay content in the sand, or from inadequate ramming adjacent to the pattern. Too rapid pouring and an incorrectly proportioned in-gate can also result in the scouring away of the walls of the mould.

Uneven wall thickness

An uneven wall thickness can be caused by the moulder not assembling the core correctly or by accidentally displacing the core when reassembling the mould. Alternatively, the core may move in the mould due to ill-fitting or inadequately proportioned core prints. The buoyancy of the core sand can make it try to rise in the molten metal.

Fins

Fins are due to badly-fitting mould parts and cores allowing a thin film of metal to leak past the joints. Although fins can be removed by fettling after casting, this is an added and unnecessary expense and detracts from the appearance of the casting.

Cold shuts

Cold shuts are usually caused by casting intricate components with thin sections from metal that is lacking in fluidity and which is at too low a temperature. Consequently, individual streams of metal may flow too sluggishly and either solidify before the detail in the mould is filled, or fail to merge when the streams of metal meet.

Drawing

Drawing results from a lack of risers or the incorrect positioning of risers so that thick sections are not adequately fed with metal as they cool and shrink. Since thick sections cool slowly compared with thin sections, they solidify last. This allows other parts of the casting to draw molten metal from thick sections rather than from the risers. The result is that the thick sections may have unsightly hollows which may not clean up when machined.

1.6 Shell-moulding

The use of phenolic (Bakelite) resins as a core sand binder has already been introduced in Section 1.4. In shell-moulding, the resin-bonded sand is used to make the mould as well as the core. This process has a number of advantages over green sand moulding for the quantity production of repetitive components. The shells are hard, strong and relatively light so that they can be safely handled and stored over long periods before use. They can be produced by hand or by automatic machines more quickly and easily than green sand moulds. The mould has good permeability so that sound castings of high definition and freedom from porosity can be produced. Figure 1.3 shows typical shells and cores ready for assembling together to form a mould. The figure also shows a complete casting and a section through a casting made from such a mould. It can be seen that the traditional sand mould, rammed up in a moulding box, is replaced by a relatively thin, rigid shell of uniform wall thickness. Also, cores produced for the shell-moulding process are hollow and provide much-improved venting compared with the solid rammed cores used when green sand moulding.

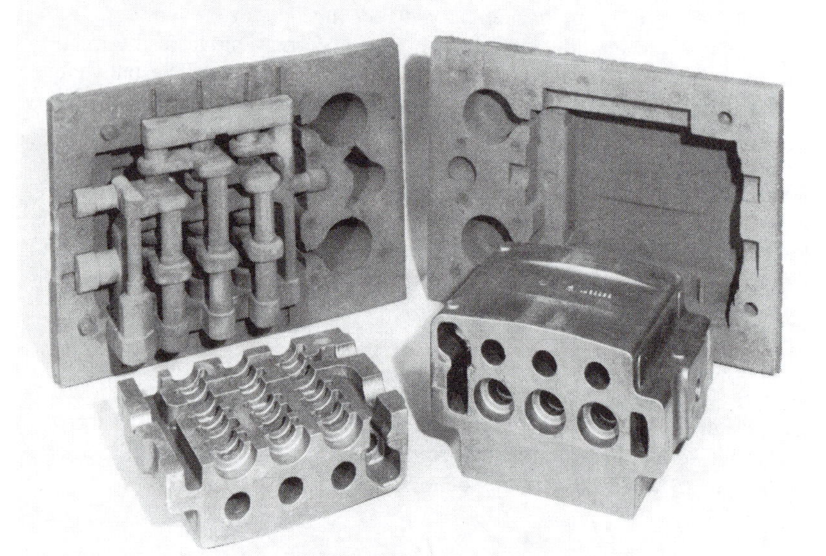

Fig. 1.3 Shells ready for assembly

The shell is made in two or more parts so that it can be stripped from the pattern. These shell parts are then assembled together with their associated cores to make a complete mould. High strength, synthetic adhesives are used during assembly to bond the individual components of the mould together permanently. The making of a shell-mould will now be described.

A metal pattern having the profile of the required casting, but enlarged to allow for contraction on cooling, is heated in an oven to between 200°C and 250°C. The pattern is removed from the oven, sprayed with a release agent and clamped over a trunnion-mounted *dump box* as shown in Fig. 1.4(*a*). The dump box contains the sand and resin-binder mixture

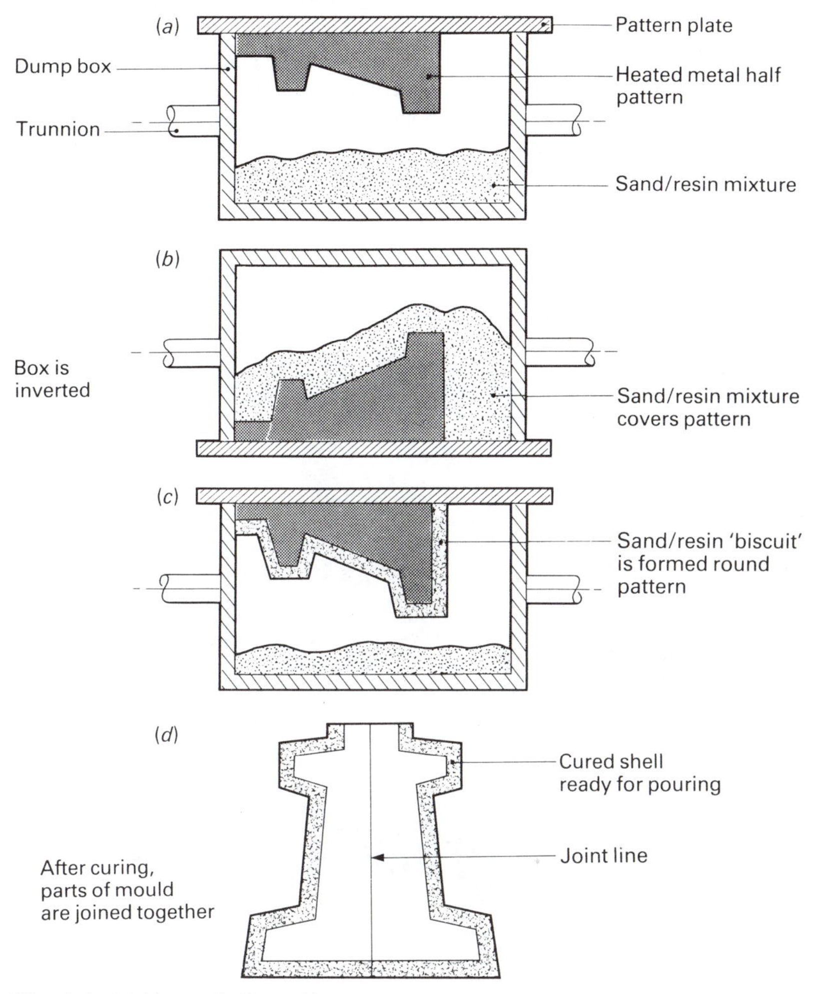

Fig. 1.4 Making a shell-mould

that will eventually form the shell mould. The dump box is rotated on its trunnions so that it becomes inverted. This allows the resin–sand mixture to fall over the pre-heated half-pattern as shown in Fig. 1.4(*b*). After some 30 s the resin–sand mixture becomes tacky and forms a soft '*biscuit*' about 6 mm thick. The dump box is rotated back to its original position so that the surplus resin–sand mixture falls back to the bottom of the box ready for re-use, as shown in Fig. 1.4(*c*). The pattern with the 'biscuit' still adhering to it is transferred to an oven for about 2 min to *cure* the resin binder. Alternatively, a silicate binder can be used and this is cured by the use of carbon dioxide gas. Either of these processes converts the 'biscuit' into a rigid shell. If the thickness of the shell needs to be increased, it is returned to the dump box and a further layer of 'biscuit' is built up on it, after which it is returned to the oven to be cured.

When the desired shell thickness has been reached, it is stripped from the pattern and placed on one side ready for assembly. The shells may be assembled by clamping, bolting or bonding the various component parts together to form the complete mould. Bonding with synthetic resin is the most usual method as it gives a clean casting, free from flash lines. To strengthen large moulds, metal reinforcing rods are introduced into the 'biscuit' when the shell is built up in several layers. Figure 1.5 shows some shell-moulds assembled ready for pouring.

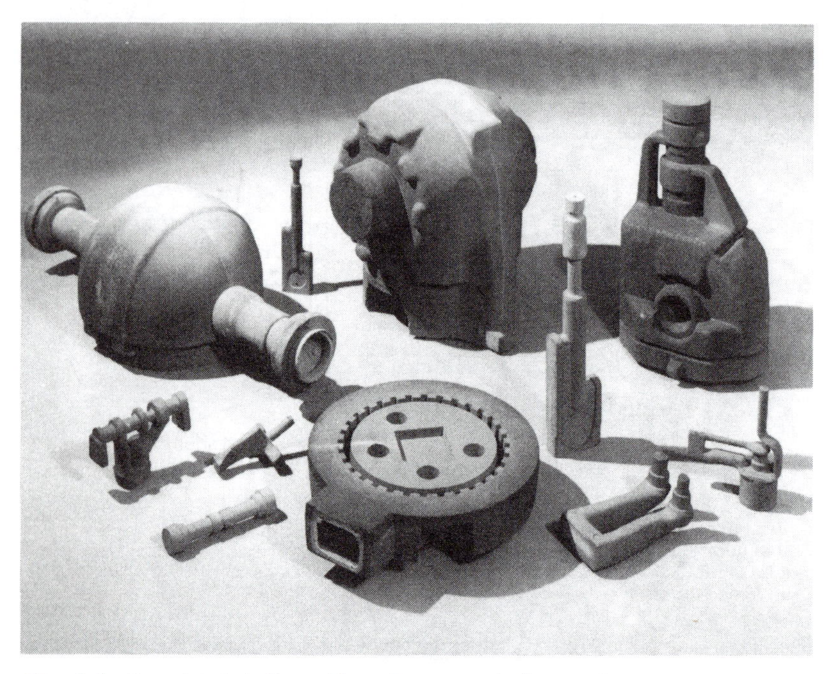

Fig. 1.5 Completed shell-moulds and cores ready for pouring

Shell cores are made in split core boxes which are similar in appearance to those used for conventional core-making except that they are made from metal. This is so that they can be heated to the correct temperature to form the core 'biscuit'. Figure 1.6 shows the stages in making a shell core.

Finally, the assembled shell-moulds are placed in a rack to prevent them from falling over during pouring. They are usually placed over a sand bed for safety in case a mould fails and molten metal is released accidentally. Large moulds are often placed in strong metal boxes and packed with a coarse sand to provide reinforcement while not interfering with the venting of the mould. Alternatively, steel shot can be packed round the shell where greater rigidity is required and where some chilling

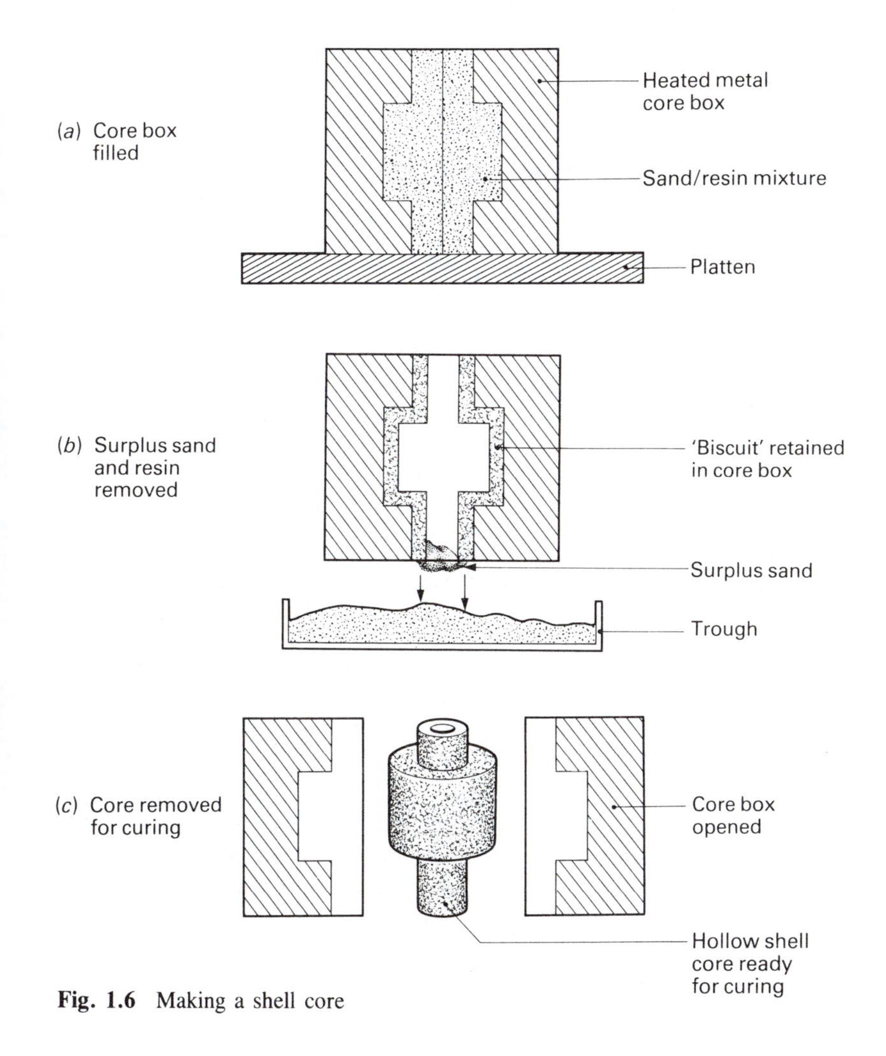

Fig. 1.6 Making a shell core

of the mould is desirable for metallurgical reasons. A pouring cup is usually built into the mould, but risers are not generally provided as the mould is sufficiently permeable to allow the escape of any trapped gasses. The metal is poured in rapidly until the mould is full, because the heat of the metal starts to burn away the resin binder. By the time the metal has solidified, the resin binder should have completely burned away, leaving only a deposit of relatively loose sand around the casting. On tapping the residual shell with a metal bar it disintegrates and the mould and the core are easily removed, leaving a clean and accurate casting requiring a minimum of finishing. For quantity production a *casting wheel* may be used. This is a horizontal carousel holding a number of shells which are filled as they pass under the pouring station. By the time a filled shell reaches the loading station, the metal is sufficiently solidified to allow the filled mould to be removed and replaced with an empty shell. The casting wheel (carousel) rotates continuously.

1.7 Advantages of shell-moulding

The shell-moulding process is widely used where castings with thin sections and high definition are required, for example the finned cylinder of an air-cooled engine as shown in Fig. 1.7. On a larger scale, the water-cooled cylinder block castings for car and truck engines are also made by the shell-moulding process. The sand particles in shell-moulding are only bonded at their points of contact. This gives the mould high permeability, allowing castings of great complexity and varying wall thickness to be made without venting difficulties and without the necessity for a complex system of risers. The low mass of the shell allows it to heat up rapidly so that it does not chill the metal as it is poured. This results in:

- rapid metal flow through the runners, allowing complete filling of complex moulds with thin sections and reducing the possibility of cold shuts (see Section 1.5);
- the absence of surface chilling and hardening of the cast metal. This improves the machining characteristics of the casting, resulting in improved tool life;
- the metal cooling more slowly, resulting in improved grain structures and strength in thin sections.

Sections may be as thin as 1.5 mm or as thick as 50 mm. Thick sections are not recommended due to the tendency for the shells to burst under the pressure of the molten metal. The surface finish of the castings is excellent and free from the fins and blemishes associated with a green sand mould. Dimensional tolerances can be as close as ≤ 0.075 mm for small and medium sized castings.

Despite the fact that the shell and the resin–sand mix from which it is

Fig. 1.7 Typical shell-mould showing the fine detail that can be achieved

made are destroyed every time a casting is poured, this cost is more than offset by the rate of production of the mould and the casting and the reduced finishing time required. Automatic moulding equipment for making shell-moulds is lower in cost than that required for the automatic production of green sand moulds. Unfortunately, the metal pattern equipment required for shell-moulding is much more costly than the wooden patterns used with green sand moulding, and it is not economical to use shell-moulding for small-quantity production and 'one-off' castings.

Both ferrous metals and non-ferrous metals can be cast in shell-moulds provided suitable sand–resin mixtures are used. Excellent results can be obtained when casting aluminium and magnesium alloys provided an inhibitor is used in the sand–resin mixture so that the resin does not react with the molten metal. Copper alloys can be used provided the tin and lead content is not too high, otherwise 'sweating' can occur at the surface of the casting. Grey and malleable irons cast very well and, with care, finishes approaching that of aluminium die-castings can be achieved. Because of the effect of their high melting temperatures on the resin bond, plain carbon and alloy steels are not recommended for castings made by the shell-moulding process.

1.8 Investment casting

The process is unique through its ability to produce castings of high integrity and surface finish combined with virtually unlimited freedom of design since the pattern does not have to be removed bodily from the finished mould but is melted out. There are two essential stages in producing the mould.

Production of the wax pattern

An expendable wax pattern is first produced in a pattern die, the wax pattern conforming exactly to the shape of the desired casting but its size including allowances both for the contraction of the wax pattern and the contraction of the metal casting. Pattern dies are constructed from various materials by a variety of manufacturing techniques. These are dictated by the shape and size of the component to be cast and the quantity required, ranging from cast pattern dies in low melting point alloys and epoxy resins to multi-impression dies made from alloy tool steels for use in automatic injection machines. These multi-impression dies have a life of 50 000 injections before maintenance is required. An average die life expectancy is about 10 000 injections. An investment-casting pattern die is shown in Fig. 1.8 together with the wax pattern made from it.

The most commonly-used pattern material is a blend of specially prepared pattern waxes, although polystyrene is used on a limited scale. With few exceptions, it is normal practice to assemble together a number

Fig. 1.8 Wax pattern and pattern die for investment casting

of individual patterns around a sprue together with the necessary runners, risers (feeder heads) and pouring cup. This is referred to as a pattern assembly, all of which is in expendable wax. Investment-casting patterns are produced to a high degree of accuracy and finish, which is reflected in the mould cavity and the casting itself. Since the moulds have no natural permeability, provision for venting has to be designed and built into the pattern assembly.

Production of the mould

Two basic processes are used for producing an investment mould: the *block* or solid mould process and the *ceramic shell* process. Both techniques involve the application, to the pattern assembly, of a primary *investment* in the form of a suspension of very fine refractory ceramic particles (200 mesh size) in a binder liquid. To this wet primary coat, or investment, is applied a primary *stucco* of fine particles of dry refractory ceramic powder. This initial application of slurry and stucco constitutes the primary coat, which must be dried or chemically hardened prior to the application of secondary or 'back-up' layers.

In the *block-mould* process, a block or solid mould is produced around the coated pattern assembly, as shown in Fig. 1.9. An open-ended container or 'flask' is placed over a baseplate to which the coated pattern assembly has been attached. A coarser refractory slurry, or secondary investment, is poured into the container and the base plate is vibrated to consolidate the mould and to facilitate the escape of entrapped air bubbles. The solid mould is allowed to dry slowly over a period of several hours to avoid cracking, following which the wax pattern is melted out at a temperature of 150°C and the mould is fired at a temperature of approximately 1000°C. This final firing develops the maximum strength of the ceramic bond and also removes any remaining traces of pattern material. The mould is now ready for pouring. The wax that formed the pattern and that was melted out of the mould can be re-used to make more patterns.

The *ceramic shell-mould* process, which has largely superseded the block-mould process except for certain specialist applications, involves the application of alternate layers of slurry and stucco to the primary coated pattern assembly, as shown in Fig. 1.10; each successive coat being dried before the application of the next coat. This is similar to the building up of a resin–sand shell-mould. A completed ceramic shell-mould will normally consist of six to eight individual back-up coats, giving a final shell thickness of the order of 12 to 20 mm. The removal of the expendable pattern — *de-waxing* — may be accomplished in a steam autoclave or in a 'flash firing' furnace. Both these de-waxing processes are designed to impart a very high heat input at the pattern/mould interface. This causes immediate melting of the surface layers of the pattern, which allows for the expansion of the main body of the wax as it, in turn, heats up and melts. The process is very economical since the wax, which has been melted out of the mould, can

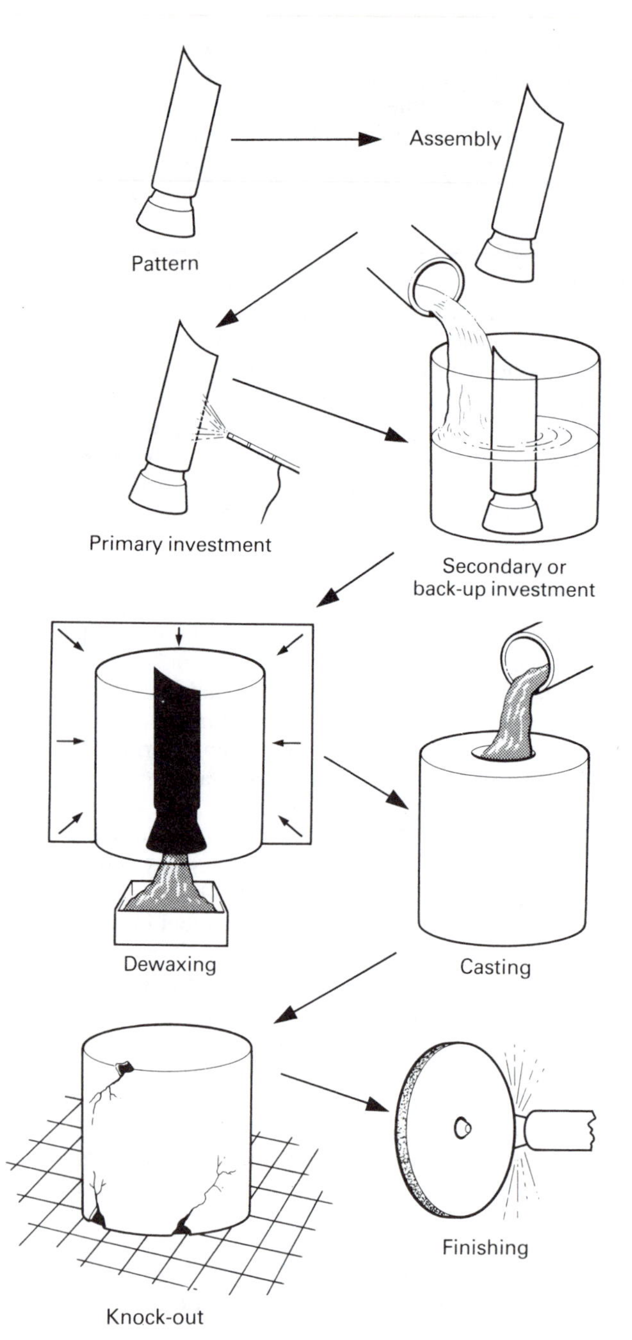

Fig. 1.9 The block-mould process

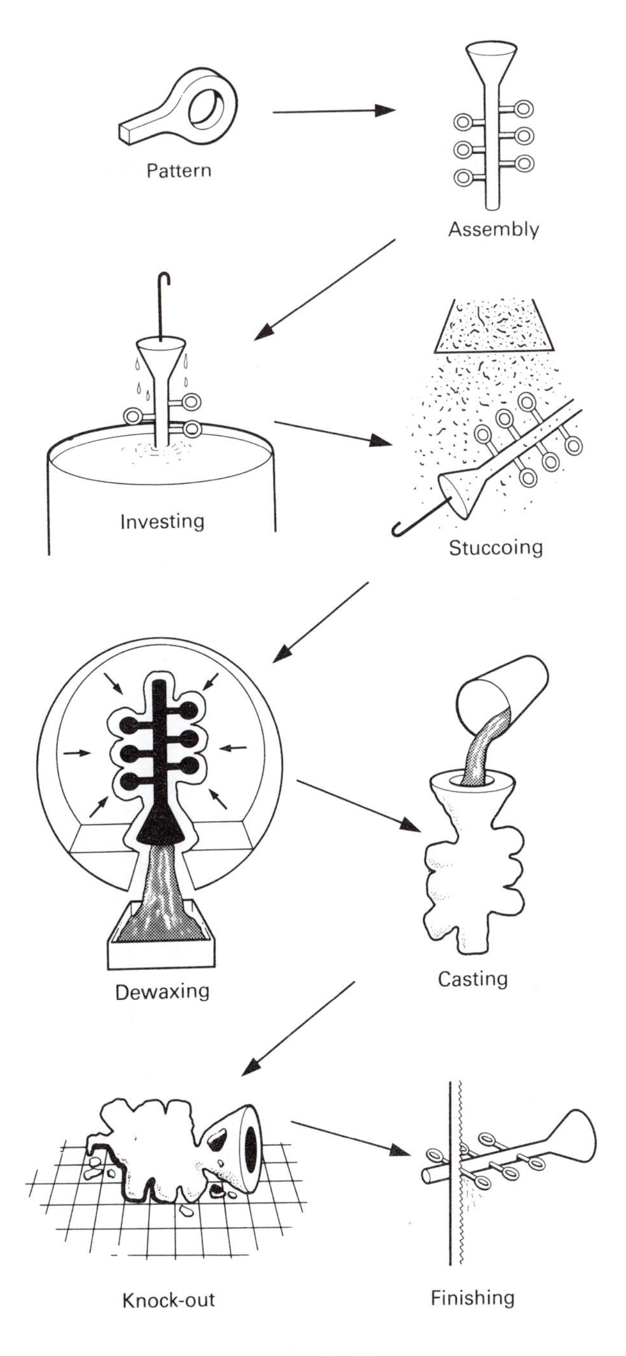

Fig. 1.10 The ceramic shell-mould process

be re-used over and over again. Following de-waxing, the mould is fired at approximately 1000°C. The metal is cast into the mould as soon as firing is complete and, preferably, while the mould is still hot. In both the block-mould process and the shell-mould process the moulds have to be destroyed to recover the castings and the broken mould material cannot be re-used.

When investment-casting ferrous metals, these are usually melted in indirect arc and induction furnaces. Sophisticated alloys whose strength relies upon high purity are often melted in vacuum conditions. The ceramic mould is clamped or bolted to the furnace and the residual air is pumped out. When the metal has melted, the furnace and mould are inverted so that the metal flows directly into the mould. Pumping continues throughout the cycle so that any gases given off during melting and casting are also exhausted. Non-ferrous metals are usually melted in induction furnaces or in crucible furnaces prior to casting and they are then poured into the moulds in the normal way.

1.9 Advantages of the investment-casting process

The investment-casting process possesses the unique ability to transform liquid metal into a cast shape with a high degree of dimensional accuracy, surface finish, and a virtually unlimited freedom of design since, no matter how complex the shape of the mould, the pattern is simply melted and poured out of it. These attributes offer to the design and value engineer the opportunity of achieving maximum material utilisation and a truly functional component. The increasingly heavy burden of machining and finishing costs necessitates a constant reappraisal of the established methods of producing both simple and complex metal shapes. The investment-cast component, requiring an absolute minimum of finishing, has rapidly emerged as a more viable proposition than the forged or the machined-from-bar component for an increasingly wide range of general and specialised engineering parts (Fig. 1.11).

The limits of design possibilities for the process have been extended by the use of preformed ceramic cores, which allow the casting of highly complex and intricate internal shapes, including small-diameter holes of considerable length. The process itself permits the casting of parts to an accuracy within a few hundredths of a millimetre in a wide range of alloys, including those which cannot be forged and which are virtually unmachinable. The advantages of investment casting may be summarised as follows:

- a high degree of dimensional accuracy and surface finish of the 'as-cast' component, coupled with virtually unlimited freedom of design. A general tolerance of $\leq$0.013 mm per 25 mm can be readily achieved on small and medium sized castings;
- the accurate reproduction of fine details, e.g. slots, holes, and lettering;

Fig. 1.11 Examples of investment casting

- maximum utilisation of the raw materials and a truly functional component;
- reduction of costly finishing operations to an absolute minimum. In many cases the component can be used 'as-cast' without further processing;
- savings in capital expenditure on plant, equipment, labour costs, space utilised, storage, and in the inter-process movement of materials.

The only limitation on the size and weight of an investment-cast component is the physical limitation of being able to handle the pattern assembly and the resultant moulds. The increasing use of mechanical handling within the industry, together with continual improvements in process materials (i.e. pattern waxes, mould binders, and refractories) has steadily extended the upper limits in terms of both volume and mass. While the majority of investment castings in current production fall within a mass range of 0.1 kg to 5 kg, castings up to 45 kg and more are in regular production. In terms of size, the maximum linear dimension in current production in the UK is of the order of 1.2 m.

The ability of the process to produce castings with extremely thin wall sections, with the dimensional, finish and reproduction of fine detail inherent in the process, is a further major attribute. Generally, the minimum wall thickness is 1.5 mm, but wall thicknesses of 0.75 mm for steel castings and 0.5 mm for light alloy castings have been achieved over small areas. The minimum economical batch size depends upon the complexity of the component and, therefore, the cost of the pattern dies. Normal batch sizes range from as few as 500 for a simple component to 50 000 for a more complex component. With automated production lines, production runs of over a million components are quite common for the automobile industry.

1.10 Designing for investment casting

The general principles of good casting design apply as much to investment casting as to other casting processes. The volumetric shrinkage of the metal as it solidifies, the creation of localised 'hot-spots' in the mould and the stresses set up in the casting as it contracts and cools in the solid state are inherent problems of any casting process. The adoption of the following design principles will greatly minimise these problems.

Directional solidification

The provision of 'feed' metal during solidification will be greatly facilitated by the avoidance of isolated, heavy sections. Changes in section thickness should be gradual, ideally allowing for the progressive feeding of a thin section by a thicker one.

Fillets and radii

The avoidance of abrupt changes in section thickness and the use of a fillet or radius of the correct size will avoid the creation of local 'hot-spots' in the mould and will help to reduce contraction stresses set up by unequal rates of cooling of different metal sections. Generous corner radii also help to eliminate planes of weakness in the crystal structure of the metal, and reduce the possibility of fatigue failure in service.

Angles and corners

Relatively sharp angles and corners can be satisfactorily investment-cast compared with other casting processes, but re-entrant angles will generally be cast more easily with the incorporation of a root radius.

Shaped holes, bosses, etc

In general, the investment casting of irregularly-shaped holes can be accomplished just as easily as casting a round shape. Holes with a high depth/diameter ratio should be designed so that preformed ceramic cores can be used.

Freedom of design

The absence of a mould joint or parting line removes the need, on the designer's part, to consider taper (draught) or the method of pattern withdrawal from the mould. This contributes to the improved dimensional accuracy of investment castings compared with other casting processes and presents the possibility of unrestricted shapes in cast form.

Metals for investment casting

Investment castings are now available in almost the complete range of engineering metals and alloys currently available, for specialised as well as general applications. These include alloys that are virtually impossible to shape by other, conventional metal-shaping techniques. Vacuum-melted and cast superalloys have become a major product of the industry, including the casting of high-purity titanium alloys.

The author is indebted to the British Investment Casters' Technical Association for their assistance in preparing the foregoing information on the process of investment casting.

1.11 Die casting (high pressure)

The principles and some process details of gravity and pressure die-casting were discussed in *Manufacturing Technology: volume 1*. Automatic high pressure die-casting machines can achieve up to 1000 'shots' per hour for small components and up to 200 'shots' per hour for large and complicated castings. When small components are being produced, multi-impression dies are frequently used to increase the rate of production and so that the machine can operate economically at its full injection capacity. Figure 1.12 shows components as ejected from a multi-impression die, together with a clipped spray to show how the die is designed to retain the spray in one piece to facilitate removal. Note how the components are arranged symmetrically so as to balance the forces acting on the dies and the ample provision for runners feeding the die cavities from the central sprue. Even after taking these precautions it is apparent that some leakage of metal has occurred between the two halves of the die. Since the dies are impervious, vents have to be machined into them along the flash line to allow the air trapped in the cavities to escape as the metal is injected. Because the molten metal is being injected under very high pressure, the machine has to be fully guarded to contain any metal which may escape if a die failure occurs or if the dies fail to close properly. The guards have to be interlocked with the machine controls so that the injection cycle cannot commence until the guards have been properly closed.

High-pressure die-casting machines and the dies themselves represent a large capital investment and it should be apparent that the process is only economic where large batches of components are involved. The minimum economic batch size varies between about 5000 components for simple parts to 20 000 components for complex parts, in order to ensure that the die costs are recovered and also the cost of setting the more complex dies in the machine. To benefit fully from the high-pressure die-casting process it is essential that the dies are expertly designed and manufactured by specialists in this field. Some of the more important factors are as follows:

- the location and profile of runners and risers sould be such that they do not lock the components into the dies;

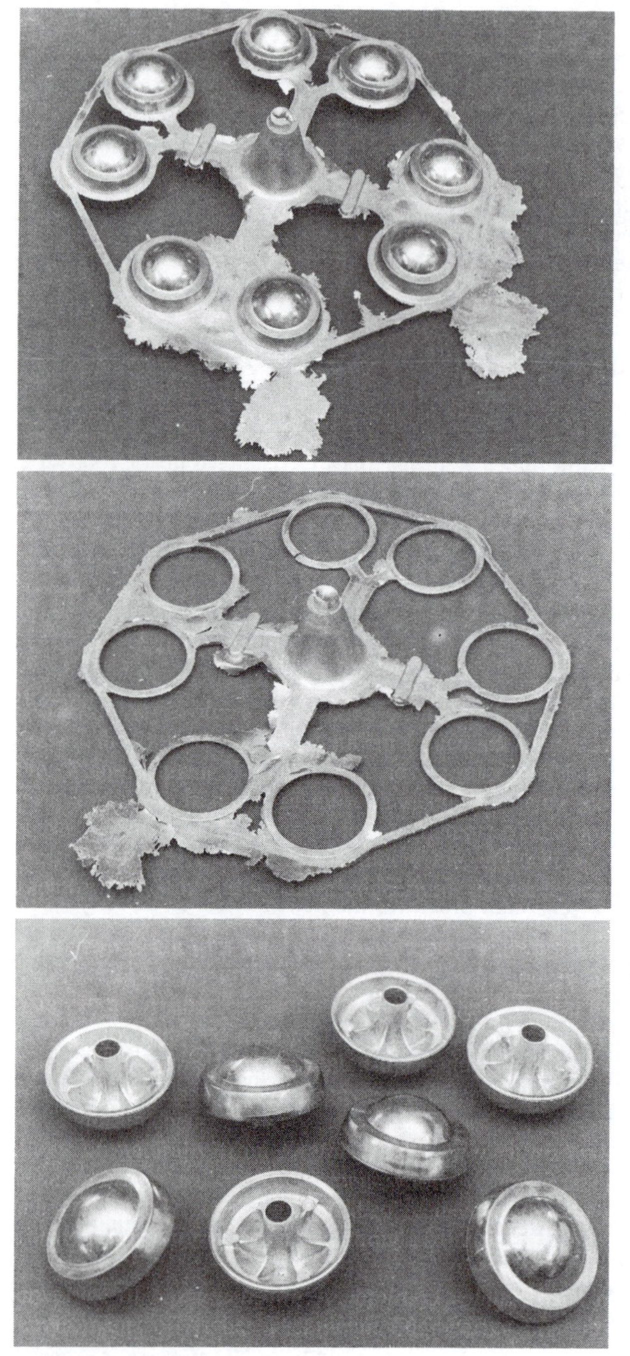

(*a*) 'Spray' of castings ready for clipping
Note: Peripheral web that improves metal flow during casting and gives the rigidity essential during flash clipping operation

(*b*) The flash after clipping. (This is melted down and used again)

(*c*) The eight components after clipping them from the spray

Fig. 1.12 Multi-impression casting

- the joint lines of the dies have to be carefully selected so that the component can be easily ejected. The component itself has to be designed with the process in mind as no undercut sections (re-entrant surfaces) can be tolerated;
- unlike sand moulds, metal dies are not self-venting and vents have to be built into the dies at the time of manufacture;
- the size of the casting is limited by the injection capacity of the machine, therefore the component design should incorporate thin walls with reinforcing ribs to give the required strength but reduce the volume of metal required, rather than using heavy sections;
- the destructible sand cores used in gravity and low-pressure die-casting are not suitable in the pressure die-casting process. Collapsible metal cores can be used but are so costly to make, maintain and use, that they are only economical in special cases. Components should be designed to obviate the need for loose pieces and internal and external cores;
- the positive knock-out pins should be sufficient in number and positioned so as to eject the casting or spray of castings from the dies without causing distortion;
- the metal from which the dies are made must be selected to resist the operating temperature and the corrosive effect of the molten metal. Since alloy die-steels are expensive and difficult to machine, die-steel inserts are usually built into a cheaper plain carbon steel yoke.

The advantages and limitations of the hot-chamber die-casting process can be summarised as follows:

Advantages

- Machining is minimised due to the high accuracy and surface finish of the castings.
- High rates of production can be achieved.
- Material consumption is low as clipped sprays and sprues can be recycled without loss of properties and without contamination.
- Complex shapes, as in decorative motifs and fine detail, can be faithfully reproduced.

Limitations

- Only low melting point alloys and metals that are chemically unreactive with the metal components of the injection equipment can be cast in hot-chamber machines. Therefore this process is usually limited to zinc-based alloys such as Mazak.
- The casting metal or alloy must have a short freezing range so that solidification is rapid, keeping the cycle time as short as possible.
- Plant costs, die costs and setting costs are sufficiently high to be prohibitive for small batches.
- The maximum size of the casting is limited by the capacity of the machine. Even the largest machines can only inject a few kilograms of metal for each ‘shot’.

1.12 Die-casting (low pressure)

This process was developed during the Second World War for the production of air-cooled cylinders in aluminium alloy for the aircraft engine industry. Up to that time, castings of this size and in alloys suitable for engine cylinders had to be sand cast or gravity die-cast. The low-pressure process is now widely used for the broad band of medium and large sized, high-quality die-castings which need to be produced in large quantities. It should be noted that, with the improved materials now available, the high-pressure die-casting process can also produce high-quality castings but generally of far less substantial dimensions and in a more limited range of metals and alloys than is possible with the low-pressure process.

Figure 1.13 shows the principle of low-pressure die-casting. It can be seen that the molten metal is retained within a crucible and furnace. The die, which is normally situated immediately above the crucible is connected to the molten metal by means of a vertical riser tube. Since the

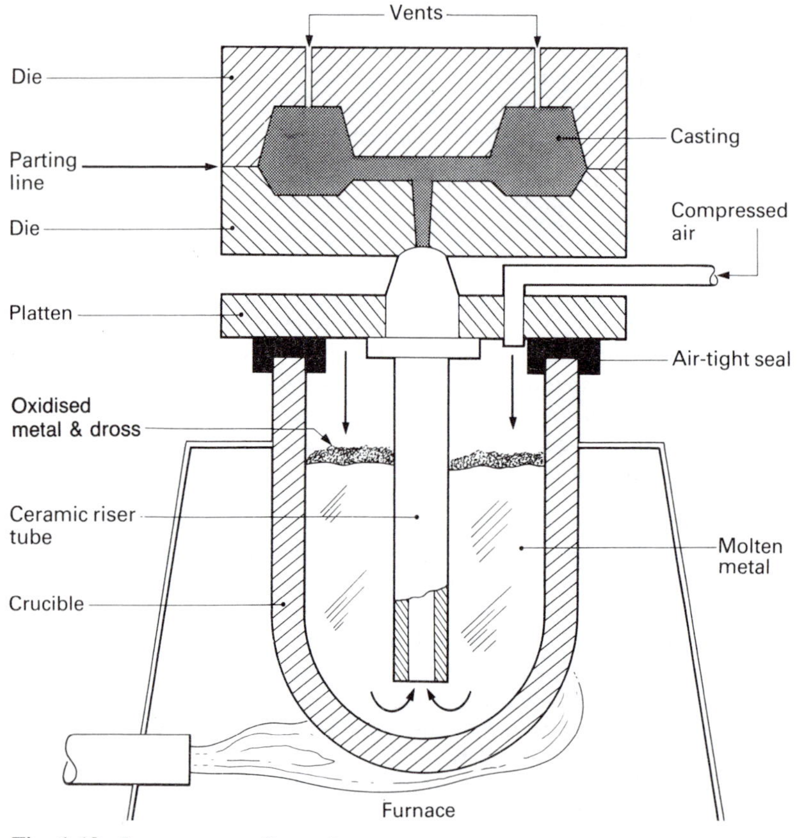

Fig. 1.13 Low-pressure die-casting

crucible and the riser are made from non-metallic refractory materials, they are completely impervious to molten aluminium and its corrosive properties.

The riser tube is immersed nearly to the bottom of the crucible in order to ensure that all metal fed to the die is taken from below any dross, metallic oxides and other impurities which always form on the surface of molten metals. The crucible is sealed against leakage of the pressure air supply by gaskets as shown. The riser is suspended from the top-plate (platten) by a flange and is connected to the gate of the die by a riser cap. After the die is closed, the crucible is pressurised by compressed air. It is this pressure on the surface of the molten metal which causes it to flow against gravity into the die cavity to form the casting. When the die is full the pressure is maintained so that feed metal is provided to avoid shrinkage voids until solidification is complete. The injection air pressure is of the order of 14 kPa to 700 kPa depending upon the size of the casting. These are substantially lower pressures than those found in high-pressure die-casting machines.

During the initial die filling, the molten metal rises smoothly up the riser and through the metal die passageways at too low a velocity to cause turbulence within the molten metal. Because the feed is from the bottom of the die, the air within the die-cavity can escape naturally upwards through the parting lines and vents. The result of this gradual die filling allows progressive solidification to start at the top extremities of the casting and slowly working its way down to the point of entry where the metal is kept under pressure until solidification is complete. After solidification, the pressure in the crucible is released, the dies are opened, and the casting is removed ready for the next cycle to commence. Little fettling of the casting is required.

A variation on this process is to use resin—sand shell moulds and shell cores where the aluminium alloy to be cast would attack a metal die material, or where the casting is too large and too complex to be produced economically in conventional metal dies. For example, motor vehicle engine cylinder blocks in light alloys are frequently produced in pressure-fed shell moulds. The more important advantages of low-pressure die-casting can be summarised as follows:

- The dimensional accuracy and surface finish is equivalent to or better than gravity die-castings.
- Production rates are higher than for any casting process other than hot-chamber pressure die-casting.
- Casting design is extremely flexible, with the choice of metal dies or shell moulds and the option to use sand cores. Heavier sections can then be used with other die-casting processes.
- Metal feed is uninterrupted without impurities since the metal is taken from the bottom of the crucible.
- The dies or moulds being filled under pressure and the pressure being maintained during cooling results in good mechanical properties.

- The full range of aluminium alloys can be cast by this process including the heat-treatable alloys. Even alloys that would normally erode the metal components of other die-casting equipment, because of their high melting temperatures and chemical affinity, can be low-pressure die-cast as all the components in contact with the molten metal can be made from non-reactive refractory materials.
- Operation and control, including ejection and removal of the finished casting can be fully automatic.
- Rapid die changing enables short runs to be justified.
- Capital and running costs are relatively low compared with high-pressure die-casting equipment.

Thus it can be seen that this process has all the attributes of good-quality gravity die-casting, but with reduced costs for fettling and finishing operations, coupled with the levels of automation usually associated with the high-pressure process. Since the process can be automated, operator fatigue and operator error can be eliminated.

Assignments

1. Outline, with the aid of diagrams, the process of shell moulding and compare the advantages and limitations of this process with the green sand moulding process (Volume 1).
2. Outline, with the aid of diagrams, the process of investment moulding and compare the advantages and limitations of this process with the green sand moulding process (Volume 1).
3. Compare the advantages and limitations of any two of the following die-casting processes:
   (a) gravity;
   (b) low pressure;
   (c) high pressure (hot chamber);
   (d) high pressure (cold chamber).
4. Describe typical components and materials associated with the following processes giving reasons for your choice:
   (a) shell mould casting;
   (b) investment casting;
   (c) high-pressure (hot-chamber) die-casting.
5. Describe the essential differences between the die-casting process and the sand casting process in terms of capital outlay and operating costs, and explain why die-casting is more econnomical for the mass production of components in low melting point alloys.
6. The dies used for pressure die-casting represent a considerable capital investment and their design is a highly specialised task. Summarise the main design factors and show, with the aid of diagrams, how the profitability of the process can be increased by the use of multiple impression moulds and describe the precautions that must be taken in the design of such moulds.

2 Flow-forming processes

2.1 Hot- and cold-working

The forming of metal by plastic flow without loss of volume was introduced in *Manufacturing Technology: volume 1*. These flow-forming processes were categorised into *hot-forming* and *cold-forming*. Hot-forming processes are those flow-forming processes carried out above the temperature of recrystallisation, while cold-forming processes are carried out below the temperature of recrystallisation. This will now be considered in greater depth.

Metals other than lead and zinc will be cold-worked if they are flow formed at normal ambient (room) temperature. During a cold-working process the grain of the metal becomes distorted and internal stresses are introduced into the grains. If the internal stresses are sufficiently great, and if the temperature of the metal is raised over a critical value, new crystals will start to grow at each stress point. The formation of these new seed crystal is called *nucleation*. The more severe the cold-working, the greater the distortion of the original grain structure will be, and the greater the internal stresses will be. The greater the internal stresses, the lower the critical temperaure at which nucleation commences for a given metal will be. If the metal is maintained at this temperature, the new crystals will continue to grow until their boundaries collide, a new grain pattern emerges, and the original distorted grain structure no longer exists. Since there are several stress points in each distorted crystal, and

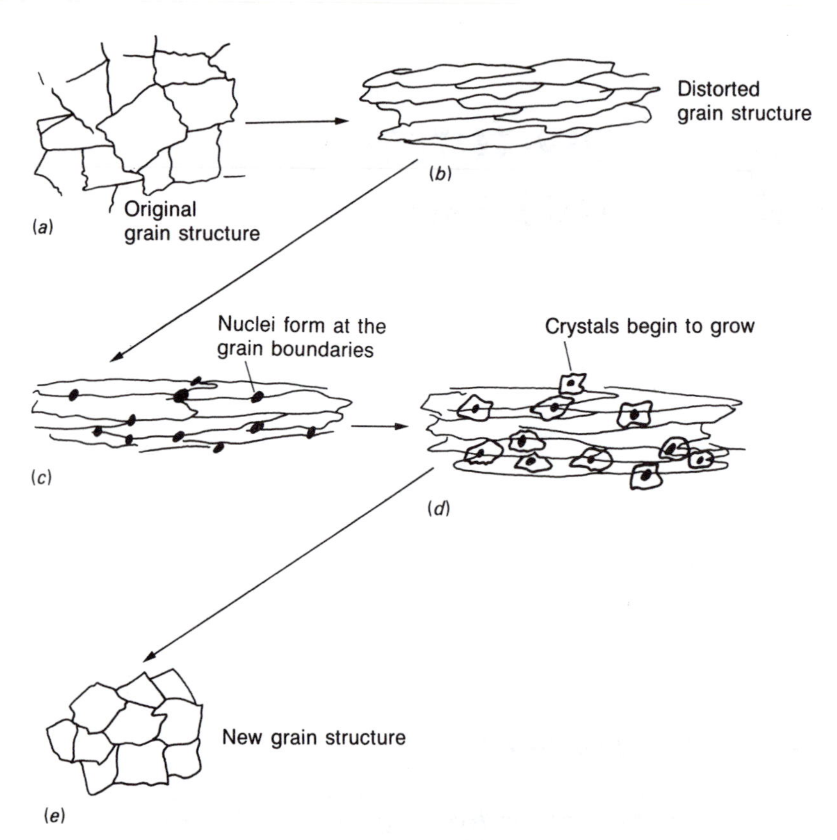

Fig. 2.1 Recrystallisation; (a) before working; (b) after cold-working; (c) nucleation commences at recrystallisation temperature; (d) crystals commence growth as atoms migrate from the original crystals and attach themselves to the nuclei; (e) after annealing is complete the grain structure is restored

since a new crystal grows from each stress point, it follows that the new grain structure will consist of a larger number of smaller grains than existed originally, i.e. the grain structure has been *refined*. The process of nucleation and crystal growth is referred to as *recrystallisation*. The principle of nucleation and recrystallisation, as outlined above, is shown in Fig. 2.1.

Cold-working

As previously stated, cold-working is when metal and alloys are flow formed *below* the temperature of recrystallisation. Since cold-working results in distortion of the crystal lattice of the grains of the metal, flow becomes increasingly difficult and the metal *work-hardens*, becoming harder and stiffer. If flow forming continues the metal will eventually fracture. Once work-hardened, the metal must be heated to its recrystallisation temperature and held at that temperature until recrystallisation is complete, before further flow forming can be undertaken. This process is variously described as sub-critical annealing, process annealing, and inter-stage annealing.

Hot-working

As previously stated, hot-working is when metals and alloys are flow-formed *above* the temperature of recrystallisation. Since the process temperature is above the temperature of recrystallisation, the grains reform as fast as they are distorted and no work-hardening takes place. If the metal could be maintained at a constant forging temperature, there would be no limit to the amount of manipulation the metal could tolerate. In practice, of course, the metal is constantly cooling down from the moment it is taken from the furnace and flow forming should not continue below the process temperature or the load on the forming equipment will become excessive, leading to structural failure, and surface cracking of the work may occur. On the other hand, the process temperature must not be exceeded or the metal may become 'burnt'. That is, oxidation of the grain boundaries occurs and the metal is seriously weakened, resulting in scrapping of the work.

2.2 Hot-forging

When metals are forged to shape, their grain structure flows to the shape of the component and this greatly increases the strength of the workpiece, as does the grain refinement which also occurs during forging. The orientation of the grain is shown in Fig. 2.2 which compares a gear

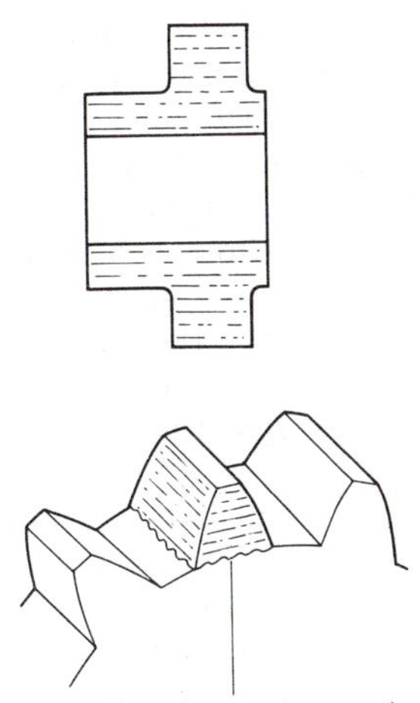

Plane of weakness where tooth will break off under load. This is due to the grain lying parallel to the tooth

(a)

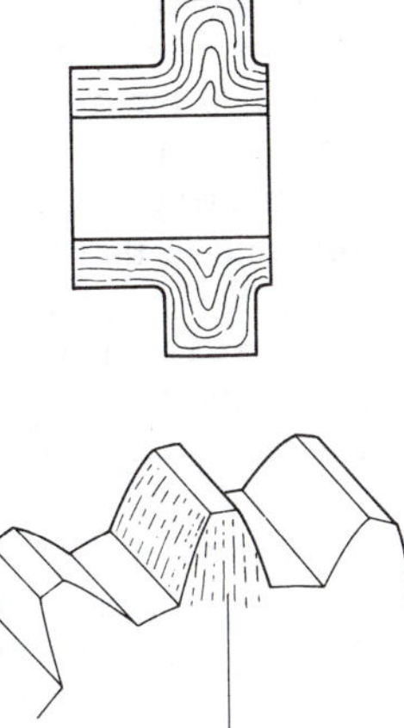

The tooth is very much stronger when the grain flows radially from the blank. This results in the grain lying at right angles to the tooth

(b)

Fig. 2.2 Grain orientation

wheel machined from the bar with a gear wheel machined from a forged blank. Since metals break more easily along the lay of the grain than across the lay of the grain, the teeth of the gear wheel machined from the forged blank will be stronger.

Forgings are widely used where a component is subjected to high impact loads, high fatigue (vibrating) loads, abrasive wear, high operating temperatures, and where high strength is required.

2.3 Forging plant

The energy required to cause the metal to flow may come from a manually-wielded hammer for small components, from various sizes of power hammer for larger components, and from hydraulic presses for the largest components. Forging machines may be classified as:

- Energy-restricted machines: pneumatic hammers, steam hammers, drop hammers and screw presses.
- Stroke-restricted machines: crank and eccentric mechanical presses.
- Load-restricted machines: hydraulic presses.

Conventionally, the capacity of a hammer is specified by the falling mass of the moving parts, or their equivalent if the down-stroke is power assisted. Figure 2.3(*a*) shows a drop-hammer or drop-stamp. These are frequently used with closed-dies for the mass production of small and medium sized components for the automotive industry but can also be used for open-die work for single components. Pneumatic hammers are made in capacities up to 4 tonnes; an example is shown in Fig. 2.3(*b*). Steam hammers are larger and are made in capacities up to 12 tonnes; an example is shown in Fig. 2.3(*c*). Hydraulic presses are the largest and can range from 200 tonnes capacity to 14 000 tonnes capacity, and an example is shown in Fig. 2.3(*d*). The scale of the machine can be assessed by comparing it with the man at the bottom of the figure.

As a useful 'rule of thumb', the falling weight required to hot-forge a low-carbon steel is 3.5 kg/cm^2. Thus a 135 kg hammer can forge stock 63 mm^2 and a 5.5 t hammer can forge stock about 380 mm^2. Hammering tends to concentrate the maximum flow towards the surface of the metal, while squeezing tends to modify the grain structure thoughout the mass of the metal.

In the past, a team of men was required to manipulate the metal and drive the hammer or press. However, modern electronic control now allows a single operator to carry out all the functions previously performed by the team from an operating console in a comfortable, soundproof control cabin. A *manipulator* holds and positions the work under the hammer, while a *charger* loads and unloads the furnace. These functions are often combined when components of under 20 t are being forged.

(*a*) Drop hammer

(*b*) Pneumatic hammer

(*c*) Steam hammer

(*d*) Hydraulic forging press

Fig. 2.3 Hot-forging plant

2.4 Open-die forging

Open-die forging is a hot-forming process, similar to blacksmithing on a large scale, in which the metal is hammered or squeezed to shape while it is above the temperature of recrystallisation. Suitable temperature ranges are shown in Fig. 2.4. The basic principles of the process were introduced in *Manufacturing Technology: volume 1*. Open-die forging refers to any forging operations performed without the metal being completely constrained. That is, the forging operations are carried out between a hammer and an anvil with only the aid of standard forging tools such as swages and fullers. These tools and their uses were

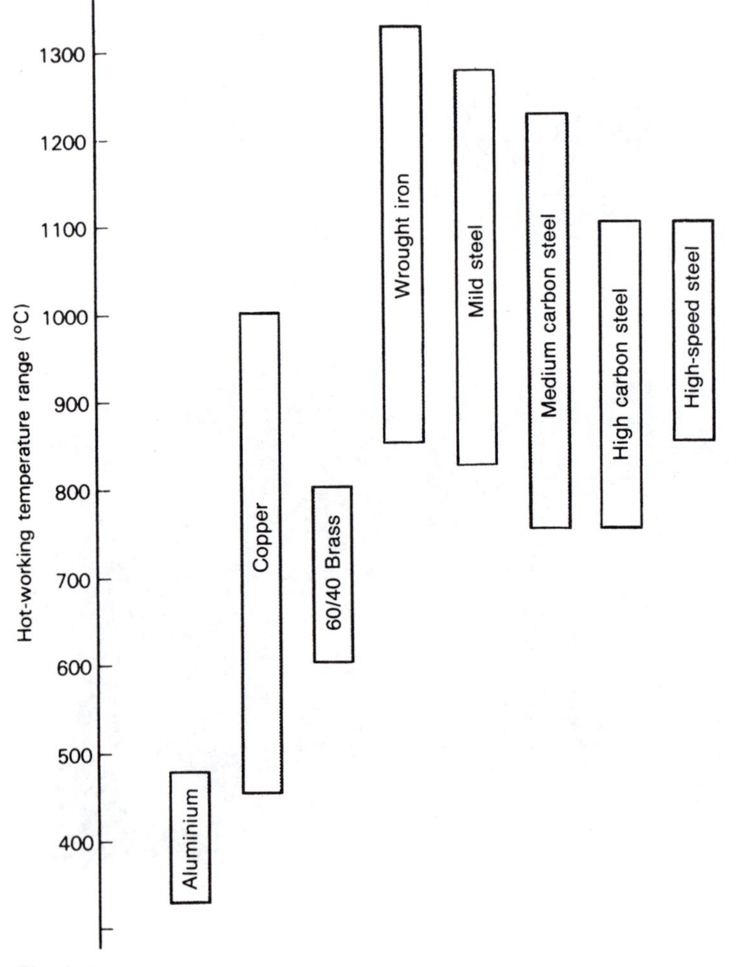

Fig. 2.4 Hot-working temperatures

introduced in volume 1. When applied to power hammers and presses, they are rigidly attached to the hammer (tup) and the anvil. The accuracy of the finished component depends upon the skill of the hammer operator and machining will be necessary to finish the component. Open-die forging requires highly-skilled labour and is a relatively slow process compared with closed-die forging. It is used for single components or for small batches of similar components of any size from the smallest to the biggest. It is a relatively expensive process and is used where the special properties imparted to forged metal need to be exploited.

2.5 Closed-die forging

Closed-die forging is also a hot-forging process in which the metal is worked above the temperature of recrystallisation. However, in closed-die forging the metal is totally constrained within 'impressioned dies' and, after clipping off the residual flash, the forgings require the minimum of finishing. Closed-die forging can be used for the finishing of components roughed out by open-die forging or it can be used for the complete manufacture of forged components. The latter is the more usual and the process is used for the mass production of parts for the automotive industry.

Figure 2.5 shows a section through through a pair of impressioned forging dies. To ensure complete filling of the die cavity, the blank is

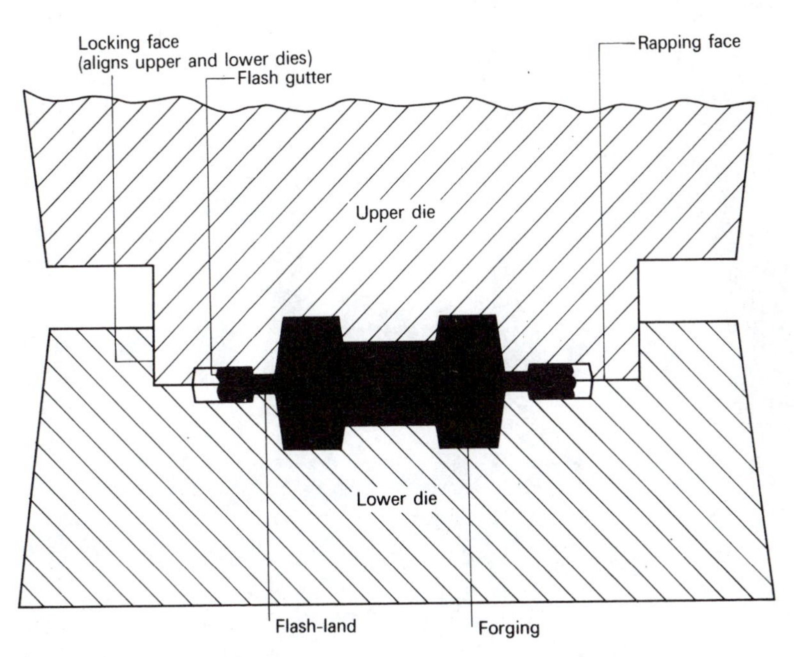

Fig. 2.5 Closed-forging die

slightly larger in volume than the finished forging. As the die closes, the surplus metal is squeezed out into the *flash gutter* through the *flash land*. The flash land offers a constriction to the flow of the surplus metal and tends to hold it back in the die cavity to ensure complete filling. It also ensures that the flash is thinnest adjacent to the component being forged, resulting in a neat, thin flash line being left after the flash has been trimmed off. The flash land should be kept as short as possible otherwise the dies may fail to close. The *rapping faces* ensure that the component is the correct thickness. The hammer driver knows when the dies are fully closed by the sound of the sharp 'rap' when the hardened surfaces meet and the operation is complete. The dies are given a 7° taper, or 'draught', on all vertical surfaces so that the forging can be easily removed.

Except for very small and simple components, most forgings have to be produced in a sequence of operations. Figure 2.6 shows a set of forging dies and the component produced at the various stages in the sequence. Instead of using separate forging dies as shown in Fig. 2.6, components are frequently produced in multi-impression dies. The preforming impressions are sunk into the outer edges of the die and the finishing

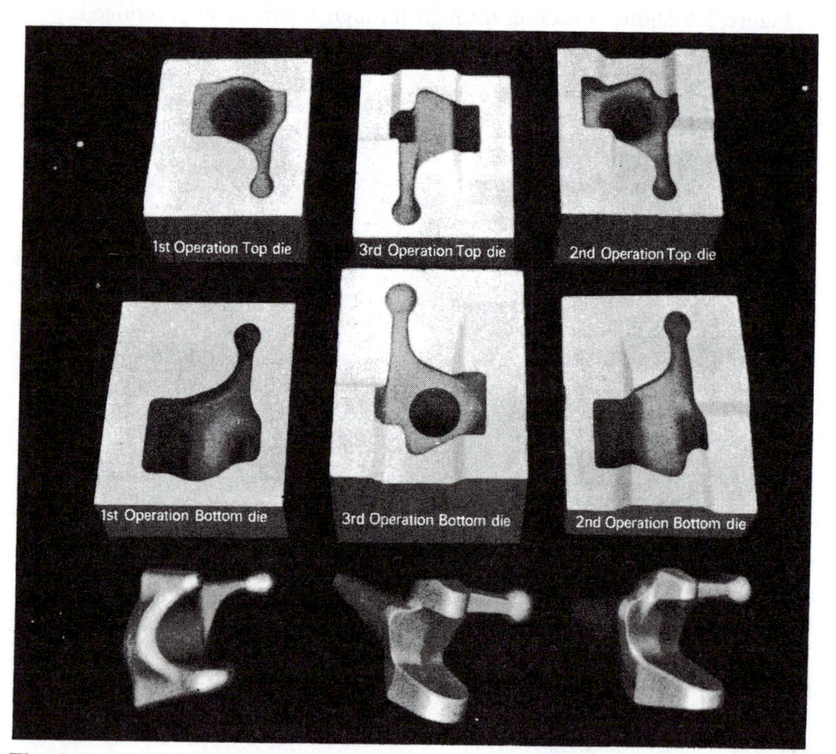

Fig. 2.6 Set of drop-forging dies

operations are sunk into the middle of the die where the blow is more solid and there is less chance of the dies tilting out of alignment.

Die life depends upon many factors and may range from only 1000 components up to 25 000 or more, depending upon the size and complexity of the forging, and the material from which it is being made. The composition of the die steel and its heat treatment also has a significant effect upon the die life. Low-alloy die steels generally provide toughness and shock resistance but wear fairly quickly, whilst high-alloy die steels have better wear resistance and thermal fatigure resistance. The usual causes of die rejection are:

- *abrasive wear* – occurs where pressures are highest and metal sliding is greatest, e.g. flash lands and other points where the reduction in vertical cross-section is greatest;
- *thermal fatigue* – heat or 'craze' cracks on the surface of dies caused by frequent and high ranges of temperature change;
- *mechanical failure* – catastrophic failure which usually occurs as a result of overloading the dies, i.e. using an oversize blank. This is the most common cause of die rejection when precision forging without a flash land and flash gutter.

2.6 Upset forging

Figure 2.7 shows an upset-forging machine and a set of dies. Unlike the processes described so far, the *header* moves horizontally and strikes the stock bar 'end on' to force it into split dies. Figure 2.8 shows the principle of an upset forging operation. The bar stock is fed into the machine after being heated to the correct temperature and gripped between the split dies. The dies are closed by a crank- or cam-operated toggle mechanism. A crank- or eccentric-operated header then moves the punch forward against the end of the bar stock and forces it into the cavity formed in the dies (or partly in the dies and partly in the punch). This process is used to manufacture high-tensile bolt blanks, internal combustion engine valve blanks, small cluster gear blanks and similar components.

2.7 Cold-forging

Figure 2.9 shows the principles of forward and backward extrusion processes as performed on mechanical and hydraulic forging presses. These cold-forging processes were developed for the manufacture of constant-velocity universal joints and similar components where close tolerances, high levels of surface finish and maximised mechanical properties are required. In fact, the tracks in constant velocity joints produced by this process require no machining or further finishing after extrusion.

The following process description is for the manufacture of a constant-velocity joint member from 527A19 induction hardening steel. The bar

Fig. 2.7 Upset-forging machine and dies

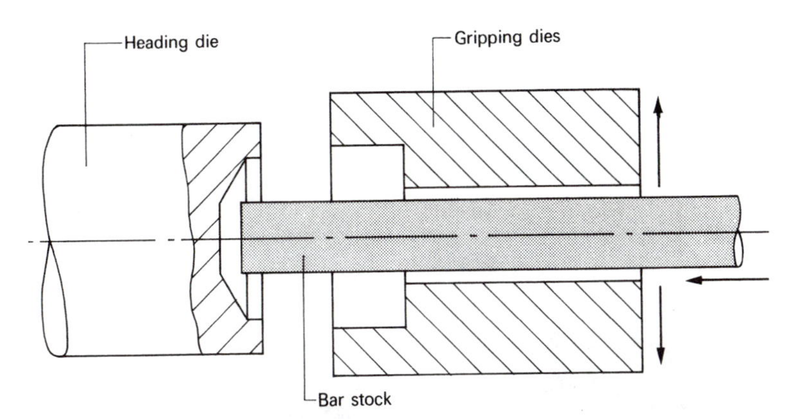

(a) **Gripping dies open whilst stock is fed in**

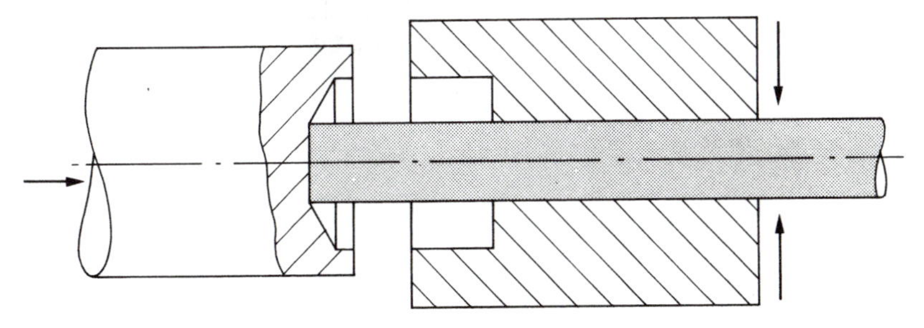

(b) **Gripping die close and heading die moves against stock**

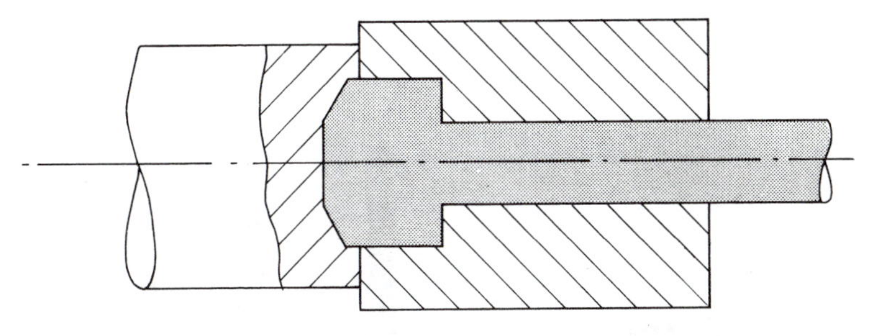

(c) **Heading dies upsets stock to fill die cavity**

Fig. 2.8 Heading dies

stock is supplied in the bright condition with the outer skin removed by a peeling operation to ensure that no surface or subsurface defects are present. Defects that are too small to be detected by commercial ultrasonic testing can still cause component failure during extrusion. The billets are cropped from the bar to a high degree of accuracy. Each blank is automatically weighed as it is cut off and the setting of the cropping

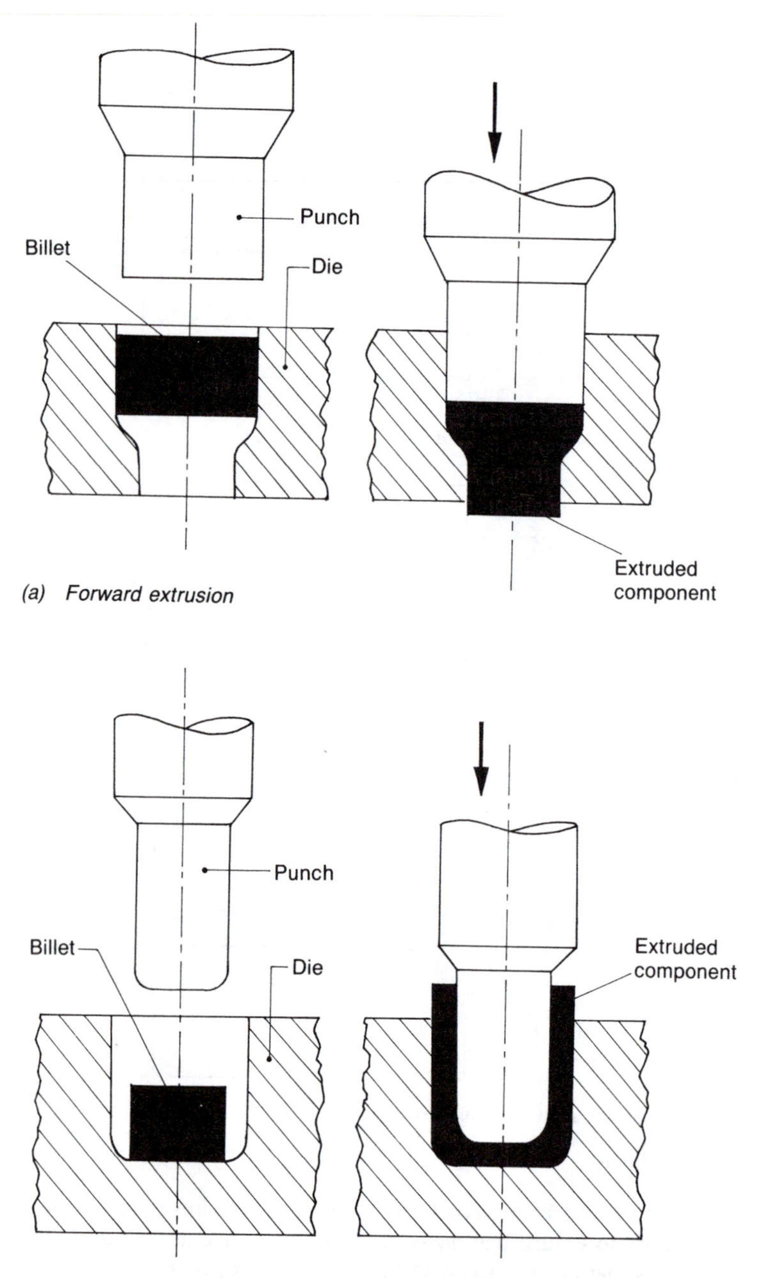

Fig. 2.9 Cold-forging (impact extrusion)

machine is automatically adjusted to keep the mass of the billets within very close tolerances. A 2.5 kg billet has to be kept within −0 +20 g. The billets are sub-critically annealed at about 710°C in a controlled-atmosphere furnace. They are then barrelled to remove any sharp corners, pickled to ensure a chemically clean surface, treated with a phosphate bonder and lubricated with a metallic stearate soap. The phosphating process ensures the lubricant adheres to the surface of the billet and also provides additional, extreme-pressure lubrication in its own right.

The billets are then extruded in mechanical or hydraulic forging presses of 750 t to 1000 t capacity depending upon the operation. The operation sequence is shown in Fig. 2.10. Inter-stage annealing is carried out at

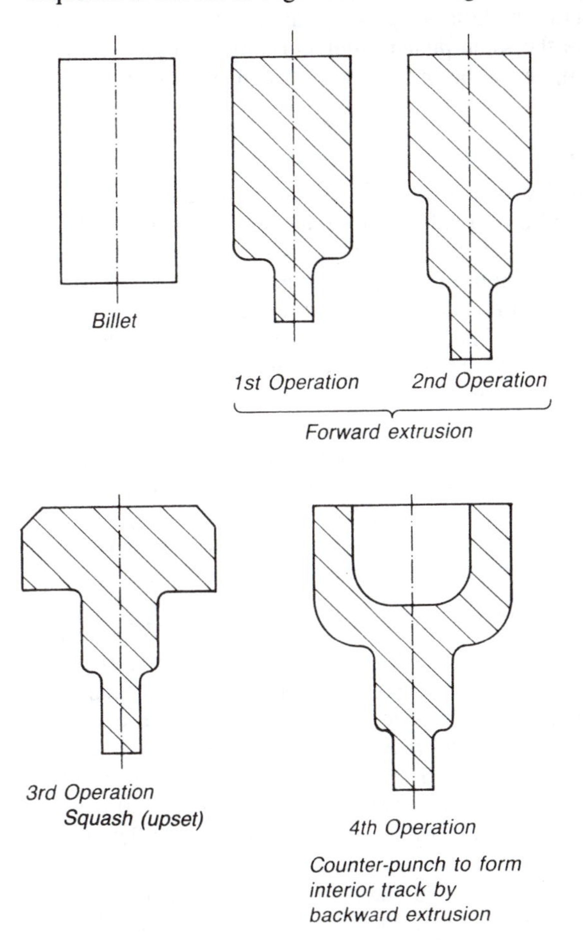

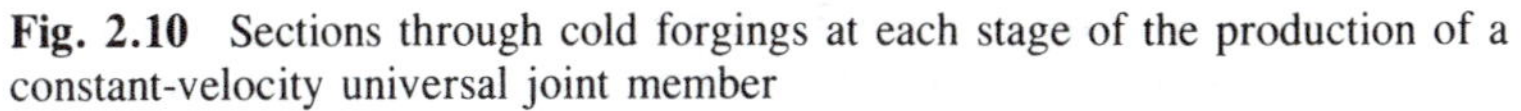

Fig. 2.10 Sections through cold forgings at each stage of the production of a constant-velocity universal joint member

sub-critical temperatures (700°C to 720°C) between the squashing and backward extrusion operations. The extrusion is then normalised at 920°C and again pickled, barrelled and lubricated with sodium strearate (PBL treatment) before the final cold calibration pressing operation which imparts an accuracy of ≤0.03 mm across the tracks. No subsequent machining of the tracks is required.

The batch size is dependent on the die life of the most severe operation in the sequence. This is normally the backward extrusion operation and the die life is limited to about 5000 components. At the other extreme, the die life of the calibration tools is about 500 000 components. Cold-extrusion tools are extremely expensive and every care has to be taken to maximise the life of the tools. For example, a tool-steel punch for the backward extrusion operation starts to 'pick up' after between 1500 and 2000 components, while the same punch, after nitriding, can have a life of 30 000 to 35 000 components before recovery becomes necessary.

2.8 Warm-forging

Warm-forging is now taking over from cold extrusion since the forces required to form the metal are greatly reduced and the life of the tools is correspondingly increased. Both these factors decrease the operating costs of the process. As the name implies, the process is carried out below normal hot-forging temperatures but above the temperature of recrystallisation for the metal. The following sequence of operations is required to produce a constant-velocity universal joint member similar to that described in the previous section.

The billets are again cropped to length and each billet is subjected to automatic weighing to ensure that any variation in mass does not exceed −0 +20 g on a billet of mass 2.5 kg. As previously, the weighing station automatically adjusts the cropping machine to compensate for any variation in mass. After cropping, the billets are barrelled and degreased but, this time, they are not subjected to a phosphate treatment nor a stearate soap lubricant treament.

The billets are then given their first induction heating to 120°C and are immediately immersed in a suspension of graphite (carbon) particles in water. The suspension flash dries and leaves a protective film of graphite on the surface of the metal. This graphite film reduces oxidation of the metal to a minimum and provides lubrication for the subsequent processing. The billets are then raised to the process temperature of 900 to 960°C by further induction heating. Temperature control is 100 per cent and is performed automatically by infrared sensors. Any billets that are not at the correct temperature are automatically rejected and re-heated. Correctly-heated billets are fed automatically to a 1600 t, four-station transfer press. The sequence of operations is the same as in Fig. 2.10 but this time operations 1 and 3 are performed at the same time and operations 2 and 4 are performed at the same time to balance the load on the press. The warm-forged (extruded) components exit the press automatically into cooling trays. When cool, the components are returned to the PBS plant for the full treatment of pickling to remove any residual oxide film, barrelled, phosphate-treated and lubricated with a sodium stearate soap. The soap reacts with the zinc phosphate coating on the component to form a film of zinc stearate at the surface of the component. This acts as an extreme pressure lubricant.

The treated components are then cold pressed to give them their final calibration. This is the same operation as that used during the cold extrusion process and the finished accuracy is equally as good so that no machining of the tracks is required. Again, the highest wear occurs during the backward extrusion process. However, when warm-forging, the die life is approximately 25 000 components compared with 5000 components for cold extrusion. Output is approximately 15 finished components per minute. The combination of full automation and extended die life makes warm-forging a much more economical process than cold extrusion. Components produced by this process have excellent mechanical properties and the microstructure and machinability can be controlled to suit the customer's specification direct from the press, thus avoiding the heat treatment processes necessary after cold extrusion. Warm-forgings are very much more accurate, have a much superior surface and superior mechanical properties compared with components produced by conventional hot-forging.

2.9 Thread rolling

Thread rolling is a flow-forming process carried out below the temperature of recrystallisation, i.e. it is a cold-working process. The thread is flow formed to shape and no cutting occurs. This results in a thread in which the grain orientation follows the thread form and this imparts greater strength to the thread than is possible with a machine-cut thread.

The workpiece blank is rolled between a pair of flat dies or roller dies upon which is formed the impression of the thread to be produced. This process is usually used for screw threads under 25 mm diameter. The pressure between the dies and the blank causes plastic deformation of the metal to take place at the surface of the blank. This results in depressions which form the thread spaces and coining up of the metal which forms the thread flanks and crests. Before rolling, the blank will be approximately the mean diameter of the thread. The rolling action results in displacement of the metal above and below this surface. In practice the blank diameter will vary slightly from the mean depending upon the properties of the metal being formed. Although the dies are expensive, they have a long life and accurate threads can be produced in large quantities.

Flat dies

Figure 2.11 shows the principle of thread rolling using flat dies. In this example only one die is moved, but in some machines there is a counter-movement of both dies. Compensation of die and machine wear is

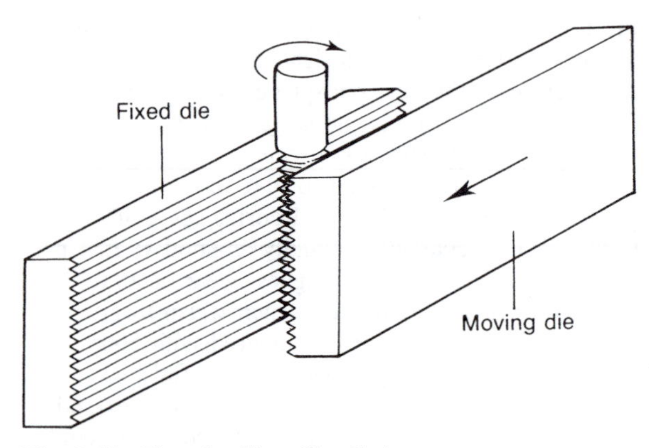

Fig. 2.11 Thread rolling (flat dies)

achieved by the use of adjustable wedges. The blanks are usually fed automatically from a hopper and the die length is arranged so that the blank rotates about four times during one pass. The dies are made from high-quality alloy tool steel and have a life of up to 500 000 components. Rolling speeds are of the order of 30 m/s to 75 m/s, depending upon the properties of the material being rolled. Use of an extreme pressure lubricant is essential to avoid early die wear.

Roller dies

Roller dies are used for rolling small-diameter threads on capstan and automatic lathes. They could also be used on CNC turning centres. Since the work is supported by three roller dies located 120° apart, slender work can be threaded without difficulty. The highly polished rollers have helical grooves and are, in effect, screw threads. They are self-feeding and no lead screw is required. The length of the screw that is rolled is not limited by the face width of the roller dies. In this, they are superior to flat dies which cannot roll threads that are longer than the width of the dies. Figure 2.12 shows the general arrangement of a thread-rolling die head, and its similarity with a thread-cutting die head is apparent.

In operation, the die head is set and this closes the roller dies to give the correct thread size. When the thread is complete, the die head opens automatically and the roller dies retract clear of the work. This enables the die head to be withdrawn over the finished screw without having to stop or reverse the machine. As for flat die rolling, a copious flow of extreme pressure lubricant must be maintained to prevent early die wear. The thread rollers themselves are mounted on needle roller bearings to keep friction to a minimum.

Tapping

Internal screw threads can also be flow formed in suitably malleable materials using lobed taps. The lobed tap can be screwed into a pre-

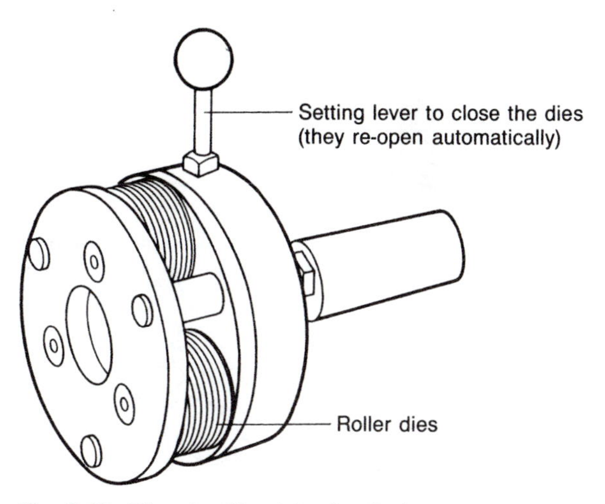

Fig. 2.12 Thread rolling (circular dies)

drilled hole in the same way as a metal-cutting tap. However, greater axial force is required to start the tap. For flow tapping, the pre-drilled hole is the mean diameter of the thread and not the core diameter as in thread cutting. Again, an extreme pressure lubricant must be used.

2.10 Press bending

Vee-bending, as shown in Fig. 2.13, is the simplest of the bending operations and was introduced in *Manufacturing Technology: volume 1*. When the metal is bent, the metal on the outside of the bend is stretched (in tension) and the metal on the inside of the bend is shortened (compressed). Somewhere between these two extremes there is a layer of metal which is neither in tension or compression; it has not changed in length. This layer lies on the *neutral axis*. It is important to understand the effects of bending on the material being processed. Initially, when bending is slight, only elastic deformation occurs and if the bending pressure is removed the material will spring back to its original shape. As the bending pressure is increased, the stress produced in the outermost layers of the material (on both the tension and the compression sides) eventually exceeds the yield strength. Once the yield strength has been exceeded plastic deformation occurs. This permanent strain only occurs in the outermost regions of the material which are furthest from the neutral axis. Nearer to the neutral axis only elastic strain will be present.

When the bending force is removed, those regions of the material that have been subjected to plastic deformation will try to retain their permanent set. However, those regions of the material adjacent to the neutral axis, which have been subjected only to elastic deformation, will try to recover. To some extent this elastic recovery will overcome the permanent set and a degree of *spring-back* will occur. To allow for this

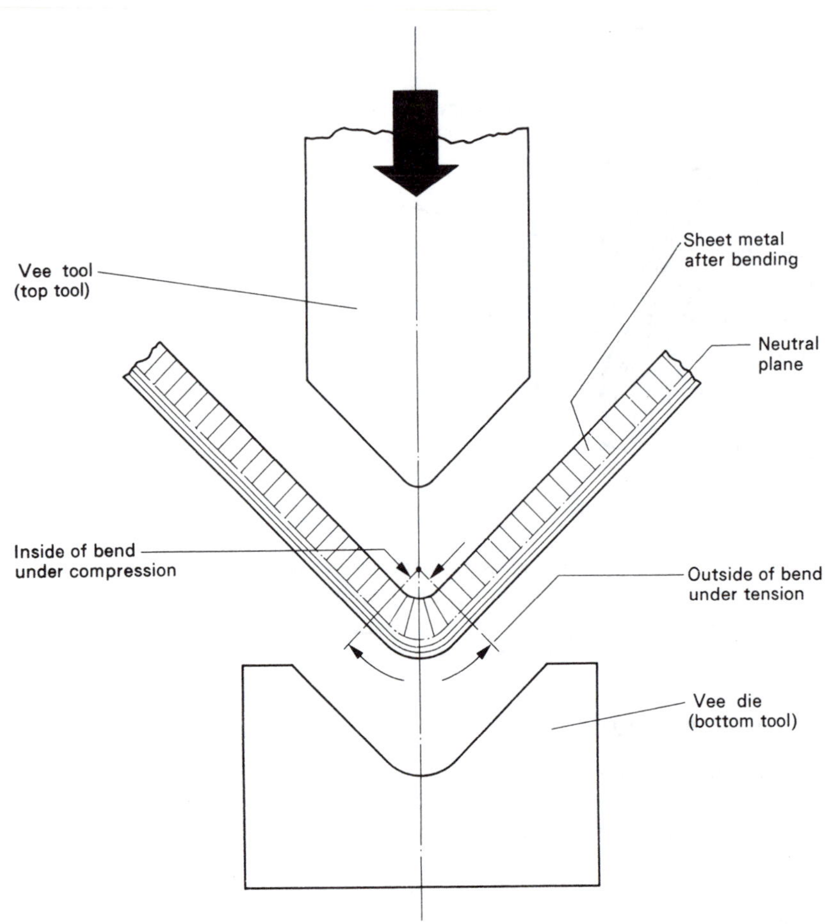

Fig. 2.13 Vee-bending

the tools have to be designed with *over-bend*. That is, if a right-angled bend is required, the tools have to be made to bend the metal to a more acute angle than 90°.

When materials are bent, allowance has to be made for the change of length which occurs either side of the neutral axis. The length of the neutral axis does not change during bending. Because there is a slight difference between the amount of compressive strain and the amount of tensile strain when a metal blank is bent, the neutral axis does not lie on the centre of the material but is slightly offset. However, the difference between the position of the neutral axis and the centre line of the material is negligible for thin sheet metal blanks and the centre-line bend allowance is used to simplify the calculation. Example 2.1 shows how the centre-line bend allowance can be calculated.

Example 2.1 Calculate the length of the blank required to form the 'U'-bracket shown in Fig. 2.14. Assume the neutral axis = 0.5 T, where the thickness of the blank, T = 12.7 mm.

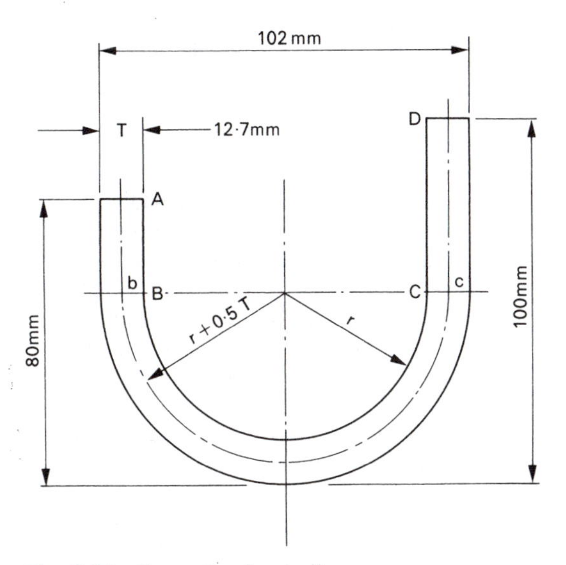

Fig. 2.14 Centre-line bend allowance

The length of the blank required is equal to the sum of the flats AB and CD and the length of the mean line bc. Thus:

L = AB + CD + bc

Now bc represents a semicircular arc whose mean radius R is equal to the *inside radius* r plus half the thickness T of the blank.

The outside diameter of the semicircle	= 102 mm	(see Fig. 2.14)
The inside diameter of the semicircle	= 102 − (2T)	(where T = 12.7 mm)
	= 102 − 25.4	
	= 76.6 mm	
From which the inside radius, r	= 76.6/2	
	= 38.3 mm	
Thus the mean radius, R	= 38.3 + (0.5 × 12.7)	
	= 38.3 + 6.35	
	= 44.65 mm	
Bend allowance bc	= πR	(for a semicircle)
	= 3.142 × 44.65	
	= 140.3 mm	
Length of flat AB	= 80 − (102/2)	
	= 80 − 51	
	= 29 mm	

Length of flat CD	= 100 − (102/2)
	= 100 − 51
	= 49 mm
Total length of blank, L	= AB + CD + bc
	= 29 + 49 + 140.3
	= 218.3 mm (before bending)

2.11 Deep drawing

The process of deep drawing is used to manufacture cup-shaped components from sheet metal. By way of introduction to the process, consider the component shown in Fig. 2.15(*a*). This shallow cup-shaped component, which is drawn from relatively thick sheet metal, can be produced in tools of the type shown in Fig. 2.15(*b*). A circular blank is forced through the die by the punch and the metal flows to shape. The punch and die clearances are arranged so that the metal is lightly 'pinched' between them. This 'irons out' any creases which tend to form during the drawing operation. This simple operation is called *cupping* and is limited to shallow components of the type shown.

Figure 2.16(*a*) shows what is likely to happen when an attempt is made to cup a component having a greater depth to diameter ratio, particularly if the metal is thinner. Puckering occurs round the edge of the blank. This may be sufficient to prevent the metal flowing smoothly through the die and the punch may then tear the bottom out of the component. Even

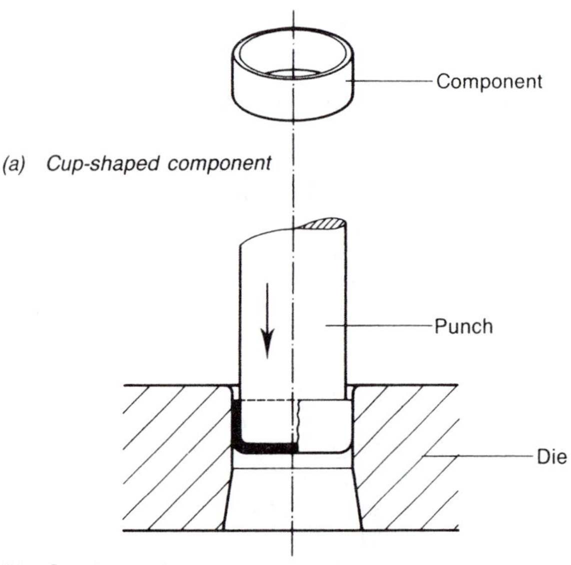

Fig. 2.15 Cupping

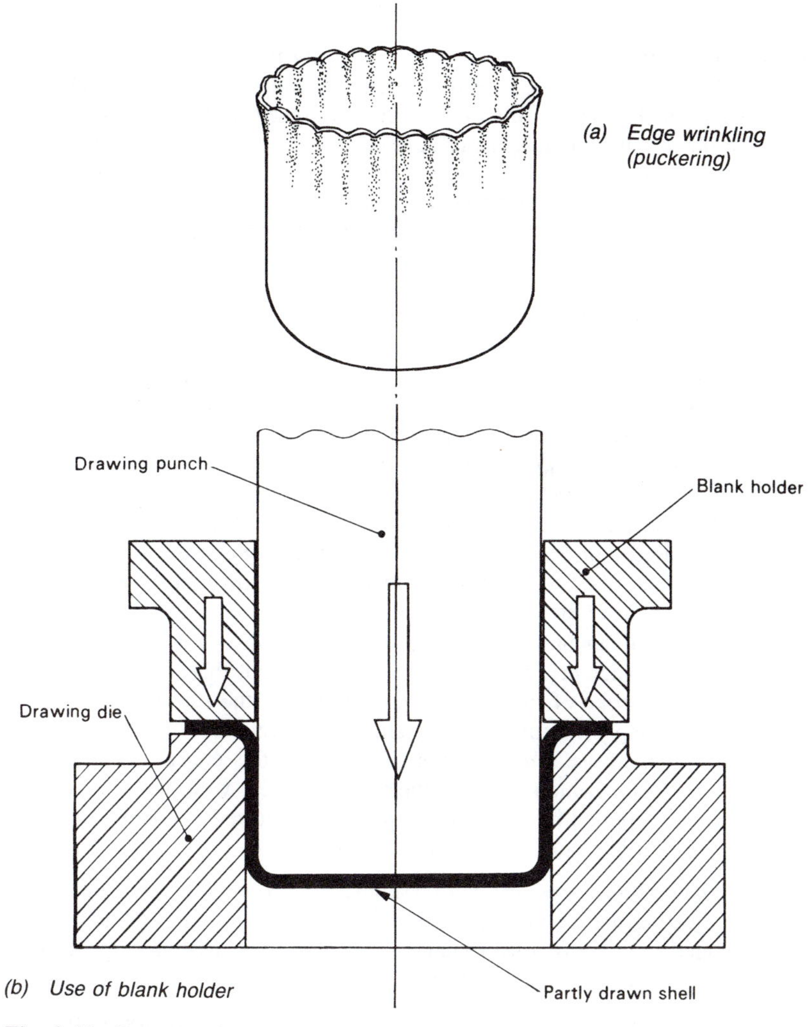

Fig. 2.16 Deep drawing

if the tensile strength of the metal allowed it to be forced through the die, it would be impossible to 'iron out' all the pucker marks. The obvious solution to this problem is to prevent the metal from puckering in the first place. This can be achieved by adding a *blank-holder* to the tools as shown in Fig. 2.16(*b*).

2.12 Deep-drawing tools

Deep-drawing tools vary widely in design, not only because of the range of components produced but also because of the variety of methods adopted for applying the necessary force to the blank-holder.

Double-action press

In addition to the centre ram which is driven by the crank and which carries the punch, there is a cam- or toggle-operated outer ram which carries the blank-holder. This outer ram is designed to descend first and grip the blank with a constant pressure while the centre ram (and punch) descends and completes the drawing operation as previously described. To control the blank-holder pressure accurately, the outer ram is counterbalanced by pneumatic cylinders. An example of this type of press is shown in Fig. 2.17.

Single-action press

It is also possible to deep draw components in single-action power presses provided they have sufficient length of stroke. Some presses have a variable throw crank so that the stroke length can be varied for diferent operations. An example of a single-action press is shown in Fig. 2.18. There are various techniques for applying blank-holder pressure when deep drawing in a single-action press, and Fig. 2.19 shows two commonly used methods. The tool shown in Fig. 2.19(*b*) is often referred to as an inverted tool because the die is attached to the ram of the press and the punch is fixed to the bottom bolster. In the inverted type of tool the punch is often referred to as the pommel.

The main disadvantages of the types of tool shown in Fig. 2.19(*b*) is that the blank-holder pressure is not only limited, but that it also varies throughout the stroke of the press as the springs or the rubber buffer become increasingly compressed. An alternative technique for use with inverted tools is to use a *die-cushion*. Essentially, this consists of a cylinder and piston which replaces the rubber buffer, with blank-holder pressure being applied by feeding compressed air to the cylinder. The use of a die cushion has the following advantages:

- the blank-holder pressure can be adjusted for optimum performance by adjusting the air pressure being fed to the piston and cylinder;
- provided the air pressure remains constant, the blank-holder pressure is constant throughout the stroke of the press.

Figure 2.20 shows how a simple die-cushion is used. The piston is fixed to the bolster by the suspension rod, and the force applied to the blank-holder is transmitted by push-rods engaging the top of the cylinder. In this instance the piston is stationary and the cylinder moves; other designs use a more conventional fixed cylinder and moving piston. The choice of die-cushion depends upon the type of press and the operation being performed.

2.13 Metal flow when drawing

Figure 2.21(*a*) shows a segment from a circular blank which is to be drawn through a circular die. The radius of the blank is R_1 and the radius of the die is R_2. During the passage of the blank across the die face, the

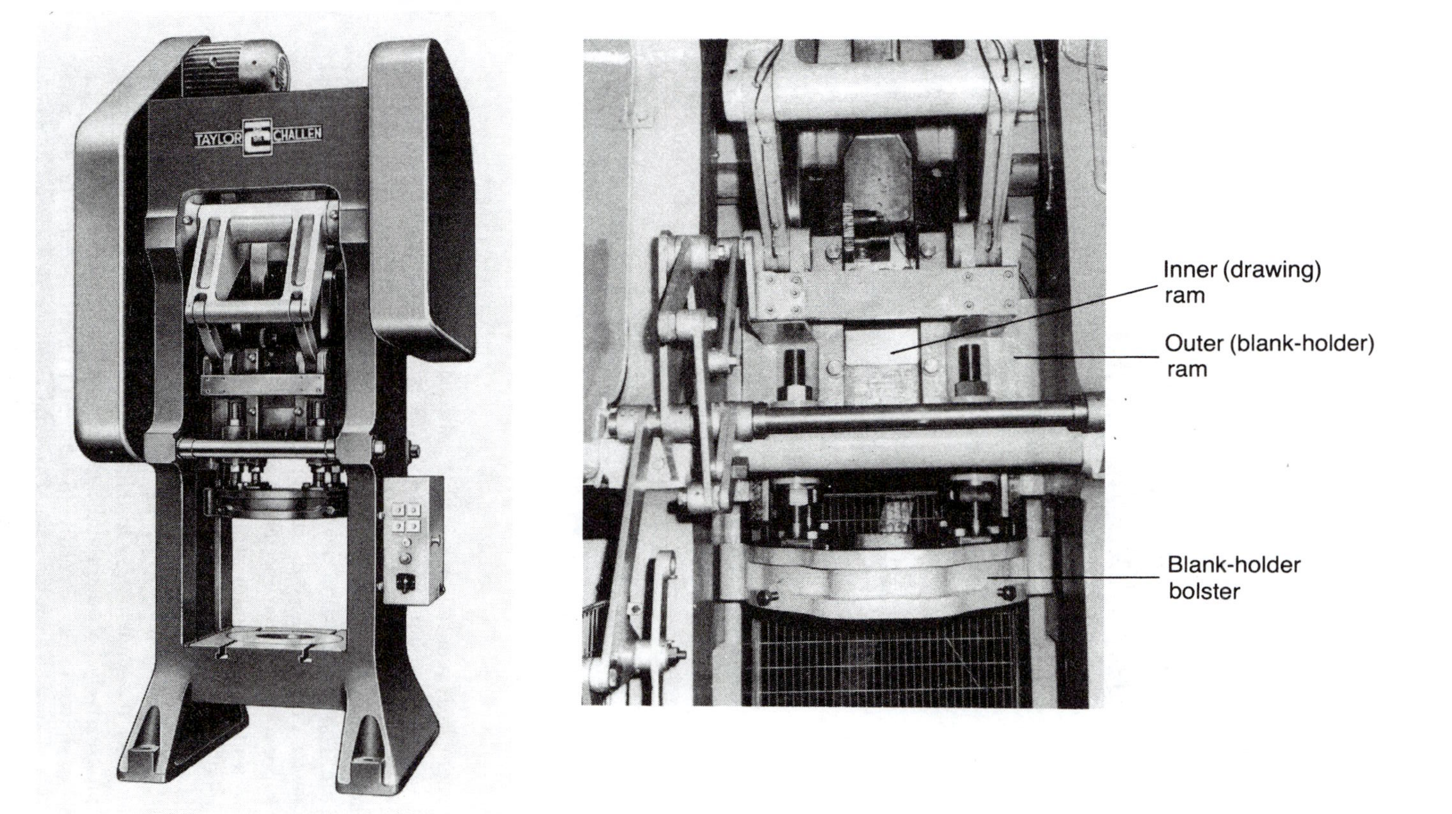

Fig. 2.17 Double-action power press (courtesy of Taylor and Challen Ltd.)

Fig. 2.18 Single-action power press

arc length L_1 is reduced to L_2 and the shaded area of the metal has to be 'lost'. Since the volume of metal does not change between the blank and the component, this loss of surface area is compensated for by local thickening of the metal in the finished component as shown in Fig. 2.21(*b*). This compression of the blank material sets up a hoop stress in the material outside the die, causing it to buckle and pucker. To prevent this happening, an opposing tensile force must be applied to the blank rim as a means of control. This tensile control force is provided by the blank-holder as previously described.

The very action of control immediately sets up a further condition that exerts an influence on the drawing operation. This is the tendency of the metal under the blank-holder to thicken uniformly around its annular rim and progressively between the die mouth and the outside edge of the blank as shown in Fig. 2.21(*b*). In addition to the thickening of the metal and the compressive hoop stress at the rim, there is also a state of tension in the walls of the component as the punch forces the blank through the

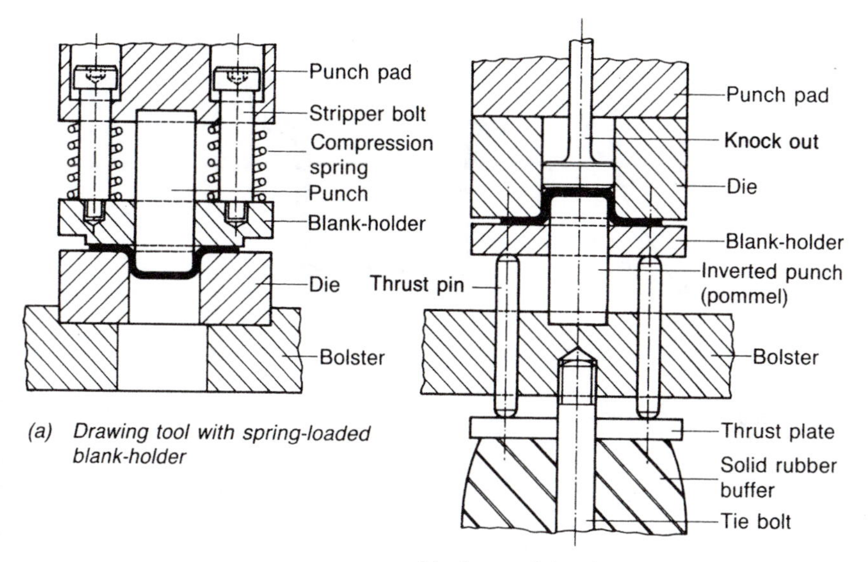

Fig. 2.19 Drawing tools for single-action press

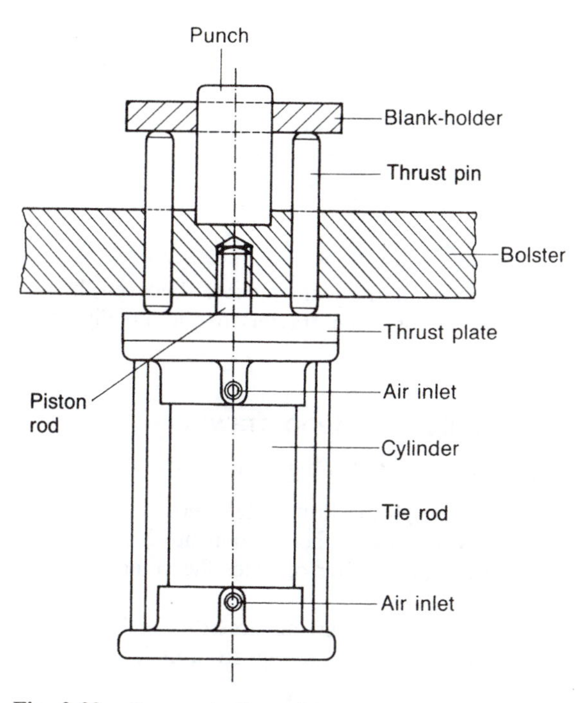

Fig. 2.20 Pneumatic die-cushion

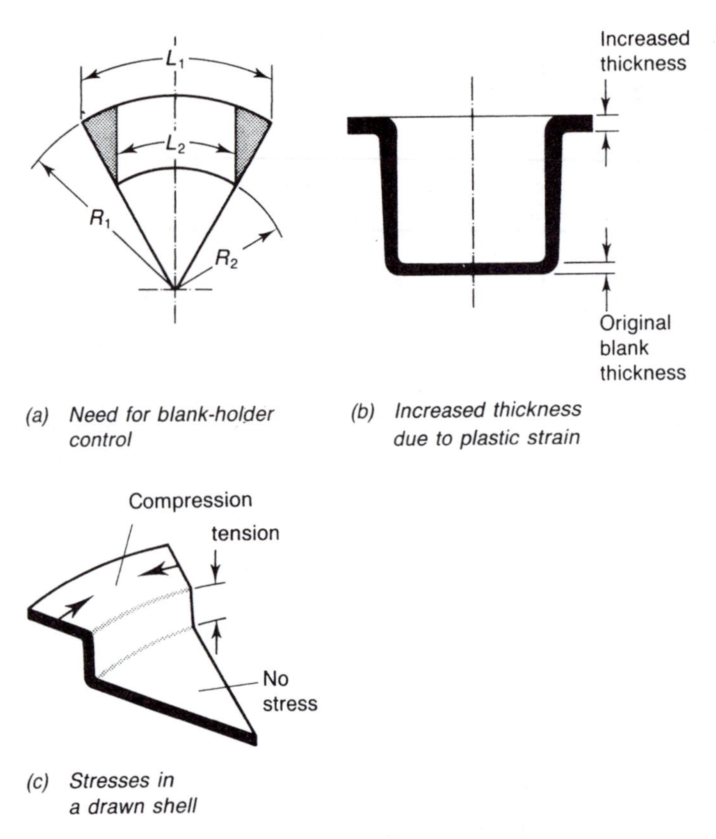

(a) *Need for blank-holder control*

(b) *Increased thickness due to plastic strain*

(c) *Stresses in a drawn shell*

Fig. 2.21 Metal flow when drawing

die. These three factors together decide the diameter ratio between the blank and the finished component in any one drawing operation. Figure 2.21(*c*) shows the stresses in a drawn shell.

2.14 Material characteristics for deep drawing

Materials for deep drawing require the following properties:

- *high ductility* to allow extensive plastic flow to take place;
- *low work-hardening capacity* to prevent the component cracking during drawing and to reduce the number of times the component has to be re-annealed (see section 2.17);
- *high tensile strength* to withstand the drawing stresses which tend to tear the walls of the shell as the metal is drawn from under the blank-holder;

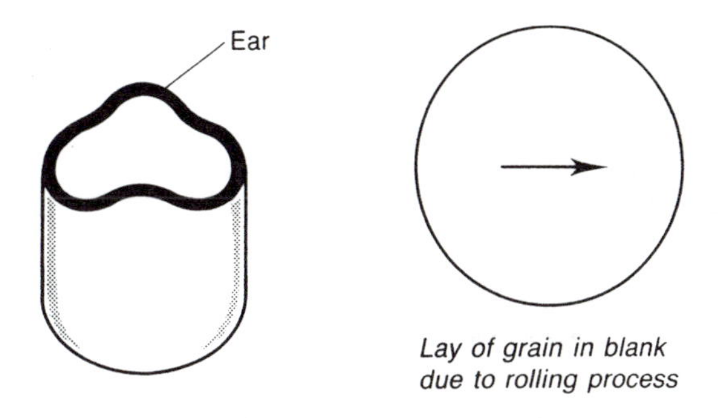

Fig. 2.22 'Earing' due to grain orientation

- *uniform grain size* and orientation to allow uniform flow to take place and to reduce *earing* to a minimum (see Fig. 2.22).

Stress–strain analysis as applied to drawing processes is still an inexact science and the ability to predict the forces involved is still largely empirical. Although the characteristics of a material can be determined from its tensile test results, a more specialised test for determining the suitability of a material for deep drawing is the *Erichson cupping test*. This test simulates a drawing operation on a sample piece of material as shown in Fig. 2.23(*a*). The circular sample is gripped between the pressure ring (blank-holder) and the die face. After the pressure ring has been tightened down on the specimen, it is slackened back by a constant amount (0.05 mm) to give the metal freedom to draw but not to pucker. The spherically-ended punch is then pressed into the specimen, drawing it into the die to form a cup. The forward movement of the punch is continued until a crack appears at the base of the cup. The depth of the cup at fracture is read from the scale of the machine and is a measure of the drawing quality of the material. Figure 2.23(*b*) shows Erichson's standard curves, giving the depth of cup for various thicknesses of sheet and various materials. Cupping tests are not absolute but merely compare a sample from an untried batch of material with a sample from a batch of material known to be satisfactory.

Account must also be taken of the critical relationship between the grain size of the metal after annealing and the amount of cold-work it underwent before annealing. Figure 2.24 shows this relationship graphically for a low-carbon steel sheet. It can be seen that the critical value for cold-working is about 10 per cent deformation for this material. If it is annealed after receiving this amount of cold-working, a very coarse grain will result. This coarse grain not only weakens the component but it also impairs the surface finish, which is said to have an 'orange peel' appearance. When heat treatment is to follow deep drawing, this critical amount of cold-working must be avoided.

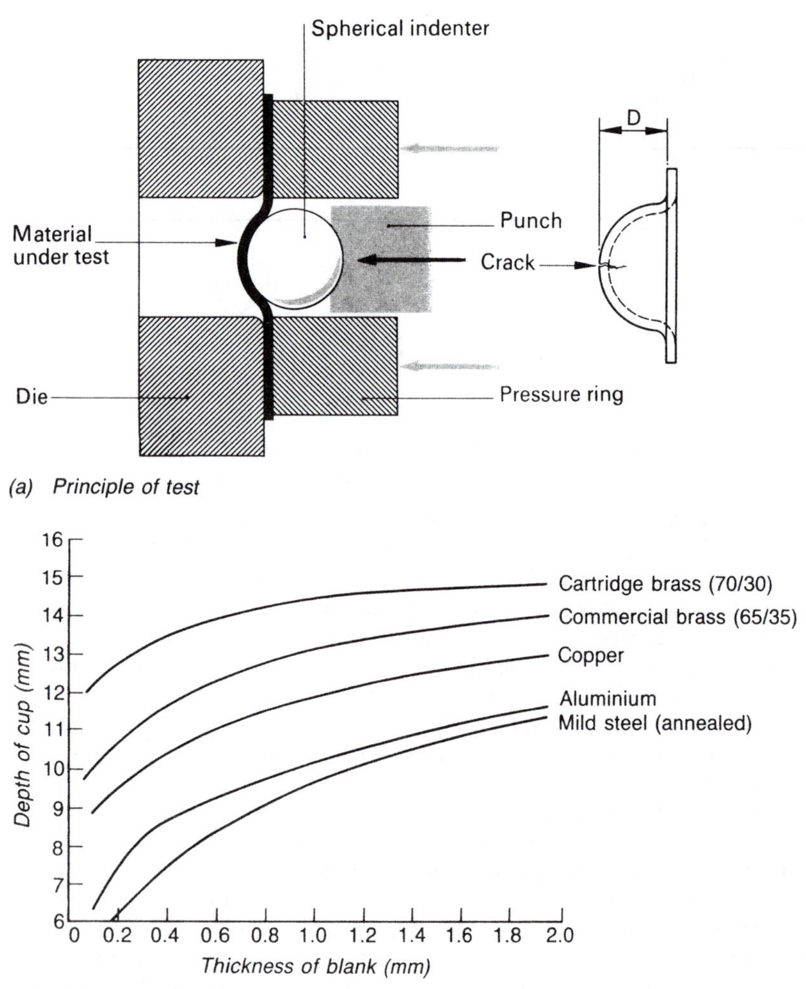

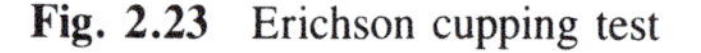
Fig. 2.23 Erichson cupping test

2.15 Blank-holder force

The blank-holder force is critical not only for the avoidance of wrinkles and puckering, but also for the extent to which it affects the elongation of the surface regions of the drawn component. As the force exerted by the blank-holder ring increases, so also does its braking effect on the material being drawn between the blank-holder and the die face. If the braking effect is insufficient, wrinkling occurs but if the braking effect is too great, the bottom is torn out of the drawn component by the punch. When using double-action presses having mechanically-operated blank-holders the setting of the blank-holder force is usually determined

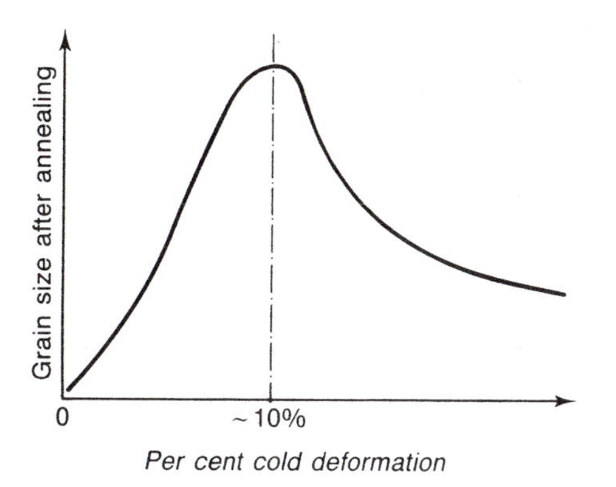

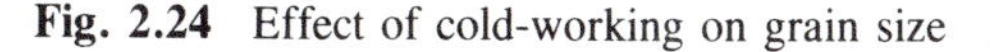
Fig. 2.24 Effect of cold-working on grain size

empirically. This is done by drawing the shell progressively deeper using blanks of the same diameter, while progressively lowering the blank-holder until the surface of the drawn shell is smooth and free from wrinkles. The pressure of the blank-holder on the blank increases during the drawing operation as the edge thickness of the blank increases under the effect of the hoop compression forces set up (see section 2.13). The blank-holder force may be calculated using the following empirical formula:

$$P_B = \pi/4[D^2 - (d_1 + 2r)^2].p$$

where: P_B = blank-holder force (N)
D = blank diameter (mm)
d_1 = diameter of die orifice (mm)
r = drawing edge radius (mm)
p = specific blank-holder pressure (MPa)

Since the drawing edge radius and metal thickness is small in relation to the blank diameter and the punch diameter (d), an approximate formula may be used:

$$P_B \simeq \pi/4(D^2 + d^2).p$$

The magnitude of the specific blank-holder pressure (p) varies with the material used, the punch diameter and the blank material thickness. The smallest blank-holder pressure that should just prevent wrinkling is given by the empirical formula:

$$p \simeq 0.0025[(\beta_0 - 1)^2 + 0.005(d/s)].\sigma_B$$

Where: p = specific blank-holder pressure (MPa)
β_0 = drawing ratio (see section 2.18)
d = punch diameter (mm)
s = material thickness (mm)
σ_B = tensile strength (MPa)

Example 2.2 Calculate the blank-holder force for a drawing operation, given the following data:

Tensile strength of material (σ_BB)	= 370 MPa
Blank thickness	= 0.8 mm
Punch diameter	= 240 mm
Drawing ratio	= 1.75 : 1
Blank diameter	= 420 mm

$$
\begin{aligned}
p &\simeq 0.0025\,(\beta_0 - 1)^2 + 0.005\,(d/s)]\,.\,\sigma_B \\
&\simeq 0.0025\,[(1.75 - 1)^2 + 0.005\,(240/0.8)] \times 370 \\
&\simeq 0.0025 \times 2.0625 \times 370 \\
&\simeq \underline{1.9\ \text{MPa}}
\end{aligned}
$$

$$
\begin{aligned}
P_B &\simeq \pi/4\,(D^2 - d^2).p \\
&\simeq \pi/4\,(420^2 - 240^2) \times 1.9 \\
&\simeq \pi/4 \times 118\,800 \times 1.9 \\
&\simeq \underline{177\ \text{kN}}
\end{aligned}
$$

Sometimes it is not possible for the press to exert sufficient force on the blank-holder to prevent wrinkling, or there may be problems with lubrication of the blank itself. Under these circumstances the tool may be designed with an entry bead to the die as shown in Fig. 2.25.

2.16 Die edge and punch radius

The die edge radius and the punch corner radius are important factors in the design of the drawing tools. If the die edge radius is too small, the blank is elongated and stressed too severely as it is drawn into the die.

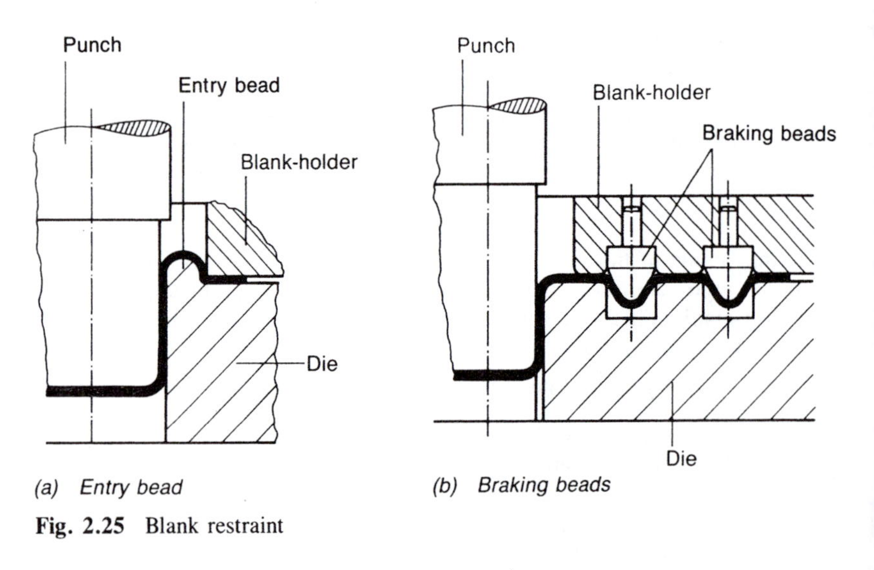

Fig. 2.25 Blank restraint

This results in rupture of the component wall and/or excessive thinning of the component. Alternatively, if the radius is too large, wrinkling can occur as the metal flow is uncontrolled for too great a distance between leaving the blank-holder and becoming pinched between the punch and the die, particularly at the end of the draw. The die radius should normally lie between 5 and 10 times the blank thickness. When deep-drawn shells have comparatively thin walls (large d/s ratio) the higher value should be used.

The corner radius of the punch is also important and should be given careful consideration at the component design stage, especially when the drawing ratio is large. A generous punch radius helps to maintain the thickness of the component wall and, hence, maintain its strength. The punch corner radius should never be less than the die edge radius and, wherever possible, it should be between 0.1 and 0.3 times the punch diameter.

2.17 Multi-stage drawing operations

Where the depth to diameter ratio of the drawn cup is too great to be produced in one operation, it is necessary to produce the component in a series of draws. Figure 2.26(*a*) shows a first draw and a re-draw tool. It can be seen that the punch in the first drawing operation is shaped so as to leave a taper of approximately 15° in the bottom of the cup. The reason for this is shown in Fig 2.26(*b*). Not only does the taper provide location in the re-draw tool, but it allows the metal to flow smoothly past

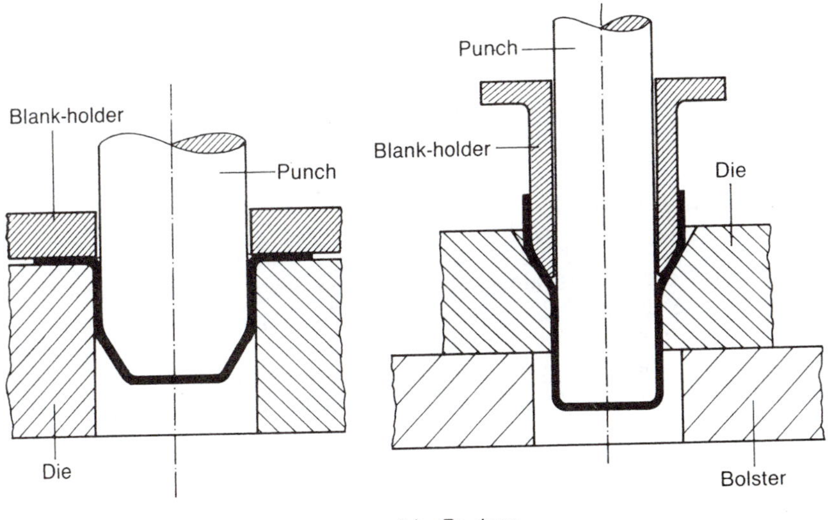

Fig. 2.26 Multi-stage drawing tools

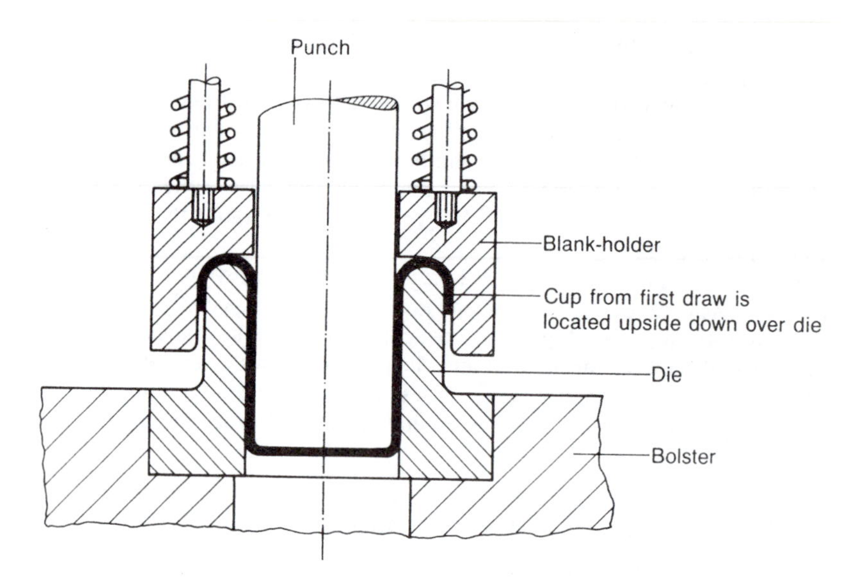

Fig. 2.27 Reverse drawing tool

the pressure sleeve which provides the necessary braking to prevent wrinkling.

Alternatively, reverse drawing can be used as shown in Fig. 2.27. In this instance the cup is turned inside out. The amount of cold-working is greatly increased without undue pressure from the blank-holder. This technique is often used with single-action presses when drawing thin-walled components. Further, the additional cold-working of a reverse draw tool reduces the likelihood of coarse grain developing during subsequent annealing. The cold-working will have exceeded the critical value (see section 2.14, Fig. 2.24).

2.18 Drawing ratio

Where multi-stage drawing is required, the percentage reduction at each stage must be carefully calculated as the allowable reduction becomes less at each stage. Figure 2.28 shows details of the stages in drawing and re-drawing a cup from a given blank. The changes in percentage reduction and depth to diameter ratio (drawing ratio) at the successive stages can be clearly seen. The drawing ratio (β_0) was introduced in the formulae used in section 2.15 where:

$$\beta_0 = D/d$$

D = blank diameter (mm)

d = punch diameter (mm)

Stage	Per cent reduction	*D/d* ratio	Component (dimensions in mm)
Blank	—	—	250
First draw	34	$\beta_0 = 1.51$	165
First re-draw	24	$\beta_1 = 1.32$	125
Second re-draw	20	$\beta_2 = 1.25$	100
Third re-draw	17.5	$\beta_3 = 1.21$	82.5

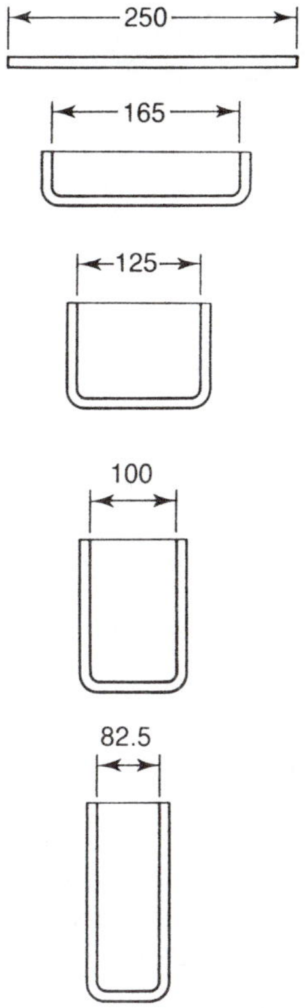

Fig. 2.28 Stages in drawing and re-drawing

This expression refers only to the first drawing operation and, for subsequent drawing operations β_1, β_2, etc., the ratios of the cup diameters before and after drawing are taken. That is:

$$\beta_1 = d/d_1 \quad \text{and} \quad \beta_2 = d_1/d_2 \quad \text{etc.}$$

The largest ratio that can be drawn in a single operation is limited by the following factors.

- the tensile strength of the material being drawn;
- tool dimensions and design;
- blank-holder pressure;
- material thickness;
- friction affecting the flow of the material being drawn between the faces of the blank-holder and the die. Also referred to as the 'braking

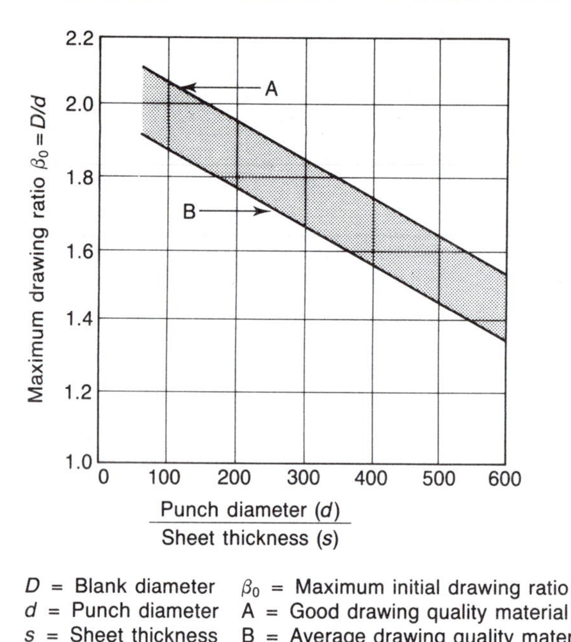

D = Blank diameter β_0 = Maximum initial drawing ratio
d = Punch diameter A = Good drawing quality material
s = Sheet thickness B = Average drawing quality material

Fig. 2.29 Deep-drawing ratios for cylindrical shells

effect'. This will depend upon the lubrication of the blank, the surface finish of the blank material, the hardness and surface finish of the blank-holder and die faces.

Figure 2.29 shows the larger drawing ratio for cylindrical forms as a function of the punch diameter and the material thickness. This figure may be used for different sheet materials drawn in steel tools of good design, using normal lubrication and appropriate blank-holder pressure.

The drawing quality of the blank material also plays a significant part in determining the drawing ratio. As a rough guide, materials with good drawing characteristics follow the upper line (A), while general purpose materials with poorer drawing characteristics follow the lower line (B). Materials with a rough surface finish need a reduced drawing ratio, particularly in the case of thin-walled components having a high d/s ratio. The use of a good drawing lubricant can increase the drawing ratio by up to 15 per cent. These specialist lubricants not only allow an increased drawing ratio but reduce die wear and operating costs. The phosphate coating of low-carbon steel sheets, as used in the drawing of car body panels, has produced outstanding results; the phosphate coating not only providing a key for conventional lubricants but also acting as an extreme pressure lubricant in its own right. Heavily sulphurised extreme pressure lubricants are also used in deep drawing; however, they tend to attack and discolour copper and copper-based alloys. More recently, use has been made of plastic spray coatings containing Polytetrafluroethylene (PTFE) which has the lowest known coefficient of friction. This coating can be stripped off after processing by the use of suitable chemical solvents.

2.19 Drawing force

It is not possible to determine a precise drawing force since this varies as the drawing operation proceeds. Since the drawing force increases proportionally to the drawing ratio, the force having to be exerted by the punch is greatest at the start of the draw when the blank diameter is at its maximum. However, as the blank is drawn into the die, the diameter of the blank remaining under the blank-holder progressively decreases and the drawing ratio also decreases. Both these effects result in the force on the punch decreasing as the operation proceeds. Unfortunately this is the reverse of what happens in a mechanical press where the force on the punch increases as the crank approaches bottom dead centre.

Empirical formulae have been developed to estimate the maximum drawing force for a given set of circumstances. One which is relatively simple yet gives a reasonable result is:

$$F\text{max} \simeq n.\pi.d.s.\sigma_u$$

where: F_{max} = punch force (N)
s = material thickness (mm)
d = punch diameter (mm)
σ_u = UTS (ultimate tensile strength) of blank material (MPa)
n = a coefficient based upon the drawing ratio and lying between unity for general purpose materials and 1.2 for very good drawing materials.

Since this is only an approximate estimation of the punch force required, the coefficient n can be ignored for all practical purposes.

Example 2.3 A cylindrical component is to be drawn to a diameter of 150 mm from 1-mm-thick material. Calculate the approximate drawing force, given that the ultimate tensile strength of the material is 370 MPa.

$F_{max} \cdot \pi.d.s.\sigma_u$

where: π = 3.142
d = 150 mm
s = 1.0 mm
σ_u = 370 MPa

$\simeq 3.142 \times 150 \times 1.0 \times 370$
$\simeq$ 174 kN

2.20 Blank size

The determination of the blank size and shape for drawing operations is largely a matter of trial and error based upon experience, particularly for non-circular and asymmetrical components. The drawing tools are made first and trial blanks are cut out by hand. These are gradually adjusted until the desired component shape, free from 'earing', is arrived at. A copy of this final blank is used as a template for making the blanking tool. Figure 2.30 shows a rectangular, drawn component with corner radii. The 'earing' which occurs with the blank shape shown in Fig. 2.30(*b*) is due to crowding of the metal at the corners where it will be in compression. To obtain a reasonably flat top to the component and to reduce the amount of 'ironing' required between the punch and the die, the blank shape should be modified to that shown in Fig. 2.30(*c*). It should be remembered that the need for excessive 'ironing' of the component results in premature and unnecessary die wear. Where even slight 'earing' is unacceptable, the drawn shell may have to be clipped or machined.

The approximate blank size for cylindrical components can be calculated by equating the surface areas of the circular blank and the finished, cup-shaped component. Assuming that no change in metal thickness occurs and that the corner radius of the bottom of the cup is minimal, the following expression may be used:

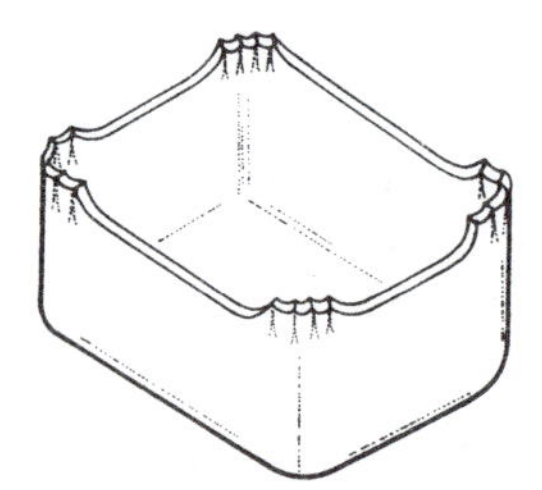

(a) Drawn rectangular box showing earing

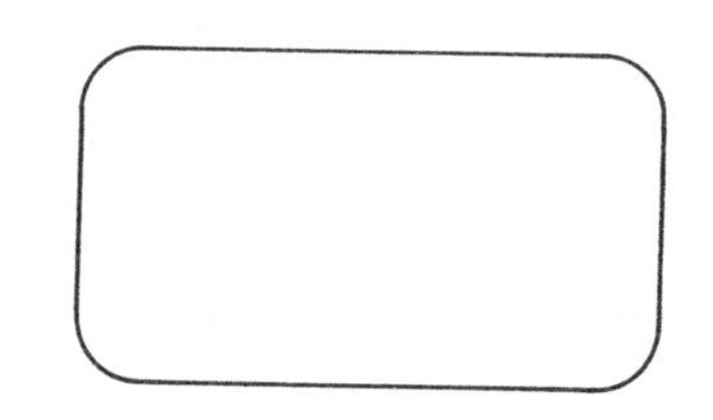

(b) Incorrect blank profile causing crowding at the corners and earing

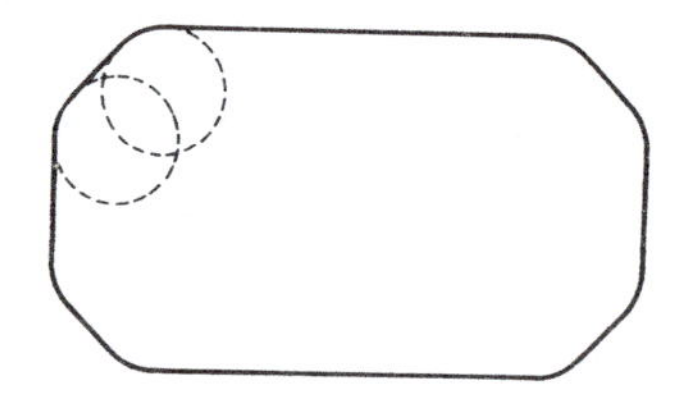

(c) Correct blank profile with corner relief

Fig. 2.30 Blank profile for rectangular box

Surface area of blank $= \pi . D^2/4$ (1)
Surface area of cup $\simeq (\pi . d^2/4) + (\pi , d . h)$ (2)
Equating (1) and (2): $\pi . D^2/4 \simeq (\pi . d^2/4) + (\pi . d . h)$
$\therefore$ Blank diameter $D \simeq \sqrt{(d^2 + 4d.h)}$

where: D = blank diameter
d = punch diameter
h = height of cup

Example 2.4 Calculate the approximate blank diameter for a cup-shaped component 250 mm in diameter and 200 mm high.

$$D \simeq \sqrt{(d^2 + 4d.h)}$$
$$\simeq \sqrt{(250^2 + 4 \times 250 \times 200)}$$
$$\simeq \sqrt{(62\,500 + 200\,000)}$$
$$\simeq \sqrt{262\,500}$$
$$\simeq \underline{512 \text{ mm}}$$

where: d = 250 mm
h = 200 mm
D = blank diameter

Press tool design is largely a matter of experience and the calculations are largely for guidance and verification. No matter how carefully the tools are designed and set, faults frequently show up on 'try-out' while the tools are being 'proved' and later, during service, as wear sets in.

2.21 Flexible die pressing

Figure 2.31(*a*) shows the principle of forming sheet-metal pressings using a rigid punch and a flexible, rubber block die. This technique can be used when:

- the production run is small (under 1000 components);
- only shallow drawing or forming is required;
- the material is soft and thin (e.g. aluminium panels for aircraft skins);
- tooling costs must be kept to a minimum — the punch may be cast from reinforced epoxy resin.

When confined as shown in Fig. 2.31(*a*), the rubber acts like a fluid and transmits the pressure uniformly in all directions. Alternatively, a rubber diaphragm may be used, as shown in Fig. 2.31(*b*). The pressing pressure is transmitted to the diaphragm by fluid pressure. It is essential that the fluid pressure is kept constant throughout the operation and that the deformation of the diaphragm is restricted to prevent excessive wear leading to bursting. With both these techniques it is essential to avoid sharp corners and sudden changes of section.

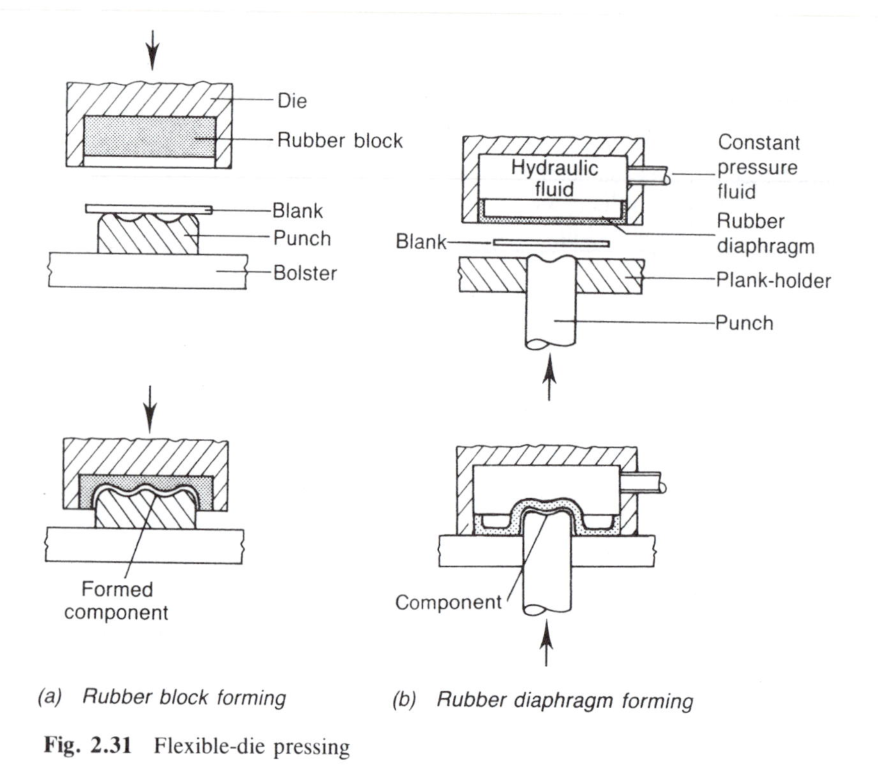

Fig. 2.31 Flexible-die pressing

2.22 Stretch forming

The principle of stretch forming is shown in Fig. 2.32. The sheet is placed in a state of tension and wrapped around a former. Normally, when a sheet is bent, one side is in tension and the other in compression. It is on the compression side where the material starts to buckle and pucker. In stretch forming, pre-stressing keeps both sides of the sheet in tension and prevents puckering from occurring. This process was developed so that large thin sheets could be formed economically in small quantities.

2.23 Spinning and flow turning

Metal spinning is used to produce symmetrical, round, hollow components from circular blanks with a surface area similar to that of the finished component. Unlike deep drawing, which requires very expensive tooling, spinning only requires a simple former. The blank is spun over the former by a skilled operator using specially-shaped stick tools. An example of a basic set-up is shown in Fig. 2.33. Since the metal flow depends upon the pressure exerted by the operator, the process is limited

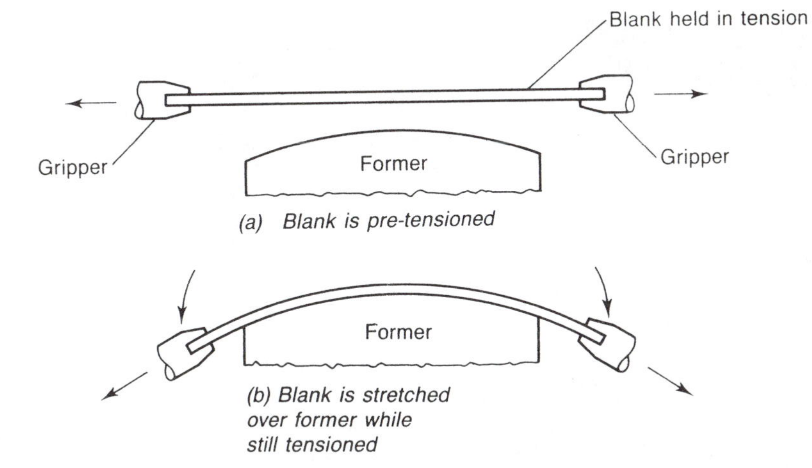

Fig. 2.32 Stretch forming

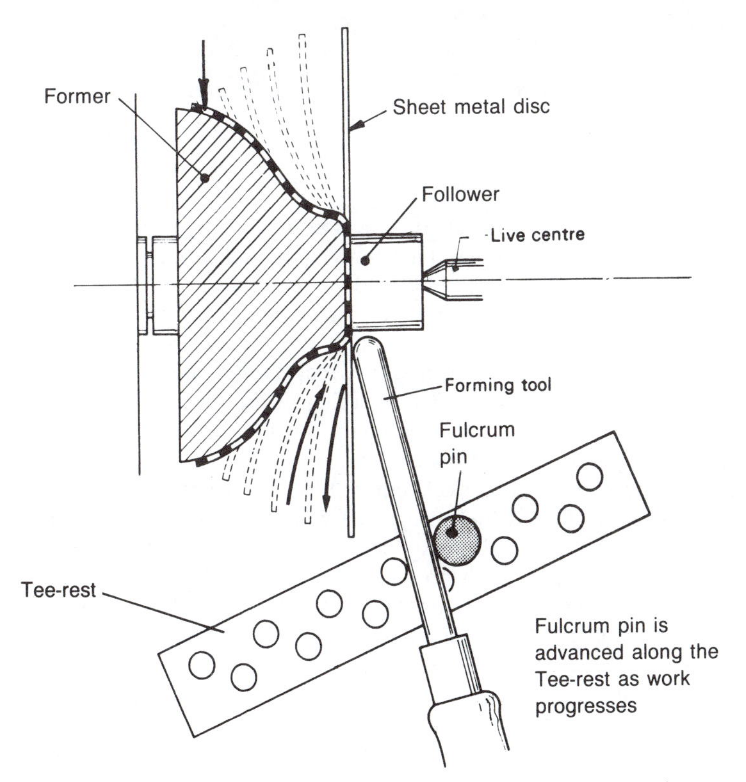

Fig. 2.33 The metal spinning process

Fig. 2.34 Shear spinning

to thin metal which is soft and ductile. It is also possible to internally spin from preformed cups on special internal spinning formers. Considerable material savings can be made using these methods as a skilled operator is capable of thinning or thickening up the material as and when required. The burnishing action of the spinning tools results in an exceptionally good surface finish which reduces the need for such finishing operations as polishing. The working of the metal also improves its mechanical properties and the high quality of the surface finish improves the fatigue life of the component.

Because of the high level of skill required by the operator, the unit labour cost of production is high compared with pressing. However, due to the very low tool costs and negligible setting costs involved in the spinning process compared with pressing, the overall unit cost of production becomes competitive for the production of single components and for small quantity batches. Spinning is also useful for producing components where the material strength and/or the blank thickness is inadequate to resist the forces encountered when deep drawing.

The principle of *shear spinning* is shown in Fig. 2.34 It can be seen that the stick tool used for manual spinning is replaced by a roller and that some extrusion of the metal, resulting in wall thinning, takes place. The products manufactured by shear spinning are as diverse as: hollow explosive charge units for underwater seismic exploration, rocket motor

Fig. 2.35 Flow turning

tubes, parabolic nose cones, pressure cylinders, brake cylinders, funnels, champagne buckets and wine coolers, thick-based thin-walled hollow-ware, accurate thin gauge seamless tube, and cylinder liners. Traditionally, shear-spun components were limited to cylindrical and conical shapes but with the advent of computer-controlled flow-turning lathes, more complex shapes can be produced. The principle of *flow turning* is shown in Fig. 2.35.

Both flow turning and shear spinning allow a greater degree of deformation of the material to be achieved than is possible by any other process. This is because deformation of the material only occurs at the point of contact between the rollers and the blank, and the remaining material remains stress free. Therefore, in many instances, components can be produced in one operation when they would require a number of operations with inter-stage annealing when produced by pressing. Due to the severe working of the material the crystal structure is elongated, increasing the strength and hardness of the material in the finished component. Thus the possibility of manufacturing a given component from a thinner blank can be considered, with a corresponding reduction in the cost of material required and a reduction in the mass of the component. Components up to 685 mm in diameter and up to 600 mm in length can be spun. The benefits of flow turning are: low tooling costs and short lead times, close dimensional accuracy, considerable savings in materials, and the effect of the process on the microstructure of the metal results in improved strength and hardness and a surface finish that requires little if any further finishing.

2.24 Advantages and limitations and the justification of flow-forming processes

Advantages

- Economical use of materials: there is no loss of volume, except for conventional hot-forging where the flash has to be clipped from the forged component.
- Improved mechanical properties due to controlled grain refinement and orientation.
- Cold-forging and warm-forging operations produce components with a high geometrical and dimensional accuracy and a good surface finish. Such components require little further finishing or machining to complete them.
- Many flow-forming processes lend themselves to full automation where large quantities of components are required.

Limitations

- Compared with casting, the stock material is in a highly processed condition and, therefore, more costly.
- Flow-forming processes are capital intensive as they have to be performed on large and expensive machines.
- Flow-forming processes generally require complex and expensive tooling which can only be justified for large batch or continuous production.
- Because of the complexity of the tooling, the production lead time can be lengthy.
- The down time for tool changing and setting can be relatively lengthy, thus frequent tool changing is uneconomical and flow forming is generally restricted to medium and large quantity batch production. The exceptions to this are open-die forging using standard smithying tools and spinning; however, these processes require the use of highly skilled and expensive operators. Note that the use of pre-set and modularised tooling substantially reduces the down time for tool changing.

Justification

When comparing flow-forming processes with such alternatives as casting or machining from the solid, a number of factors have to be considered. Flow-formed components have better mechanical properties than castings and components machined from the bar. Compared with machining, there is much less waste of material. For example, when turning a bolt from a hexagonal bar, roughly half the material is wasted as swarf. Compared with casting, the finish and accuracy of flow-formed components is superior with the exception of components produced by die-casting and investment casting. Further, die-casting can only be used with relatively low strength metals and alloys. Therefore, on an overall cost and performance basis, flow forming can be justified as follows despite the high cost of plant and tooling:

- Low loss of volume — economical use of stock material.
- Improved mechanical properties: for some components, such as wing spars for aircraft, precision forging is the only viable process that can provide components with the necessary strength/weight ratio and service reliability.
- High accuracy and a good finish reduces or eliminates machining and finishing operations, particularly when precision forging, so that the overall component cost is competitive with, say, a lower-cost sand casting that has to be fettled by sand or shot blasting and extensively machined.
- High rates of production — especially in automated plants — leads to low unit production costs providing that the costs can be defrayed over an economically-large number of components.

Assignments

1. Compare and contrast the advantages and limitations of *hot working* and *cold working* metals in terms of finish, properties, accuracy, grain structure, process costs.
2. With the aid of sketches, describe the processes of *open-die* forging and *closed-die* forging and compare and contrast the advantages and limitations of these two processes.
3. Compare and contrast the advantages of *cold forging* and conventional *hot forging* in terms of finish, accuracy, properties, grain structure, and process costs.
4. With the aid of sketches, describe the process of *warm forging* and explain the advantages of this process compared with *hot forging* and *cold forging*.
5. With the aid of sketches, describe the process of *thread rolling* and discuss its advantages compared with conventional thread cutting.
6. (a) Sketch a section through a 'cut and cup' deep drawing press tool to produce a cup 50 mm diameter and 20 mm deep from annealed brass strip 1.00 mm thick.
   (b) Calculate the approximate blank diameter.
7. Compare and contrast the advantages and limitations of flexible-die forming sheet metal components with rigid-die forming.
8. Describe, with the aid of sketches, the process of stretch forming sheet metal components and compare the advantages and limitations of this process with flexible-die forming.
9. Compare and contrast the advantages of metal spinning with deep

drawing in terms of finish, accuracy, process costs, and the properties of the finished component.

10. With the aid of sketches explain the essential differences between shear spinning and flow turning and give examples of components where each process would be appropriate, with reasons for your choice.

3 Sintered particulate processes

3.1 Basic Principles

As the name implies, sintered particulate processes are those processes by which components are formed by the compaction and sintering of particles of metals and metallic compounds. The study of these processes and the materials so formed are also referred to as *powder metallurgy*. Basically, the manufacture of components by means of sintered particulate processes involves a sequence of three operations.

(1) The production of a controlled blend of metals, non-metals and metallic compounds, in powder form, which will provide the required properties when formed and sintered into the finished product.
(2) The consolidation, or compacting, of the powder mixture in a die to form the unsintered or 'green' shape of the component. This will have the same shape as the finished component, but will allow for between 1 and 2 per cent shrinkage, by volume, for most processes.
(3) The sintering of the 'green' compact at high temperatures in a reducing-atmosphere furnace for several hours. Sintering is a heat treatment process carried out at between 60 per cent and 90 per cent of the melting point of the majority constituent powders. It causes the bonding together of the individual powder particles at their points of contact to provide a homogeneous solid. This results in a slight reduction in volume and a corresponding consolidation (increase in density) of the completed compact. Sintering is discussed more fully in section 3.9.

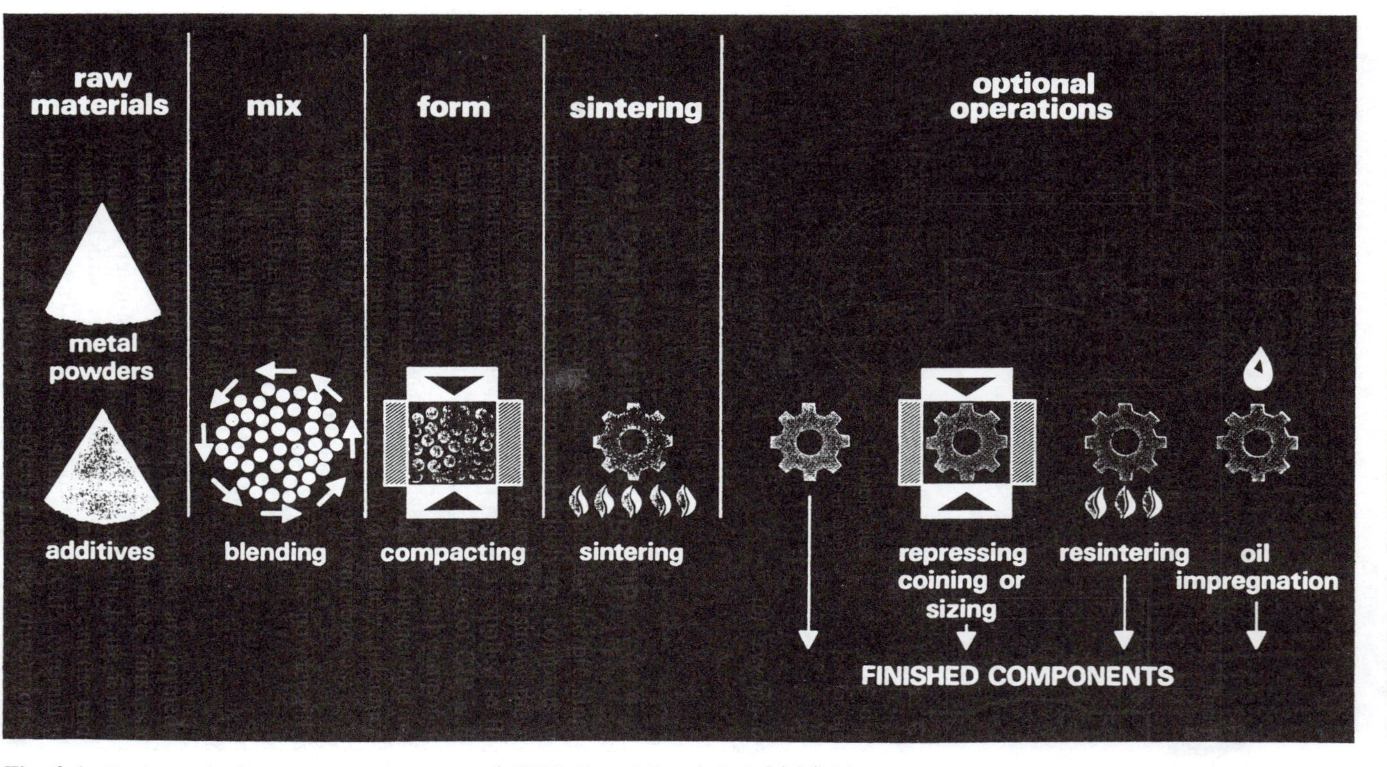

Fig. 3.1 Basic production processes (courtesy of GKN, Bound Brook Ltd, Lichfield)

Note that where close control of accuracy and mechanical properties are required, some post-sintering operations may have to be performed. These are discussed in sections 3.11 and 3.12. Figure 3.1 shows the basic production processes for making sintered components.

From the foregoing outline, it might well seem that the process of manufacturing components by the sintering of metal powders must be complex and costly, and only applicable to highly specialised components. Nothing could be further from the truth. Sintered components range from gears for washing machine components and for adjusting the tilt of car seats to highly stressed aircraft components; from small spacing washers and oil-impregnated bearing bushes to aircraft undercarriage components. The process is also used for producing hard 'tips' for cutting tools. Since the process can be readily and highly automated both for the actual manufacture and for in-process quality control, the common factor between all the examples quoted is batch size. Providing the batch size warrants the cost of making and setting the compaction tools, the process can compete on a cost basis with casting, forging, pressing and machining, because the accuracy and finish of the sintered component often removes the necessity for further processing. As a rough guide, a minimum batch size for a small and simple component would be 5000 components per batch with a minimum of four batches per year.

The limitation of component size is set by the maximum capacity of compaction press availability. For example, a 150 mm diameter sprocket wheel made from an iron–carbon powder mixture would require a compacting press capable of exerting a pressure of 925 MN/m^2, compared with a press capable of exerting 1.850 GN/m^2, which would be required if the wheel was made from an alloy steel powder. In the UK, the largest diameter component normally produced is approximately 250 mm in diameter with a mass not exceeding 2.5 kg. By introducing 'cut-outs' (holes) into the component, as shown in Fig. 3.2, the projected area of the component can be reduced. This, in turn, reduces the compaction pressure required for a given component diameter.

It has already been mentioned that one of the main cost advantages of making components by a sintered particulate process, lies not in the process itself but in the fact that the finish and accuracy is so good that subsequent machining can be largely eliminated, as the following example shows. Figure 3.3 shows a conveyor chain link that was originally made from a forging. After the forging had been clipped to remove the flash and sand-blasted to remove the scale, it had to be milled on faces A and B and then cross-drilled. The time came when the milling machine and tooling required replacement, but the value of the component was insufficient to warrant such an investment. Fortunately, the links were — and still are — required in sufficient quantities to warrant manufacture by a sintered powder process. After slight modifications to the component to aid compaction, the links are now made to an accuracy and finish that has eliminated the need for milling faces A and B. The only post-sintering machining required is the cross-drilling of the holes. Although

Fig. 3.2 Reduction of projected area (courtesy of GKN, Bound Brook Ltd, Lichfield)

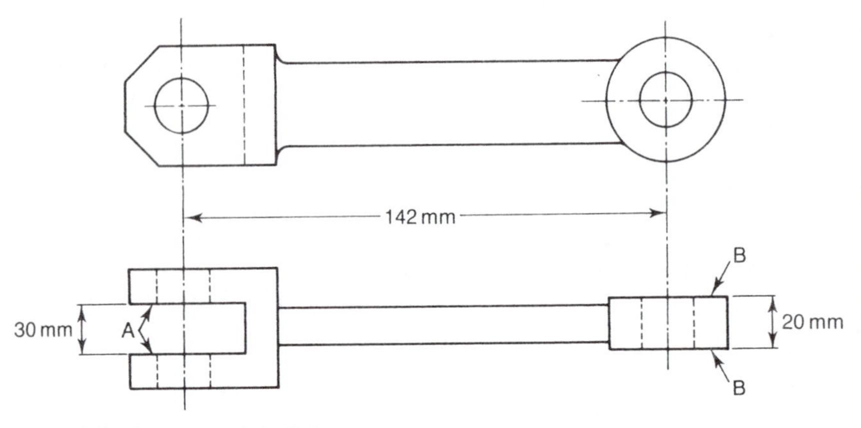

Fig. 3.3 Conveyor chain link

the cost of the sintered links is higher than the cost of the original forgings, the elimination of the milling and surface finishing processes has resulted in the overall unit cost being lower. Further, the capital outlay on new milling equipment was avoided.

3.2 Powder production

Sintered particulate processes can be applied to most metallic elements and many metallic compounds. The most notable exception being aluminium which has an affinity for steel and, therefore, tends to cold-weld itself to the die walls during the compaction process. If aluminium or aluminium alloys have to be produced by powder particulate processes then the compaction dies have to be manufactured from materials such as tungsten carbide for which the aluminium does not have an affinity. Such tools are very much more costly than tools made from alloy die-steels and the batch size has to be sufficiently large to warrant such high tooling costs. The basic properties of the individual powder particles depend not only upon their composition but also upon the process by which they were manufactured. These basic properties affect their behaviour during consolidation. Powders are produced by the following processes:

- The chemical reduction of metal oxides by carbon or by hydrogen gas results in a 'spongy' type of powder with good compaction and sintering properties. For example, sponge iron powder is widely used for the manufacture of ferrous components. The purity depends upon the quality of the metal oxides and the reducing agent used.
- Metal powders can also be produced by atomisation of liquid metal using water or an inert gas as the quenching (cooling) agent. Atomisation is akin to the metal spraying technique used for building up worn components. However, the atomised metal is sprayed into the quenching agent instead of onto a worn component. Atomisation can be used with most metals and is the most widely used commercial process for producing alloy powders.
- Metallic compound particles are usually produced by grinding these materials in atritor mills.

3.3 Selection and blending of powders

High compressibility is a desirable feature in all powders, since it allows more efficient use of the pressing (compacting) equipment with corresponding reductions in pressing forces and die wear. A die lubricant such as a metal stearate is usually added to improve the particle flow in the die and reduce die wear. Unfortunately, the lubricant has to be removed before sintering to prevent contamination and this additional operation adds a slight cost to the process. Under uni-directional compaction forces — which exist in press and die compaction — the addition of a lubricant improves the uniformity of the compaction throughout the mass of the component, reduces friction between the particles and the die wall and eases ejection of the compacted component.

Single powders

Refractory metals such as molybdenum, tantalum and tungsten are very expensive and difficult to melt and cast by conventional means due to their high melting points. It is substantially cheaper to produce components from these materials by sintering powder compacts, provided there is sufficient demand for any given product to warrant the relatively high tooling costs. In these instances, single powders of a given metal are used and no blending is required. Sintered bronze bearings and bearing shells have the advantage of being porous and capable of being impregnated with lubricant. Such bearings are invaluable in 'sealed for life' installations and other applications where the regular lubrication of conventional bronze bearings is not convenient.

Metal/metal-compound two-phase systems

A typical example of a metal/metal-compound two-phase system is the cobalt/tungsten carbide mixture widely used for cutting tool tips. When these powders are sintered, hard brittle particles of tungsten carbide form, dispersed within a softer but tougher matrix of metallic cobalt, as shown in Fig. 3.4.

Alloy mixtures

Sintered particulate processes permit alloys to be achieved between metals that are completely or partially immiscible in the liquid and solid phases. If such metals are melted and cast together they will not give a satisfactory structure but will separate out into individual layers. However, by mixing together particles of the alloying elements, and sintering them, the required alloy will be formed by dispersion, as shown in Fig. 3.5.

Manufacturers of sintered metal products buy in the powder in drums ready for blending and mixing. Weighed quantities of the powders are loaded into double-cone mixing machines together with a lubricant and any other necessary additives. Mixing must be sufficiently thorough to

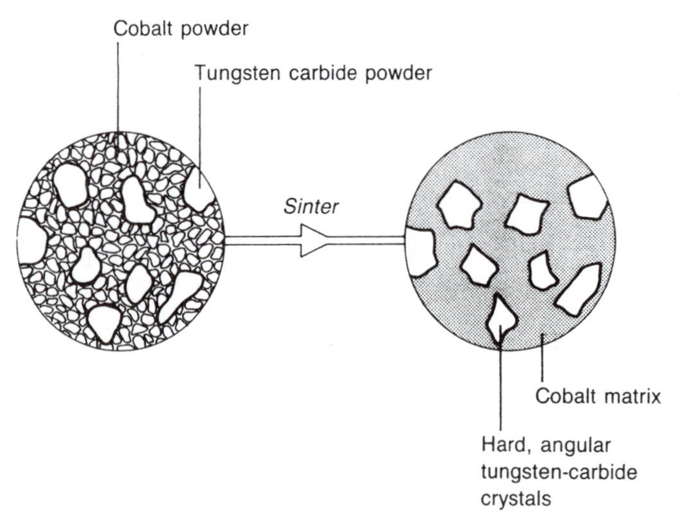

Fig. 3.4 Metal/metal-compound system

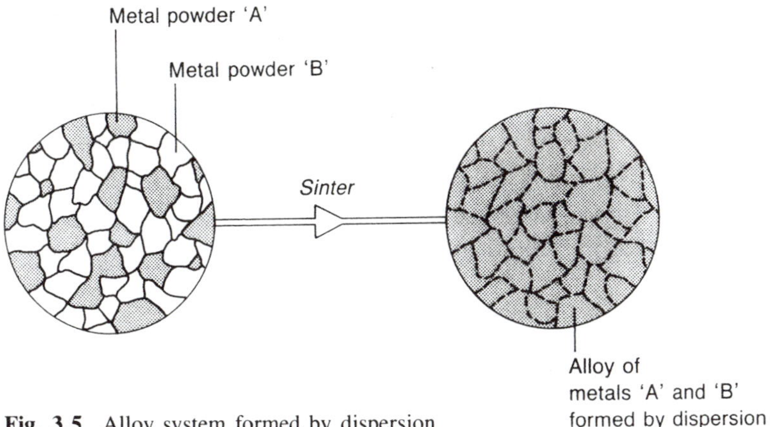

Fig. 3.5 Alloy system formed by dispersion

ensure homogeneity but, at the same time, mixing must not be excessive or the particle size and compacting characteristics of the powder will be adversely affected, as will be the mechanical properties of the component produced from the powder.

When the blend has been thoroughly mixed it is tested for two important criteria. The apparent density (AD) and the flow rate (FR). The processing and properties of a sintered component are dependent upon the characteristics of the powder, particle size, size distribution, and particle structure, shape and surface condition. The apparent density of the powder (mass of a given volume) strongly influences the strength of the compact prior to sintering. A powder with irregular particles and a porous texture will have a low apparent density and this will give rise to a large volume reduction during pressing. This, in turn, gives rise to a greater degree of cold-welding during compaction and greater green strength. It also leads to more efficient sintering.

However, irregular particles with a low apparent density also have a low flow rate and will not readily fill the die cavity and this may lead to bridging-density variations and slower rates of production. The greater reduction in volume associated with irreguler particle shapes usually requires greater compaction pressures, resulting in greater tool loads and the need for more powerful presses.

The ease and efficiency of packing the powder in the dies depends largely upon the range of particle sizes in the powder, with the voids between the large particles being filled with progressively finer particles. While an excess of 'fines' increases the density and improves the mechanical properties after sintering, too many fines reduce the flow rate and increase the compaction pressure necessary for a given size of component. Excessive mixing tends to break up the particles and increases the percentage of fines present in the blend.

The purity of the particles in a powder is also critical, particularly where the mechanical properties of the sintered component have to be closely controlled. Most powder grains have a thin oxide film on their surfaces, but these films are ruptured during compaction to leave clean and chemically-active metal surfaces which are easily cold-welded.

3.4 Compaction of the preform

Compaction of the particles into the required preform shapes can be performed by:

- closed-die pressing,
- cold isostatic pressing (CIP),
- hot isostatic pressing (HIP),
- roll compaction,
- extrusion,
- slip casting, and
- injection moulding.

The majority of preforms are compacted in closed dies using mechanical or hydraulic presses. Of the other processes listed above, only isostatic pressing is used to a significant extent commercially. Closed-die pressing and isostatic pressing techniques will now be considered in greater detail.

3.5 Closed-die pressing

The principle of compaction by closed-die pressing is shown in Fig. 3.6. A controlled amount of powder is automatically fed into the die cavity as shown in Fig. 3.6(*a*). Counter-punches close on the powder and compact it into the required shape, as shown in Fig. 3.6(*b*). Finally, the lower punch ejects the preform from the die, as shown in Fig. 3.6(*c*). As the punches close on the powder, compaction proceeds through a number of overlapping stages. These are:

(1) slippage of the powder particles with little deformation,
(2) elastic compression of the particle contact points,
(3) plastic deformation of the particle contact points to form contact areas of greater size, and
(4) bulk elastic and plastic deformation of the entire mass of the preform.

During die compaction, high shearing stresses are generated on those powder particles adjacent to the punch faces and the die walls. This results in more efficient compaction at the surfaces of the preform than within its body. Therefore counter-punches are used, as previously mentioned, to give a better density distribution to the preform. Figure 3.7 shows a comparison of the densification mechanism of a single punch compared with a counter-punch system. Under good conditions, up to 90 per cent of the theoretical maximum density can be achieved in closed-die compaction. However, the die-wall friction restricts the length to width ratio of a closed-die preform to 2.5:1 even when a lubricant is used.

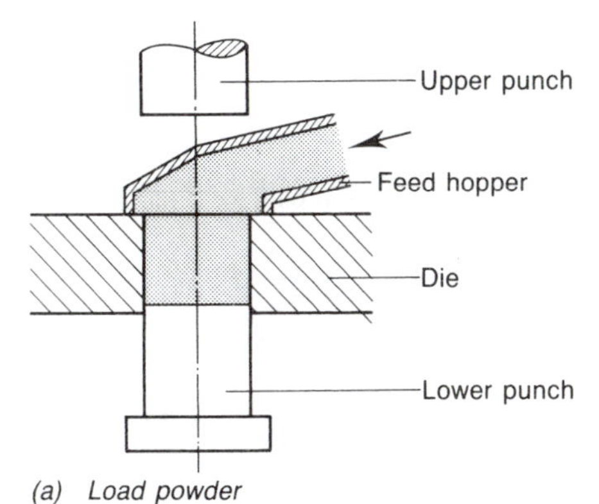

(a) Load powder

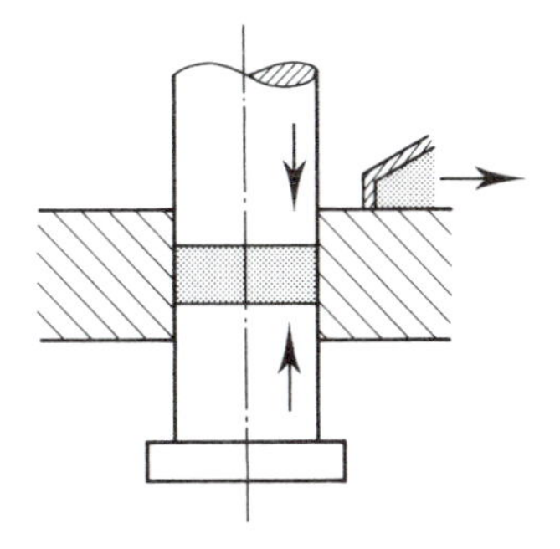

(b) Powder compaction

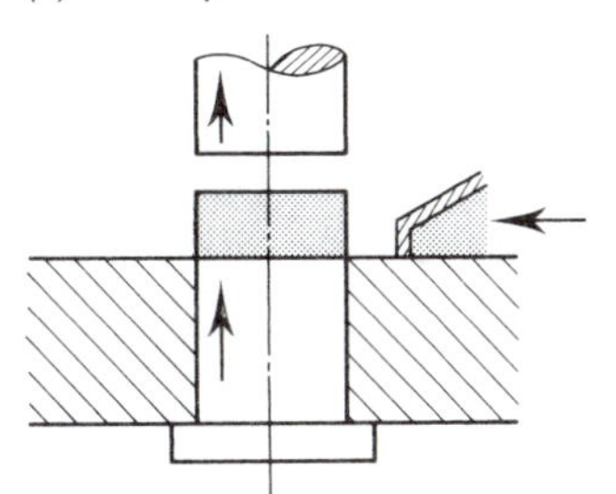

(c) Component ejection

Fig. 3.6 Closed-die pressing

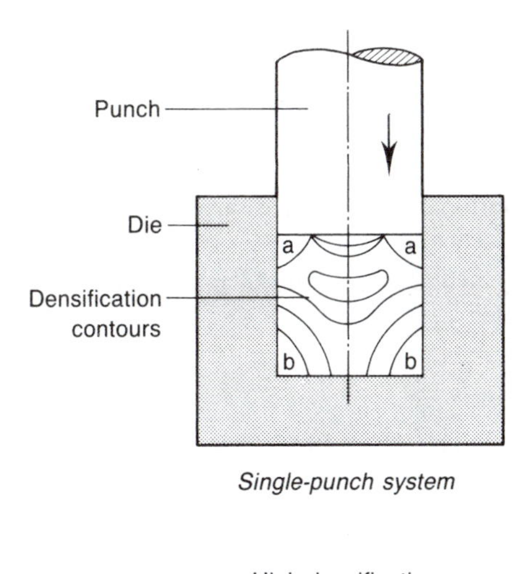

Single-punch system

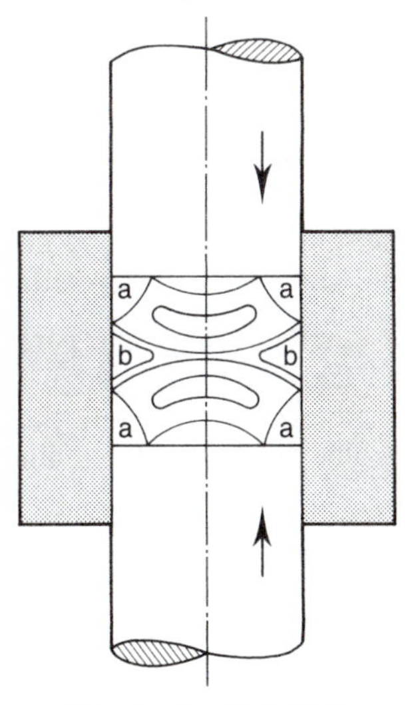

Counter-punch system

a = High-densification zone
b = Low-densification zone

Fig. 3.7 Densification

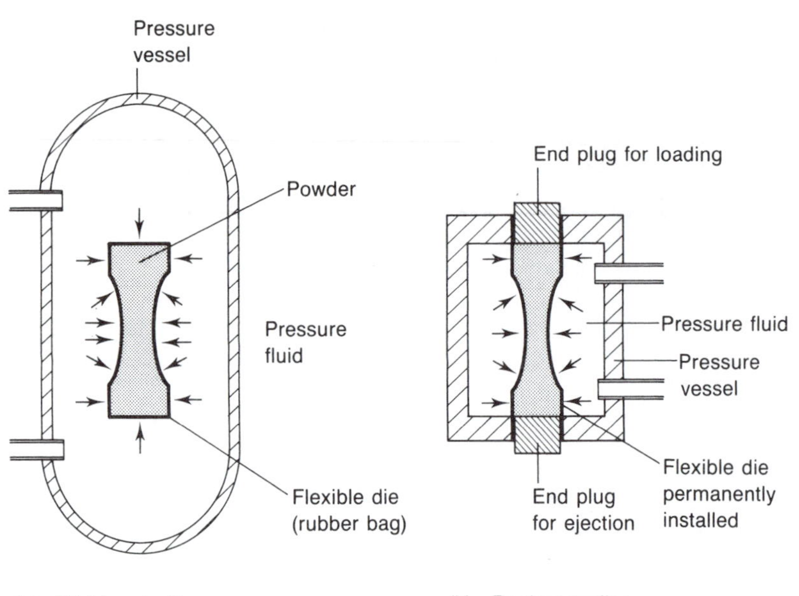

Fig. 3.8 Cold isostatic pressing

3.6 Cold isostatic pressing (CIP)

Cold isostatic pressing was developed for the pressing of fine, non-ductile particles and is now being used widely for compacts whose shape precludes the use of rigid dies. Figure 3.8 shows the principle of cold isostatic pressing (CIP) where pressure is applied to flexible tooling at ambient temperatures. The main advantage of CIP is that more complex component design geometry can be used and components of increased size can be produced. Density levels are higher and more uniformly distributed than for rigid dies. Unfortunately, tolerances are less tightly held with CIP and depend upon such variables as:

- the uniformity and particle size of the powder fill in the tooling,
- the type of powder.
- the compaction ratio, and
- the rigidity of the tooling.

Since the external surfaces of the compact are formed by a flexible membrane, close tolerances and good surface finishes cannot be produced. However, internal surfaces which are formed by a rigid mandrel can be held to tight tolerances, as can the ends of the compacts under certain conditions. Usually, with CIP, external tolerances cannot be held to better than 3 per cent of the linear dimensions and closer control requires machining after compaction or sintering.

With 'wet-bag' tooling, as shown in Fig. 3.8(*a*), flexible rubber, neoprene, PVC, or polyeurethane bags may be used. After filling, the tooling (i.e. the bags) is sealed and placed in a pressure chamber. Several bags may be pressed in a single pressure chamber at the same time using a high-pressure, incompressible fluid such as water. With 'dry-bag' tooling, as shown in Fig. 3.8(*b*), the bag or tool is fixed in a pressure vessel and only one component is produced at a time. However, tool design is simpler, loading and ejection is easier and, with a high durability membrane, high volume production runs are possible with outputs of up to 1500 components per hour.

3.7 Hot isostatic pressing (HIP)

Hot isostatic pressing combines the processes of pressing and sintering. Although the associated equipment involves high capital expenditure, HIP has many advantages in terms of improved metallurgical control of structure and quality. Full theoretical densities can be achieved and, since the compacting pressure is applied at lower temperatures than those associated with conventional sintering, the components are free from massive secondary phases. Densification by HIP not only eliminates porosity, which improves the mechanical properties of the finished component; it can reduce scatter in the material properties, resulting in closer quality control. Closer control of grain size in the microstructure of the finished compact allows subsequent processing operations such as rolling, hot-forging, and heat treatment to be successfully carried out. In such cases the particulate process is used to create alloys between immiscible elements, and the compact is only a blank for further processing.

Hot isostatic pressing is successfully applied to a wide range of high-alloy systems where the high temperatures that would be required for conventional sintering would not only be uneconomical but would result in degradation of the grain structure of the finished compact. HIP is used to process tool steels, nickel-based superalloys, 'hard metals' and some complex alloys based upon titanium and beryllium which cannot be processed economically by other and more conventional techniques.

3.8 Injection moulding

Injection moulding is more widely used in the USA than in the UK. The metal powder blend is added to an organic binder to form a material suitable for plastic injection moulding. Plastic injection moulding is discussed in section 4.8. The powder/polymer mixture is then injected into split dies, pre-heated to remove the binder and, finally, sintered. Volumetric shrinkage during sintering is very high, some 20 per cent, but this is consistent and predictable and can easily be allowed for in the design of the moulding dies. This process allows shapes to be produced which are impossible with conventional compaction in rigid dies.

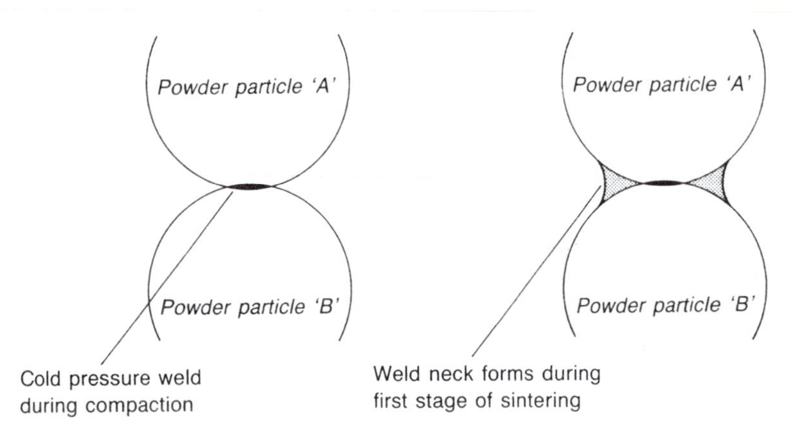

Fig. 3.9 Formation of a 'weld-neck'

3.9 Sintering

This is a heat treatment process carried out at between 60 per cent and 90 per cent of the melting temperature of the majority powder present. However, in a polyphase system, one or more of the minority materials may become molten and will infiltrate between the particles of the solid material by capillary attraction. The sintering process occurs in three stages.

(1) The sintering process begins with rapid neck growth between particles which remain essentially discrete. This is shown in Fig. 3.9 where it can be seen that the initial cold-weld caused by powder compaction is extended in area. As the weld neck grows there is little change in the density of the compact. Thus, at this stage in the sintering process, it is possible to control the porosity of the component. For example, porous bearings would receive no further sintering beyond this stage.
(2) During the second stage, optimum densification with recrystallisation and particle diffusion occurs. This is the most important stage in the sintering process for the majority of components. At the same time, the volume of the compact is reduced by some 1 to 2 per cent with a corresponding increase in density and reduction in permeability of the structure. The structure, and therefore the properties of the sintered compact, can be significantly affected by the temperature and atmosphere of the furnace and the process time. The properties achieved during sintering can also be modified by the physical and chemical treatment of the powder prior to pressing and by the introduction of reactive gases into the furnace atmosphere during sintering.
(3) During the final stage, the spheroidisation of isolated pores occurs with some further increase in densification. These are the result of

surface tension effects and continuing diffusion which, although slower than at stage 2, produces components that are homogeneous. Such components have mechanical and physical properties equal to, or in some instances superior to, similar components produced by conventional melting and casting processes.

3.10 Sintering furnaces

Sintering furnaces may be heated by gas, oil or electricity. The components may be carried through the furnace on a wire mesh belt made from a non-reactive, high melting point alloy such as inconel. Sintering is a relatively slow process measured in hours rather than minutes. The furnace atmosphere is given reducing characteristics in order to avoid the possibility of oxidation. For example, metal carbides are usually sintered in an atmosphere of hydrogen at a temperature of 1350°C. Alternatively, 'cracked' ammonia gas can be used. Although very much cheaper than hydrogen, it contains 25 per cent nitrogen and, therefore, can only be used with compact materials that are not susceptible to the formation of nitrides which would degrade their properties. The least costly and most widely used atmospheres for the commercial-quantity production of components made from the less exotic materials are made from hydrocarbon gases such as methane (natural gas). The hydrocarbon gas is 'cracked' by heating it at a high temperature with insufficient oxygen to complete combustion. This gives an atmosphere containing up to 45 per cent hydrogen. Unfortunately, contaminants such as nitrogen, carbon dioxide and water vapour will be present. The furnace has a number of chambers.

Pre-heating chamber

The pre-heating chamber is used to burn off any lubricant that has been added to the powder as an aid to pressing. This 'de-waxing' stage is critical for, should any residual lubricant be present when the compact is raised to the full sintering temperature, the lubricant will boil off causing porosity and blistering the surface of the component as well as contaminating the furnace atmosphere.

Sintering chamber

In the sintering chamber the compact is raised to the full sintering temperature in an appropriately-controlled atmosphere. The process time is controlled by the conveyor belt speed. Automatic temperature control to very fine limits is essential, as is uniformity of temperature within the sintering chamber, which must be devoid of hot or cold spots.

Carbon control chamber

When sintering compacts made from materials such as ferrous metal powders, decarburisation can occur at the high sintering temperatures met with in the previous chamber. This carbon can be restored by the

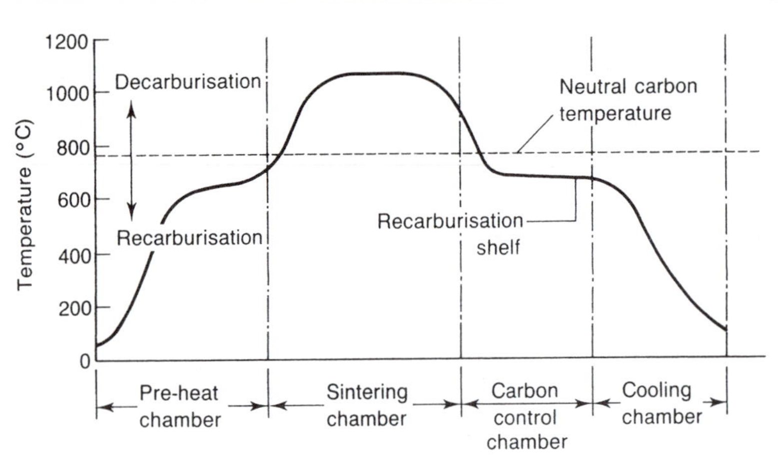

Fig. 3.10 Process temperature profile

incorporation of a carbon-control chamber operating at a lower temperature (carbon control shelf) following the sintering chamber and containing a carbon-rich atmosphere such as carbon monoxide.

Cooling chamber

It is essential to maintain atmosphere control until the components have cooled down below the temperature at which they will react with the normal atmospheric gases outside the furnace. The rate of cooling also affects the properties of the component as in any other heat treatment process. The rate of cooling can be controlled by the speed of the conveyor belt and by the rate at which the controlled-atmosphere gases for this chamber are circulated through the heat exchanger where they are cooled. A typical temperature-process chart for a ferrous powder compact is shown in Fig. 3.10.

3.11 Post-sintering treatment (sizing)

The dimensional tolerances of sintered components can be held within 0.1 mm per 25 mm in the transverse direction and within 0.2 to 0.25 mm per 25 mm in the direction of pressing when compacted by the closed-die process. Nevertheless, since sintering is carried out at high temperatures, some distortion and dimensional inaccuracy will occur. When closer tolerances are required, sizing will be required after sintering. This is a re-pressing process, similar to coining, which is performed by squeezing the sintered component in dies using mechanical or hydraulic presses. The dies are similar to those used for the initial compaction but without any allowance for densification. High rates of production are achieved for the sizing process, especially if automatically-fed presses are used. After sizing, the component tolerance for linear dimensions can be within 0.02 mm per 25 mm in the transverse direction and 0.15 mm per 25 mm in the direction of pressing.

3.12 Post-sintering treatment (forging)

For this post-sintering treatment the preform can be compacted by closed-die or isostatic processes. The compact is then sintered to a density of 85 to 90 per cent of the theoretical maximum. The sintering process both improves the strength of the compact and reduces the included oxygen content from about 2000 ppm to below 200 ppm, thus improving the ductility and the impact strength of the compact to that approaching wrought metal.

The preform is then used as a 'billet' in a conventional hot-forging operation in closed dies. This improves the densification of the sintered compact up to the theoretical maximum and a high-quality forging of great precision is produced. The mechanical properties achieved are comparable with conventional forging but the weight control and distribution is very much better, eliminating balancing operations for engine components used in the automobile industry. Other advantages of this process are:

- more economical use of material since no flash is produced;
- increased productivity since flash clipping is eliminated;
- tolerances can be held to closer limits than with conventional hot-forging;
- forging flow patterns are not introduced into the materials resulting in isotropic properties.

Unfortunately, the process is capital intensive and only economical for the continuous production of large numbers of the same component, or where the strength and quality achieved outways considerations of cost as in some highly-stressed, key components in the aircraft and automotive industries.

3.13 Components for sintered particulate processes

The most important consideration in component design geometry is that the component must be capable of ejection from a solid die. Thus conventional casting features such as undercuts, reverse tapers, cross-holes, threads, and re-entrant angles are not feasible when compaction takes place in closed dies. However, a more flexible approach to component design is possible when isostatic pressing is used. Unfortunately, production costs are higher when using isostatic pressing. Figure 3.11 shows some typical components that may be produced by powder particulate processes. It can be seen that such features as holes, slots and teeth can be produced readily in the direction of pressing, but that a minimum size limit is set by the rigidity and strength of the tool

Fig. 3.11 Typical powder metal components (courtesty GKN, Bound Brook Ltd, Lichfield)

elements. Similarly, while parts with multiple steps can be produced, thin walls must be also avoided and, where changes in the profile occur, feather edges in the tooling must be avoided.

Structural materials

Structural materials such as bronze, carbon steels and alloy steels have already been introduced. These are the most widely used materials and are found in components where mechanical strength is the main requirement.

Porous bearings and filters

The controlled porosity that can be achieved in sintered components is put to good use in porous, oil-impregnated bearings and filters. Mixtures of tin and copper powders form bronze alloys by diffusion during processing. Alternatively, pre-alloyed bronze powders may be used. Tin–copper mixtures tend to expand during sintering, while pre-alloyed powders tend to shrink. By using a combination of the two techniques, volume changes during sintering can be avoided. Sometimes graphite is added to help control the porosity and act as a lubricant in its own right. The porosity of the bearing is approximately 20 to 25 per cent by volume. The rate of oil supply automatically increases with temperature and therefore, with speed of rotation of the shaft. Thus an impregnated bearing system is self-adjusting. For heavy-duty applications, non-porous steel-backed bearings are also made by sintering.

Sintered metal filters and diaphragms are also made. These are stronger than their ceramic counterparts. As well as separating solids from liquids, they can be used to separate liquids having different surface tensions. In this way they can separate water from jet engine fuel as well as filtering out solid impurities.

Friction materials

Sintered metal friction materials are used for heavy duty brake and clutch linings and disc brake pads. The metal matrix contains non-metallic friction materials such as carborundum, silica, alumina and emery. Asbestos is no longer included for safety reasons. Such friction materials can operate at temperature up to 800°C where a copper matrix is employed, and up to 1000°C where an iron matrix is employed. These are well above the operating temperature for resin-bonded friction materials, and even at lower operating temperatures, the wear rate of sintered friction materials is lower than for resin-bonded materials.

Electrical and magnetic components

Electrical and magnetic components often require materials with composite properties that can only be achieved by using particulate processes. For example, commutator or slip-ring brushes require a combination of the high contact conductivity of silver or copper with the strength, heat resistance and resistance to arc-erosion of tungsten, and the lubricating properties of graphite.

High performance 'hard' and 'soft' magnetic materials can be made by sintering. Such magnetic components have the equivalent magnetic properties of cast alloys but with a finer grain and better mechanical properties. Further, the high finish and dimensional accuracy of sintered magnets and cores, compared with similar cast components, eliminates much of the need for machining. This is important since some of the high performance 'hard' magnetic materials are almost impossible to machine — some can only be ground.

3.14 Die design

The high wall pressures generated during compaction (approximately 1500 MPa) necessitate the use of solid dies and punches for closed-die techniques, as shown in Fig. 3.12. As has already been mentioned, counter-punches are used to achieve uniform compaction. The powdered

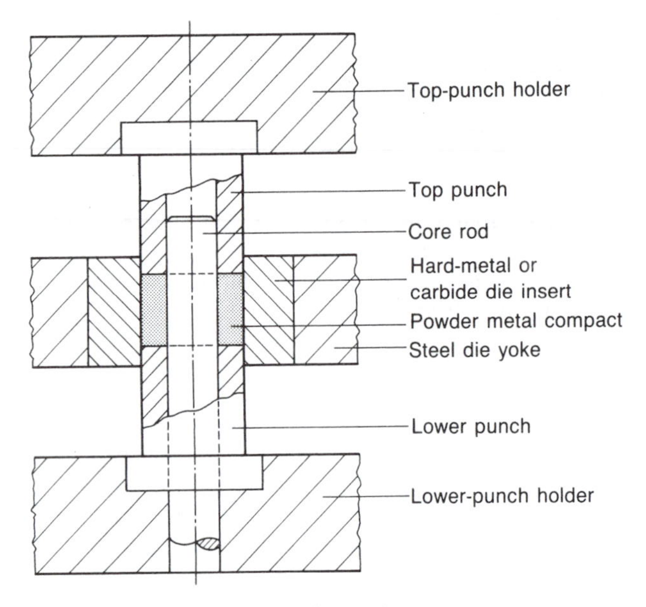

Fig. 3.12 Solid-die compaction tool

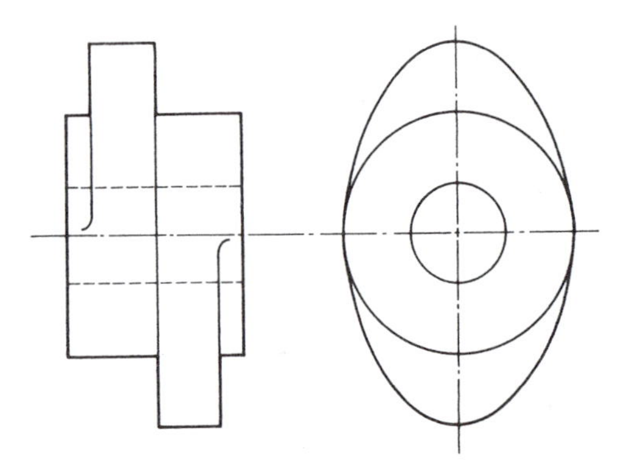

Fig. 3.13 Component requiring split-die tooling

materials are very abrasive and frictional conditions in the die are very severe. For this reason the die is often made from tungsten carbide set in a high-tensile steel yoke to prevent it from bursting. Split die, or double-die, techniques have been developed to permit components with simple offsets and projections to be manufactured in rigid dies. Figure 3.13 shows such a component, while Fig. 3.14 shows details of the double-die required for its manufacture.

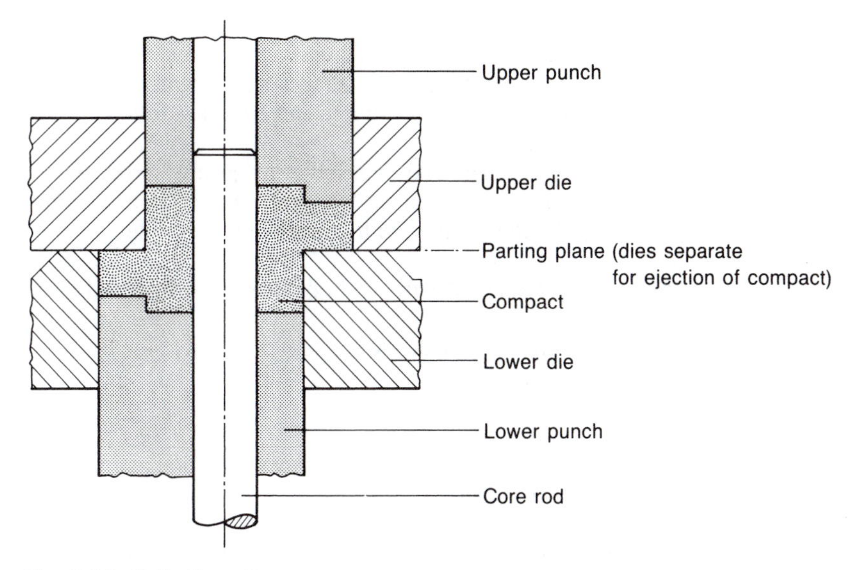

Fig. 3.14 Split-die tooling

3.15 Advantages and limitations of sintered particulate processes

Like any other processes, sintered particulate processes have a number of unique advantages and limitations which may be summarised as follows:

Advantages

- Sintered particulate processes have the ability to form components from refractory metals such as molybdenum, tantalum and tungsten, whose high melting points preclude normal casting processes.
- Materials whose phases are normally immiscible, and which will not form solutions under normal melting and casting processes, can be made to form alloy systems by the intimate mixture of particles of the materials prior to compaction and sintering.
- The porosity and permeability of components can be controlled during the compaction and sintering processes so as to produce such parts as porous metal bearings and high-pressure filter elements.
- Alloy systems can be controlled to very fine limits compared with conventional melting and casting. This is achieved by accurate mixing of the constituent particles prior to compaction and sintering. Alloys such as controlled-expansion nickel alloys, high-speed steels, superalloys, and 'hard' metals can be produced in this way.
- The microstructure of the alloy can also be accurately controlled where grain orientation is important, as in permanent magnet materials requiring high domain alignment.
- Components made by sintered particulate processes can be produced

to very close tolerances and high standards of surface finish compared with conventional melting and casting techniques. They can be made in a wide variety of sizes and shapes and, if correctly designed, need undergo little or no subsequent finishing following compaction and sintering.
- Sintered particulate processes lend themselves to quantity production and automation which helps to reduce the production costs and makes such processes highly competitive with conventional forming processes.
- The quality control of sintered particulate products can be readily automated.
- The raw material, being in powder form, is more easily stored and handled than bar stock.

Limitations
- The process is capital intensive, requiring expensive compaction equipment and sintering furnaces.
- The cost of compaction dies is high when rigid die techniques are used. Also extended lead times are involved, compared with the manufacture of casting patterns, while the dies are being made.
- The cost of the powdered materials is relatively high, although this is, to some extent, offset by economical usage. There is little waste compared with the clipped flash of forgings and the swarf produced during machining. Further, any scrap produced during setting up the compaction process can be ground down again into powder and re-used to a limited extent (a maximum of 10 per cent of the total mix) providing the compacts have not been sintered.
- The energy costs of the sintering process are high.
- There are limitations of size due to the high compaction pressures required.
- Limitations on component design are necessary to avoid problems of compaction die design and manufacture.
- Long runs are required to recover the tooling costs.

Assignments
1. By means of a flow diagram, outline the basic principles of component production by powder metallurgy techniques.
2. Describe, with the aid of sketches, the production of a plain bronze bush by a powder particulate process.
3. Describe:
   (a) how the powders are produced;
   (b) how the powders are selected and blended.
4. With the aid of sketches, describe in detail any THREE of the following compaction processes:
   (a) closed die pressing;
   (b) cold isostatic pressing;

(c) hot isostatic pressing;
(d) roll compaction;
(e) extrusion;
(f) slip casting;
(g) injection moulding.

5. With the aid of diagrams, discuss the essential requirements of a sintering furnace.
6. Explain what is meant by 'Post Sintering Treatment' and how it affects the finish, accuracy and properties of sintered components.
7. With the aid of sketches, discuss the principles of die design for a component of your choice.
8. Discuss the ecomomics of the production of a component of your choice by powder particulate techniques compared with conventional forging or casting followed by machining.

4 Plastic moulding processes

4.1 Plastic moulding materials

The moulding process selected depends upon two main factors: the geometry of the component to be moulded and the material from which it is to be made. There are two main groups of plastic moulding materials: *thermoplastic materials* which soften every time they are heated, and *thermosetting plastic materials* which undergo a chemical change during moulding and can never be softened again by heating.

Thermoplastics

Since thermoplastics soften every time they are heated, they can be recycled and re-shaped any number of times. This makes them environmentally attractive. However, some degradation occurs if they are overheated or heated too often and recycled materials should only be used for lightly-stressed components. Some typical thermoplastic materials are listed in Table 4.1.

Thermosetting plastics

Thermosetting plastics differ from the thermoplastic materials in that polymerisation (the forming of polymers — 'curing') is completed during the moulding process and, from then on, the material can never be softened again. Some typical examples are given in Table 4.2.

Table 4.1 Typical thermoplastic materials

Type	Material	Characteristics
Cellulose plastics	Nitrocellulose	Materials of the 'celluloid' type are tough and water resistant. They are available in all forms except moulding powders. They cannot be moulded because of their flammability
	Cellulose acetate	This is much less flammable than the above. It is used for tool handles and electrical goods
Vinyl plastics	Polythene	This is a simple material that is weak, easy to mould and has good electrical properties. It is used for insulation and for packaging
	Polypropylene	This is rather more complicated than polythene and has better strength
	Polystyrene	Polystyrene is cheap, and can be easily moulded. It has a good strength, but it is rigid and brittle, and crazes and yellows with age
	Polyvinyl chloride (PVC)	This is tough, rubbery, and practically non-flammable. It is cheap and can be easily manipulated; it has good electrical properties
Acrylics (made from an acrylic acid)	Polymethyl methacrylate	Materials of the Perspex type have excellent light transmission, are tough and non-splintering, and can be easily bent and shaped
Polyamides (short carbon chains that are connected by amide groups — NHCO)	Nylon	This is used as a fibre or as a wax-like moulding material. It is fluid at its moulding temperature, but tough, and has a low coefficient of friction at room temperature
Fluorine plastics	Polytetrafluoro-ethylene (PTFE)	This is a wax-like moulding material; it has an extremely low coefficient of friction. It is relatively expensive
Polyesters (when an alcohol combines with an acid, an 'ester' is produced)	Polyethylene-terephthalate	This is available as a film or as 'Terylene'. The film is an excellent electrical insulator

Table 4.2 Typical thermosetting plastic materials

Material	**Characteristics**
Phenolic resins and powders	These are used for dark-coloured parts because the basic resin tends to become discoloured. These are heat-curing materials
Amino (containing nitrogen) resins and powders	These are colourless and can be coloured if required; they can be strengthened by using paper-pulp fillers, and used in thin sections
Polyester resins	Polyester chains can be cross-linked by using a monomer such as styrene; these resins are used in the production of glass-fibre laminates
Epoxy resins	These are also used in the production of glass-fibre laminates and as adhesives

4.2 Polymerisation

The difference between thermoplastics and thermosetting plastics is the way in which polymerisation occurs. In thermoplastics, polymerisation occurs by the *addition of monomers*. For example, to manufacture such a plastic material, an alkane such as methane, ethane or propane has first to be converted into its corresponding olefin. This olefin is referred to as a 'chemical intermediate' since it lies between the monomer and the polymer. A single molecule of an olefin is referred to as a *monomer* (*mono-* means one or single), so the next stage in the process is to combine several monomers together to form a much larger molecule called a *polymer* (*poly-* means many). In the form of a polymer, the olefin takes on the properties of a plastic material. Two examples are shown in Fig. 4.1.

There are some simple basic rules governing the number of monomers that may be brought together to form a polymer. For example, at room temperature ethylene, which is made up of single molecules (monomers), is a gas. A polymer of 6 ethylene monomers is a liquid. A polymer of 36 ethylene monomers is a grease. A polymer of 140 ethylene monomers is a wax. A polymer of 500 or more ethylene monomers is the plastic polyethylene. The maximum number of monomers that can be brought together in a single polymer is limited in practice to about 2000. At this point there is little further increase in strength but a considerable increase in hardness and brittleness.

In thermosetting plastic materials, polymerisation usually occurs through *condensation*. In this process, the plastic material reacts within itself — or with some other chemical (hardener) — when heated to or

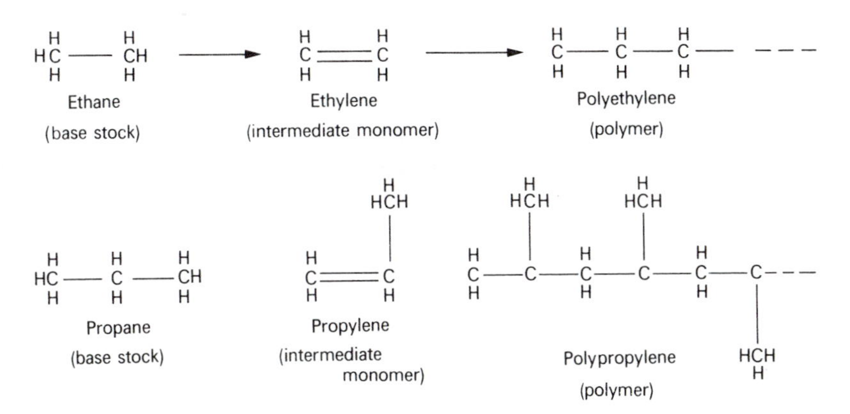

Fig. 4.1 Simple thermoplastic polymers

above a critical temperature. When this happens the plastic material releases ('condenses out') some small molecules such as water. When this happens, a permanent chemical change takes place and the plastic is said to be 'cured'. The change is irreversible and the material can never again be softened by heating. The loss of the water molecules causes shrinkage and this has to be allowed for in the moulding process. Also, the moulds have to be designed with vents so that the water vapour, generated as steam during the moulding process, can escape. The principle of condensation polymerisation is shown in Fig. 4.2.

4.3 Forms of supply

Both thermoplastic and thermosetting plastic moulding materials are normally available as powders or as granules packed in bags or in drums.

Thermoplastic powders and granules are homogeneous materials consisting of the polymer together with the colouring agent (pigment), lubricant and die release agent.

Thermosetting plastic materials are unsuitable for use by themselves, and the thermosetting plastics come in powder or granule form mixed with additives to make them more economical, to improve their mechanical properties, and to improve their moulding properties. A typical thermosetting plastic moulding material could consist of:

Resin powder or granules	38% by weight
Filler	58% by weight
Pigment	3% by weight
Mould release agent	0.5% by weight
Catalyst	0.3% by weight
Accelerator	0.2% by weight

The low-cost filler not only bulks up the powder or granules and makes the material more economical to use; it has a considerable influence on

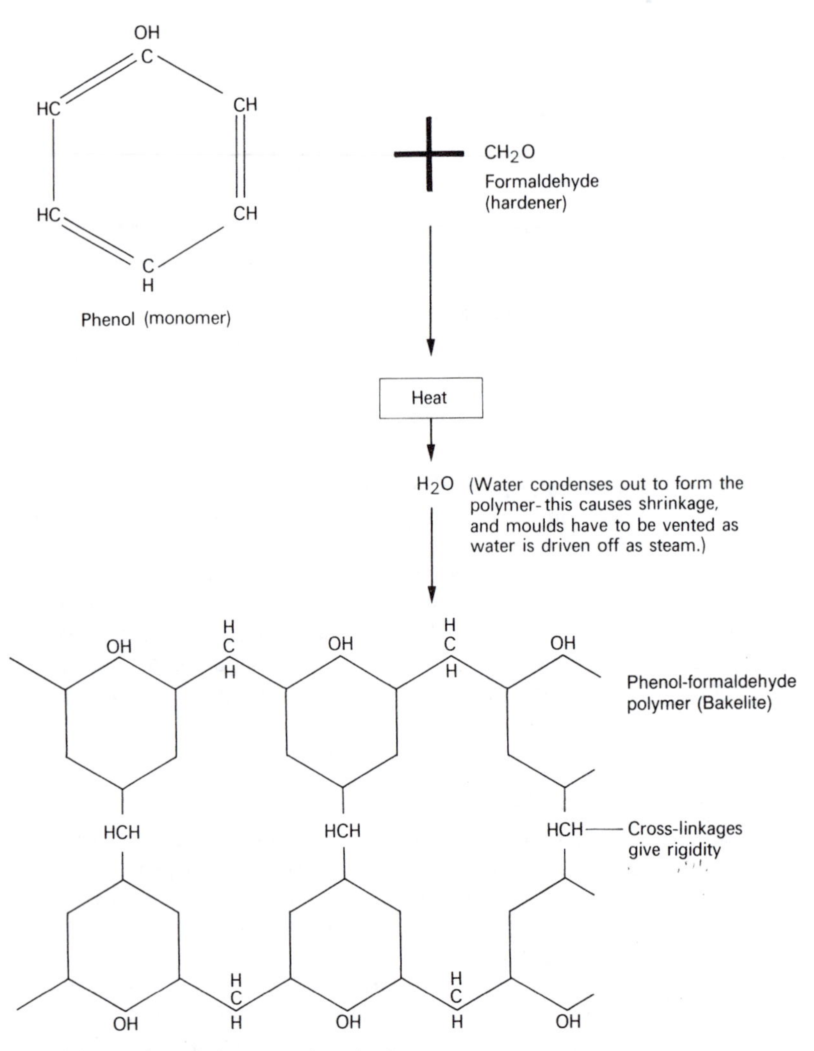

Fig. 4.2 Curing of thermosetting plastics

the properties of the mouldings produced from a given resin. Fillers improve the mechanical strength, electrical insulation properties, and heat resistance. Fillers also reduce shrinkage during moulding. Typical fillers are:

Glass fibre:	high strength and good electrical insulation properties.
Wood flour:	low cost, high bulk, low strength.
Calcium carbonate:	low cost, high bulk, low strength.

Rock wool: heat resistance (note: asbestos was used but is no longer recommended as it represents a health hazard).
Aluminium powder: wear resistance and high strength.
Shredded paper: good strength but inclined to absorb moisture.
Shredded cloth: higher strength, also inclined to absorb moisture.
Mica granules: heat resistant and good electrical insulation properties.

The pigment adds colour to the finished product. Some pigments are also used to prevent ultraviolet degradation of the polymer when it is exposed to sunlight. This also applies to thermoplastics (for example, UVPC for window frames). The mould-release agent prevents the moulding from sticking to the mould (most plastics are also good adhesives). It also acts as a lubricant and helps the moulding material to flow during the moulding operation. The catalyst promotes the curing process during moulding and ensures uniform properties throughout the moulding. The accelerator speeds up the curing time and, therefore, the moulding process, so reducing the process cost. Figure 4.3 shows the stages in manufacturing a typical thermosetting plastic material, such as paper-filled melamine formaldehyde, up to the stage where it is packed in bags or drums for delivery to the moulding company.

Polymeric (plastic) materials and adhesives are dealt with more fully in *Engineering Materials: volumes 1 and 2*.

4.4 Positive-die moulding (thermosetting plastics)

In the positive-die moulding process the two-part mould is fitted into a hydraulic press of the type shown in Fig. 4.4. The press may be of the upstroke type in which the hydraulic ram and cylinder are housed in the base (as shown) or the downstroke type with the cylinder and piston mounted above the upper bolster. In the latter type the upper bolster is free to move and the lower bolster is fixed. Large presses are usually of the upstroke type. The plattens are provided with steam or electric heating elements so that the moulds fastened to them can be heated to the curing temperature of the thermosetting plastic being processed. Modern presses are fully automatic in operation and this ensures constant moulding conditions at each stroke. This, in turn, ensures that mouldings of consistent quality are produced. The three factors which require presetting are:

- the moulding pressure (closing force on the die);
- the time that the mould is closed (curing time);
- the temperature of the mould.

In addition, feeding of the moulding powder or granules may also be automatic and this helps to ensure correct filling of the moulds and uniform product quality.

Figure 4.5 shows a section through a typical pressure mould. In this

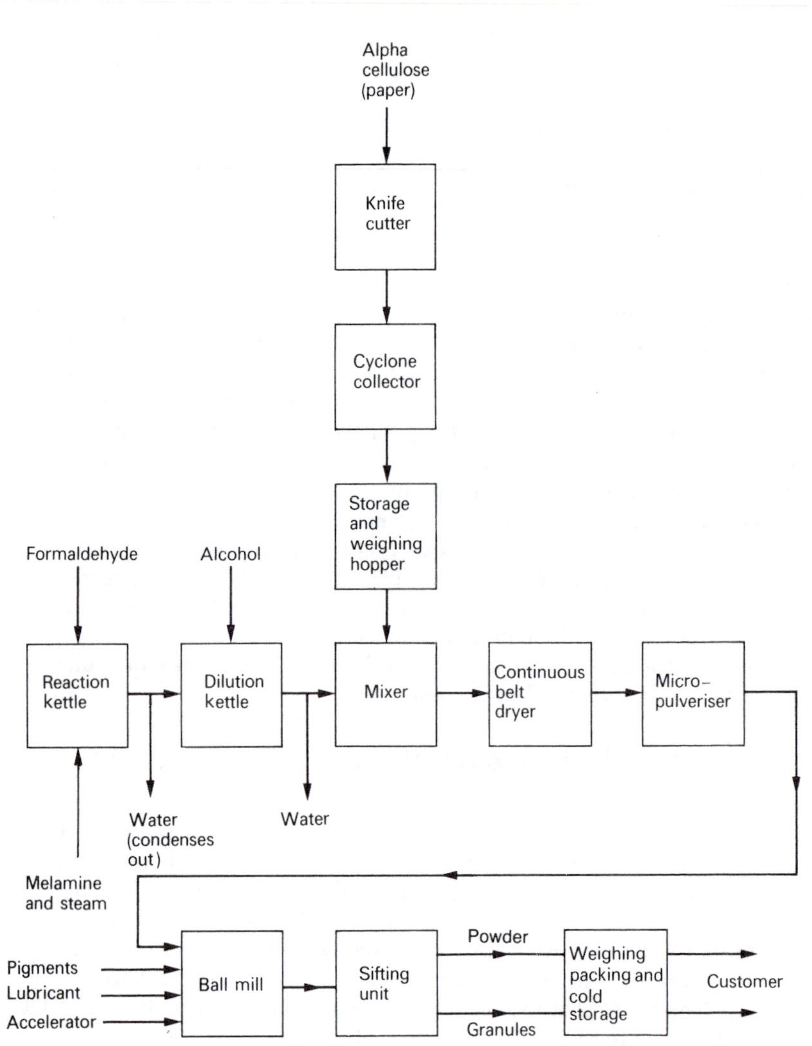

Fig. 4.3 Manufacture of a typical thermosetting plastic moulding material

type of mould a predetermined amount of thermosetting plastic material in powder or granular form is placed in the heated mould cavity. The mould is then closed by the press while the moulding material cures under pressure. The mould is then opened and the moulding is ejected. In positive moulding a vertical flash is produced in the direction of the moulding pressure. This surplus material is extruded between the upper and lower mould through the slight gap (0.01 mm to 0.03 mm) which has to be left for venting purposes (remember that steam is produced during condensation polymerisation). 'Flash' is the name given to any surplus

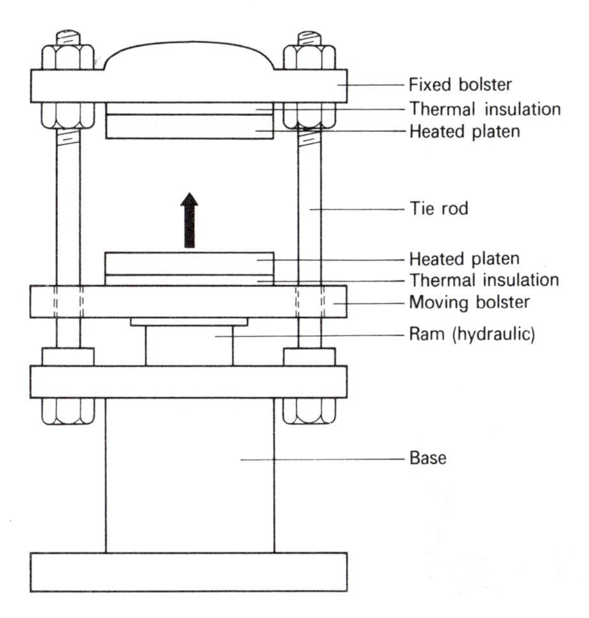

Fig. 4.4 Moulding press

material that has to be trimmed from the moulding. The mould shown in Fig. 4.5 allows for complete closure to ensure constant thickness. The disadvantage of this type of mould is that the amount of moulding powder fed into the mould has to be very accurately controlled: too little and the cavity will not be filled; too much and the mould will not close. In practice, very slight over-filling is aimed at to ensure that the mould is always filled. This very small amount of surplus material escapes through the steam vents. Where the thickness of the product is not critical there are no closing faces on the two parts of the mould, and the product thickness is controlled solely by the amount of material fed into the mould and the moulding pressure.

4.5 Flash moulding

The flash-type mould is used for simple, shallow components where compaction of the moulding powder during curing is not critical. To ensure complete filling, an excess of moulding powder is placed in the cavity. When the mould is closed, any excess material is squeezed out into the flash gutter. The 'flash land' forms a constriction which tends to hold the moulding material back into the mould cavity during curing in order to reduce shrinkage losses and uneven wall thickness. Figure 4.6 shows a section through a typical flash mould. A disadvantage of this type of mould is the heavy flash that has to be removed from the moulding and the wasted material which, being a thermosetting plastic,

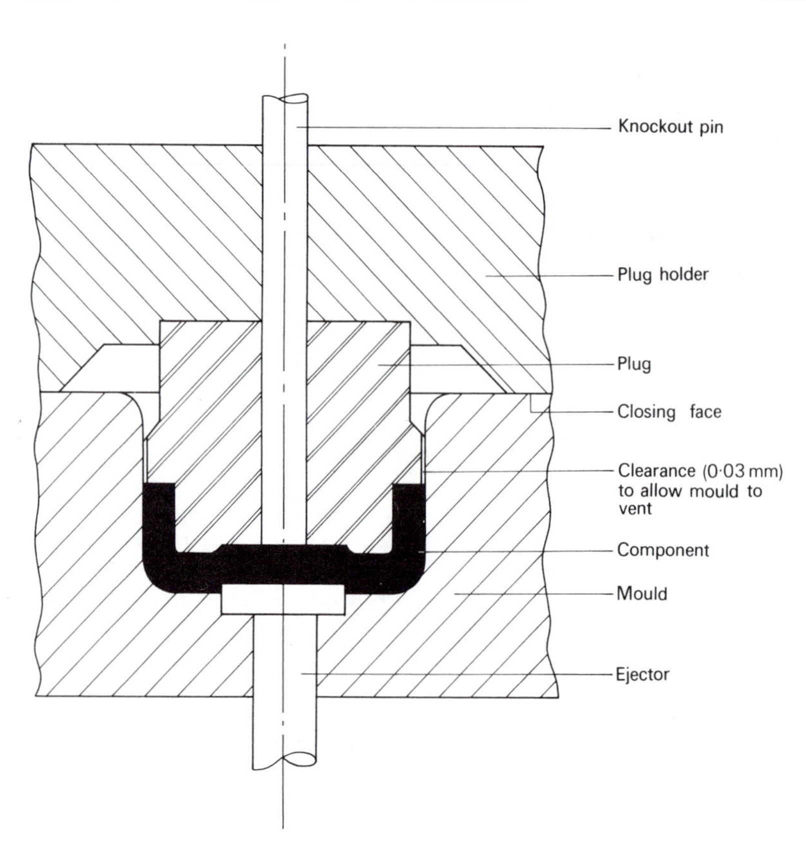

Fig. 4.5 Positive mould

cannot be recycled. However, the ease with which any surplus material can escape reduces the load on the press.

4.6 Transfer moulding

Transfer moulding uses a more complex three-part mould, as shown in Fig. 4.7. Such a mould is more costly than those previously described and transfer moulding is only used where:

- the moulding is of complex shape with many changes in wall thickness so that uniform filling of the cavity would be difficult in a more simple mould type;
- multiple impression moulds are used so that a number of components can be made at each stroke to ensure that the press is employed to its full capacity and, therefore, economically.

When transfer moulding, the powder is pre-heated and plasticised in the upper (loading) chamber of the mould. It is then forced under pressure

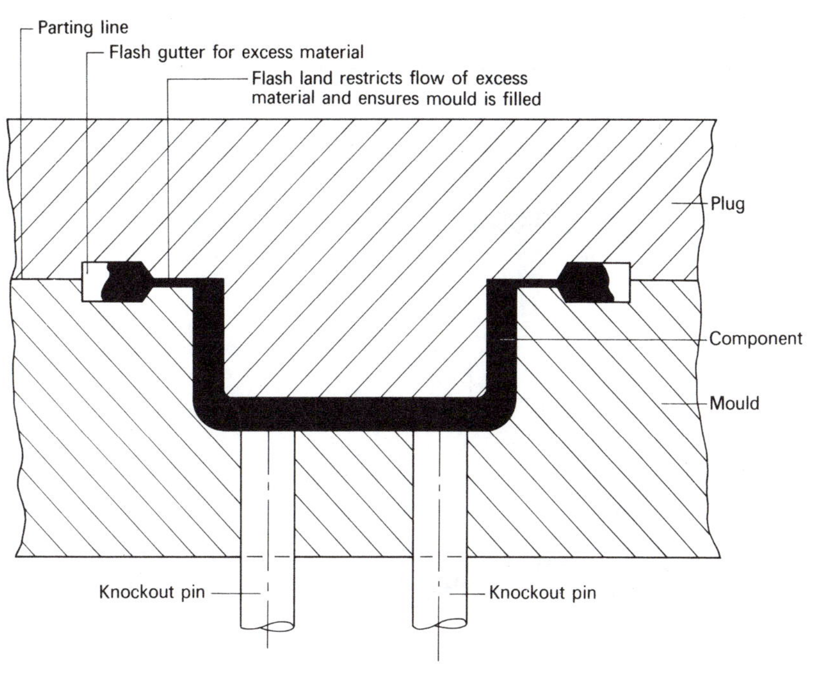

Fig. 4.6 Flash mould

into the mould cavity or cavities via the *sprue*. The sprue is removed after the mould is opened and the moulding has been ejected. Since the sprue does not form a useful part of the moulding, nor can it be recycled if a thermosetting material is used, it represents wasted material. However, this small waste is offset by the advantages of the process. Since the plasticised material is forced into the closed mould under very high pressure, complete filling of the mould cavity is ensured no matter how complex, shrinkage is reduced and improved mechanical properties are obtained from the moulding material.

4.7 Compression-moulding conditions

The moulding material may be fed into the mould as a powder, as granules or compacted into a preformed shape. The last of these is used to ensure uniform filling of the mould cavity, particularly when the cavity has a complex form. Correct loading of the mould is critical, insufficient material resulting in voids and porosity through the cavity not being properly filled. A slight excess of material is preferable as it ensures complete and uniform filling of the mould with any excess being allowed to form a 'flash'. Excessive over-charging must be avoided as the powder is incompressible and damage could be done to the mould and to the press. Automatic metering and feeding of the moulding material results in

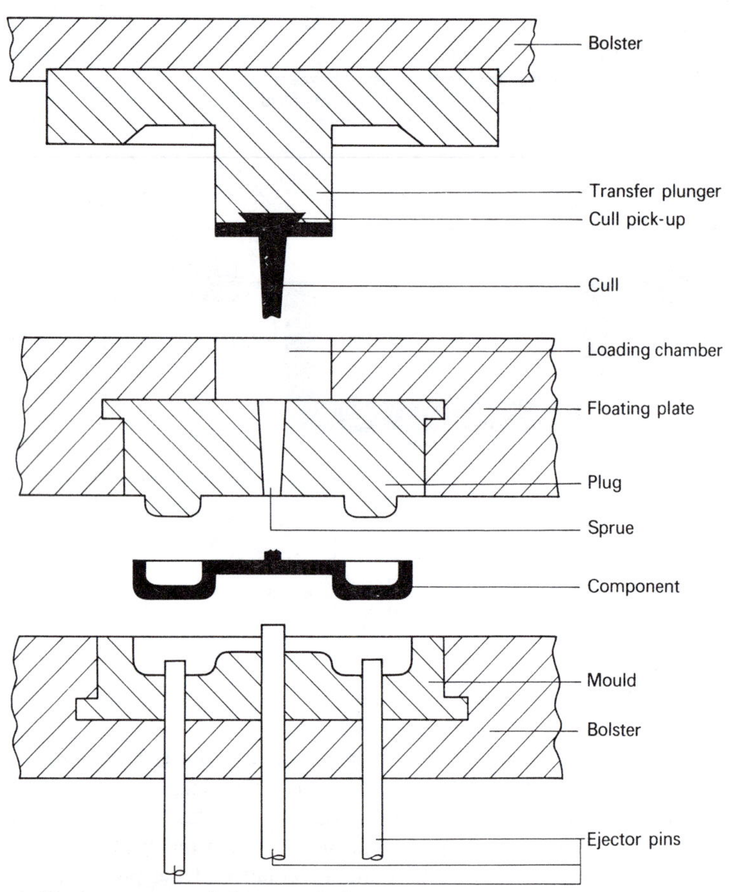

1. Die shown open ready for removal of component and cull.
2. Floating plate closes on mould and moulding powder is loaded into chamber.
3. Transfer plunger descends and forces plasticised moulding powder through sprue into mould.

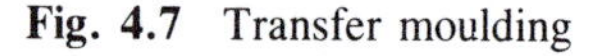
Fig. 4.7 Transfer moulding

more uniform results than hand feeding, as well as being more productive.

The moulding material can be loaded either cold or pre-heated. Pre-heating reduces the curing time and also reduces erosion of the mould cavity since the partially-plasticised material is in a less abrasive condition. During curing, volatile gases are released and these must be allowed to escape either through the mould clearances, through vents or by momentarily opening the mould part-way through the cure.

A release agent (lubricant) must be sprayed into the mould cavity, immediately prior to loading, in order to prevent sticking. The correct curing time and temperature is also critical as over-curing produces a dull

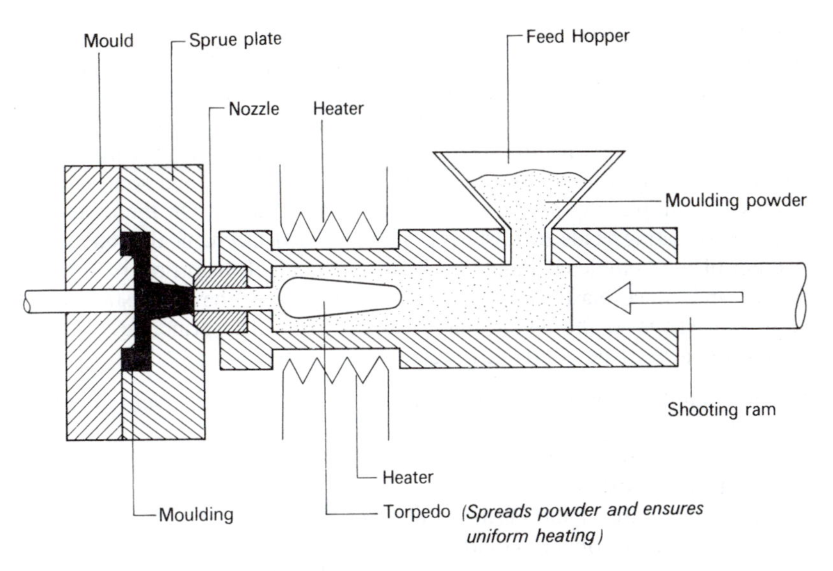

Fig. 4.8 Injection moulding

and blistered surface with some crazing, internal cracking, and poor mechanical properties. Under-curing may produce a component with the correct appearance but with poor mechanical properties. Moisture in the moulding powder can also cause blisters and porosity. The correct curing conditions are generally determined by trial and error based upon previous experience with similar mouldings.

4.8 Injection moulding

Unlike the compression-moulding processes, described in the previous section, which are usually used in conjuntion with thermosetting plastic materials, injection moulding is usually used in conjunction with thermoplastic materials. In the injection-moulding process a measured amount of thermoplastic material is heated until it becomes a viscous fluid, whereupon it is injected into the mould cavity under high pressure. In this respect it resembles transfer moulding except that, because a thermoplastic material is used, no curing has to take place and the moulds can be opened as soon as the moulding has cooled sufficiently to become rigid and self-supporting. Injection-moulding machines are generally arranged with the mould parting-line vertical and the axis of injection horizontal as shown in Fig. 4.8. As an alternative to the ram feed shown in Fig. 4.8, large-capacity machines may use a screw feed mechanism similar to the extrusion-moulding machine shown later on in Fig. 4.13.

Like so many processes that are simple in principle, the practice of injection moulding is fraught with difficulties. For instance, heating a body of thermoplastic material until it is fluid is not easy. If care is not taken, the surface becomes overheated and degraded before the interior of the plastic mass has reached its moulding temperature. Thus the injection temperature of the plastic is extremely critical as this controls its viscosity. Again, if the moulding material is not sufficiently heated, it may chill on contact with the mould and it will not necessarily fill the mould cavity. There may also be voids, sinks and shorts, and some cavities in multiple-impression moulds may not fill at all. On the other hand, over-heating leads to blistering and degraded mechanical properties. Ejection of the completed moulding is also difficult if distortion is to be avoided and, since plastic materials are also adhesive, they will stick to the mould whenever the opportunity presents itself. The principle variables that must be controlled are:

- the quantity of plastic material that is injected into the mould at each 'shot';
- the injection pressure;
- the injection speed;
- the temperature of the plastic while moulding;
- the temperature of the mould;
- the plunger-forward (pressure) time, i.e. the time the plastic material is maintained under pressure while it cools and becomes rigid enough to eject;
- the mould closed time;
- the mould clamping force;
- the mould open time.

As soon as the mould is filled, the pressure is increased and 'packing' commences. This ensures that the moulding has a high density and good mechanical properties. It also prevents sinks and shorts occurring due to shrinkage of the plastic as it cools. The injection cylinder of the moulding machine is connected to the mould by a nozzle. Figure 4.9(*a*) shows a standard nozzle, while Fig. 4.9(*b*) shows a nozzle with *breaker plates* which are used to assist mixing in colour-blending operations.

To ensure economical use of the machine, multiple-cavity moulds are used when moulding small components. This allows the machine to be used to its full capacity at each shot. Figure 4.10 shows a typical multiple-impression mould. It can be seen that the individual cavities are connected by passages, called runners, to the central sprue which connects with the injection nozzle. A 'gate' is provided at the cavity end of each runner to produce a local thinning so that the moulding may be easily separated from the runner. The combinaton of mouldings, runners and sprue produced at each shot is called a *spray*. Unlike thermosetting plastics, thermoplastic materials may be recycled. Thus the sprues and runners may be ground up and recharged into the machine to prevent waste. Since some degradation takes place each time the material is reheated, recycled material is not used when stressed components are being moulded. To ensure uniform filling of the cavities and to ensure uniform loading on the mould clamping mechanism of the injection-

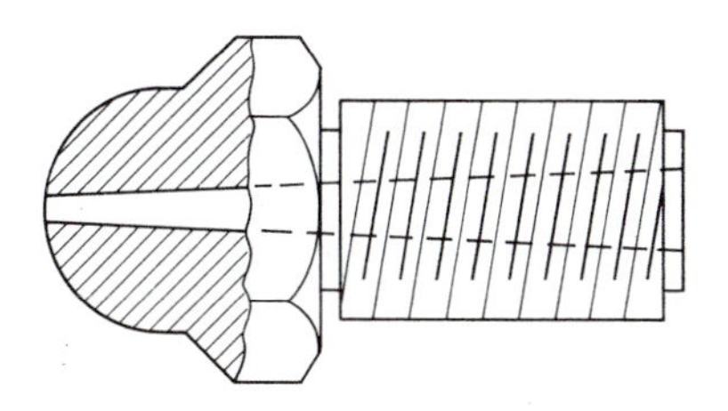

(a) **Standard nozzle**

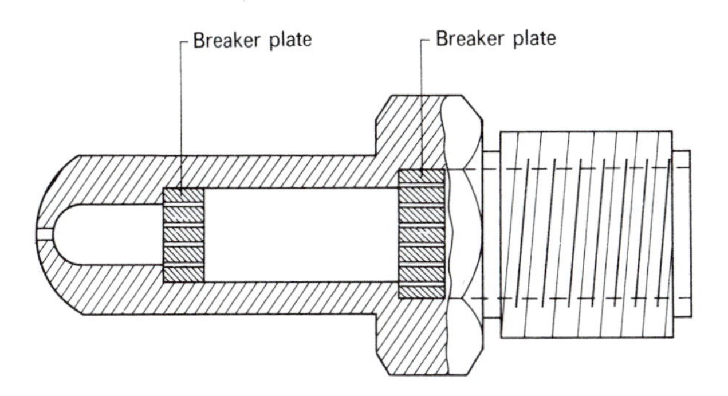

(b) **Mixer nozzle**

Fig. 4.9 Injection nozzles

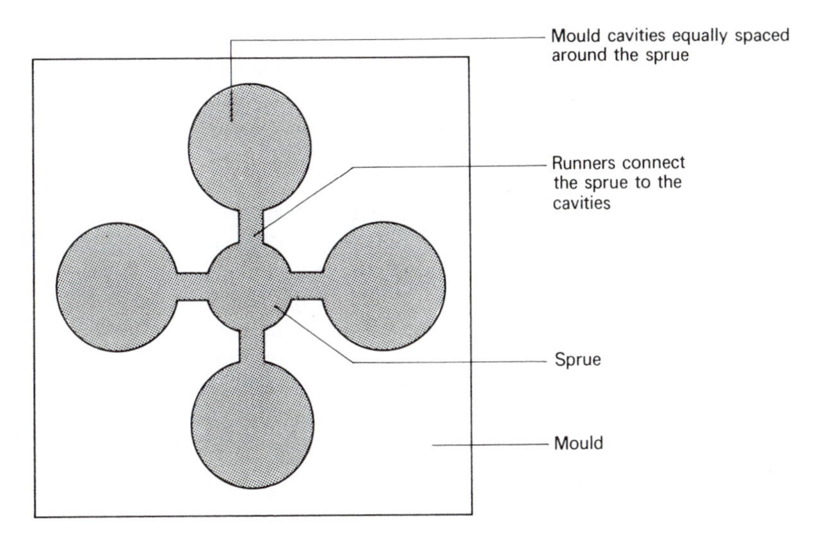

Fig. 4.10 Multiple-impression mould

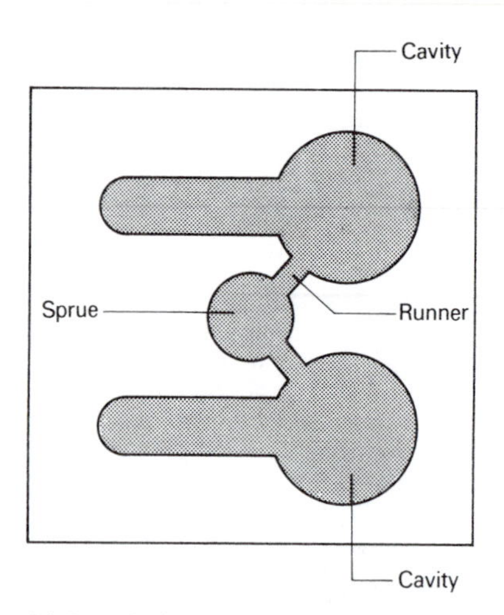

(a) **Poor design** *(Unbalanced clamping forces)*

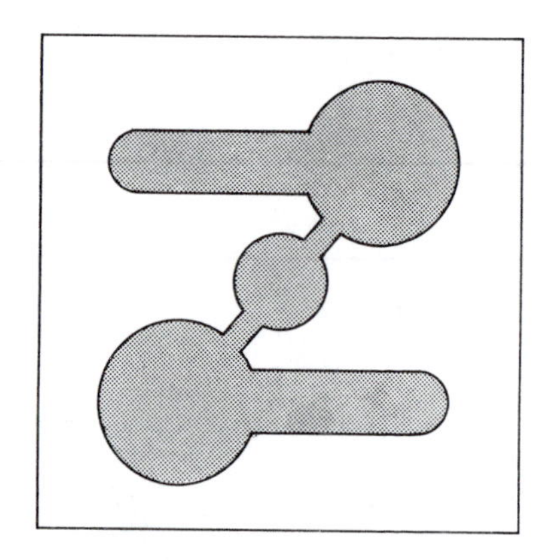

(b) **Good design** *(Cavities balanced about sprue)*

Fig. 4.11 Mould balancing

moulding machine, the mould cavities must be balanced about the central sprue as shown in Fig. 4.11.

4.9 Use of inserts

Inserts are used extensively in plastic products to provide such features as screwed anchorages, bearings and shafts, electrical terminals, and reinforcements. Metal inserts may be moulded into a component or added later by pressing them into cored or drilled holes. Inserts included during the moulding process are the most secure, but are inconvenient to use where quantity production is required as placing and locating the inserts in the mould slows down the process and increases production costs. Moulded inserts can also interfere with the ejection of mouldings from the mould cavity. Despite careful design, it is difficult to keep the moulding material out of the threads of screwed inserts, and any subsequent correction is, again, an added cost that detracts from the advantages of inserts added during the moulding process.

An ever-increasing use of fully-automated moulding processes for both thermosetting and thermoplastic materials has increased the use of inserts that are added after moulding. These inserts are pressed into cored or drilled holes. The slotted end of the insert collapses initially and then expands as the screw is driven home. This expansion locks the insert in place. Inserts are usually coarse knurled to prevent them from turning in the hole and also to prevent the insert from being withdrawn. Examples of typical inserts are shown in Fig. 4.12.

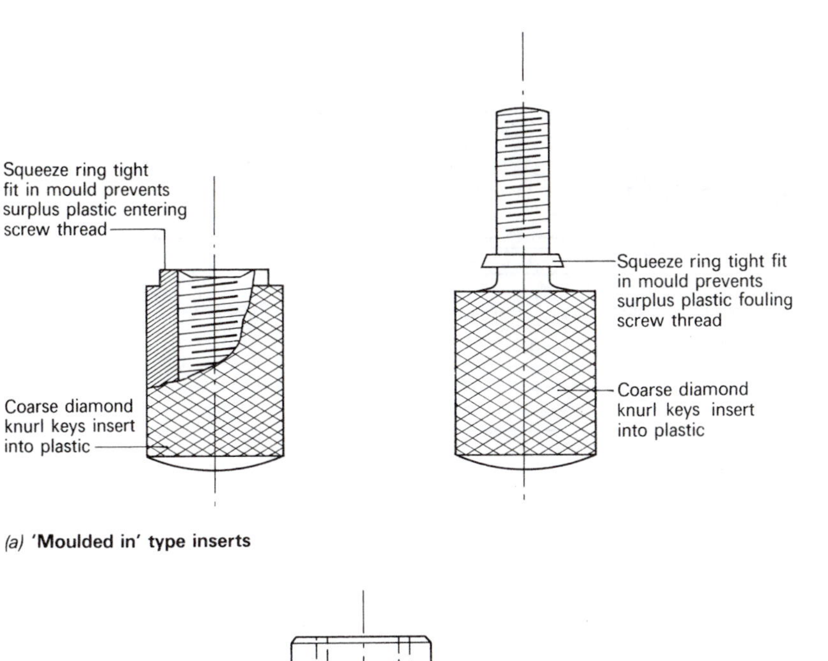

(a) **'Moulded in' type inserts**

(b) **Pressed in self-locking insert**

Fig. 4.12 Inserts

4.10 Extrusion

The extrusion of thermoplastic materials is, in principle, a continuous injection-moulding process. Any thermoplastic material can be extruded to produce lengths of uniform cross-section such as rods, tubes, sections and filaments. To produce a continuous flow of plastic material through the die, a screw conveyor is used in place of the piston and cylinder of the injection-moulding machine. The general arrangement is shown in Fig. 4.13. Unlike the injection-moulding machine, no 'torpedo' is required in the heating chamber when a screw conveyor is used. This is because the screw itself keeps the moulding powder in contact with the hot walls of

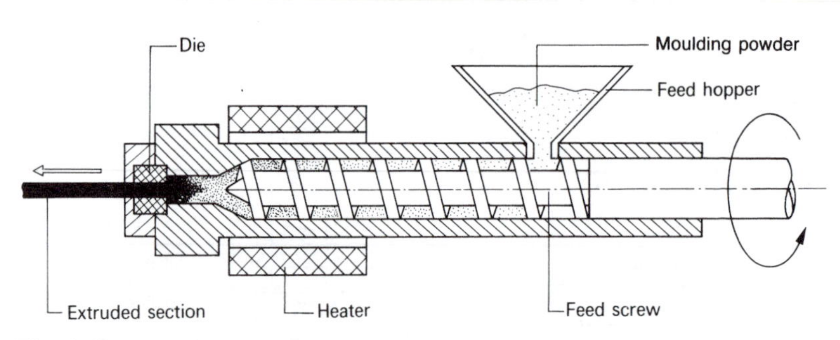

Fig. 4.13 Extrusion moulding

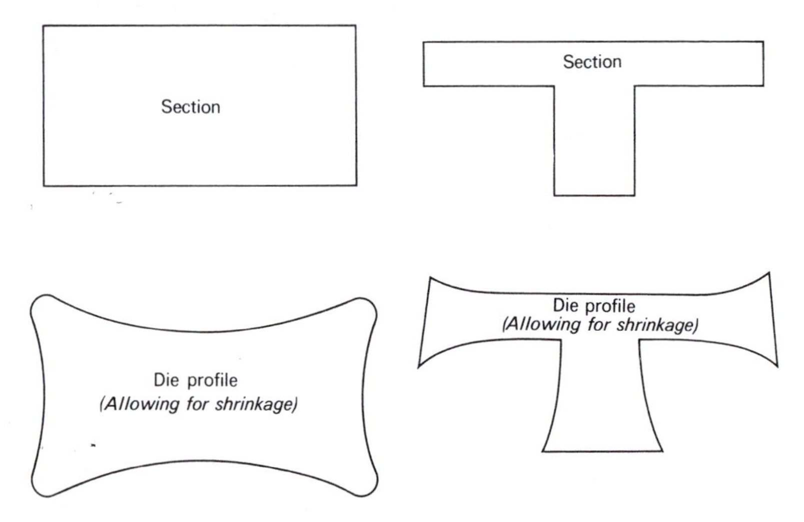

Fig. 4.14 Extrusion die (shrinkage allowance)

the chamber and no cold, central core of material can occur. To allow for shrinkage as the extruded material cools, the die profile has to vary from the finished product, as shown in Fig. 4.14. The plastic is still soft as it leaves the die and it must be supported to prevent distortion. Cooling is provided by a water tray, mist spray or air blast, depending upon the specific mass and temperature of extrusion. A conveyor draws the cooled plastic extrusion from the die ready for coiling or cutting to length.

A special application is wire coating as shown in Fig. 4.15. Since the wire runs at right angles to the axis of extrusion, this device is called a *cross-head* die. After passing through the die, the covered wire is water cooled and tested for its insulation characteristics.

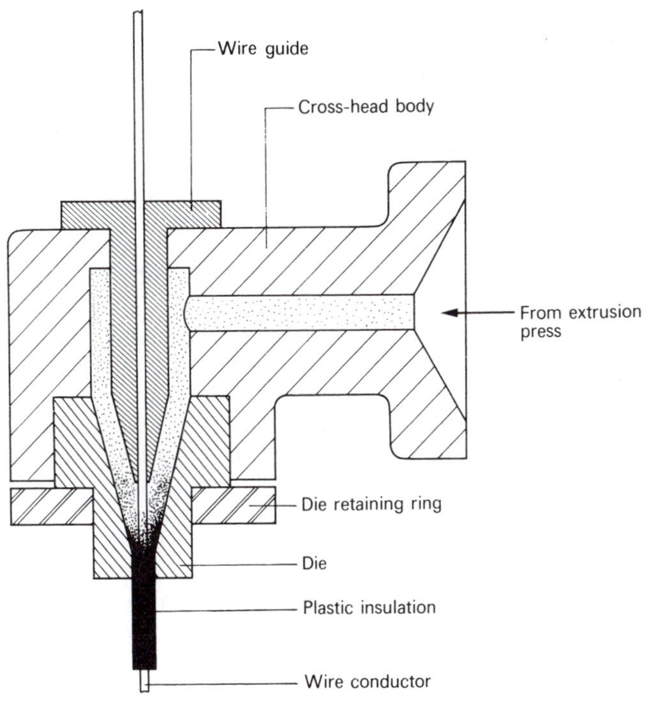

Fig. 4.15 Cross-head

4.11 Selection of moulding processes

The selection of moulding processes depends upon the component being made, the quantity required, the geometry of the component, and the materials from which it is to be made.

Pipes and sections

Since pipes and sections, such as rain-water guttering, are of uniform section and are required in lengths beyond the capacity of any mould, they are produced by the extrusion process (section 4.10). This process necessitates the use of thermoplastic materials.

Thermoplastic mouldings

Thermoplastic mouldings are usually made by the injection moulding process (section 4.8) which is suitable for the quantity production of both large and small components and is the most widely used moulding process. Small components can be made in multi-impression moulds and left on the sprue until required to prevent loss. Examples of typical components made by injection moulding can range from model kit parts made from polystyrene, and small gears for office machinery made from nylon, to rear-light clusters for motor vehicles made from transparent acrylic plastics and even complete motor vehicle bumpers moulded from impact-resistant plastics.

Thermo-setting plastic mouldings

Only under very special circumstances can thermosetting plastics be injection moulded. Almost invariably, thermosetting plastics are moulded by compression or transfer techniques (sections 4.4 and 4.5). Since the plastic resin can be readily blended with a wide variety of filler materials and pigments, mouldings made from thermosetting plastics can be given a wide range of properties and appearances. Compression mouldings are used for components such as: meter cases, electric fan bodies and blades, electrical insulators for switch gear, contactors and distribution equipment, and tableware. In all these examples rigidity and strength are required coupled with good surface finish and scratch resistance. Only thermosetting plastics have all these properties at the same time.

4.12 Justification

Providing the quantities being manufactured can offset the high cost of the moulds, the use of plastic mouldings can be justified on the following grounds:

- a wide range of materials with a correspondingly wide range of properties are available;
- polymeric materials have a lower density than metals;
- the raw materials are of relatively low cost;
- high rates of production can be achieved — ease of full automation;
- highly-polished finish is obtained straight from the moulds — little finishing required beyond trimming;
- a wide colour range is available — multi-colour mouldings can be made;
- little post-moulding machining is required — many components can be used straight from the mould.

Assignments

1. (a) Explain the essential differences between *thermoplastic* materials and *thermosetting plastic* materials.
   (b) Explain the mechanism of polymerisation by condensation and how this affects mould design.
2. Describe the composition of a typical thermosetting moulding powder and explain the purpose(s) of the various ingredients.
3. With the aid of diagrams, describe any TWO of the following compression moulding processes and compare the advantages and limitations of the processes chosen:

(a) Positive-die moulding.
(b) Flash-die moulding.
(c) Transfer moulding .

4. Describe the compression moulding cycle, paying particular attention to the precautions which must be taken to ensure successful production.
5. With the aid of sketches, compare the principles of compression moulding, injection moulding and extrusion moulding. Describe where each would be used in terms of materials and components.
6. Explain what are meant by *inserts*, and where they are of use in plastic mouldings.
7. Justify the selection of a plastic material and moulding process for a component of your choice.

5 Cutting processes

5.1 Sheet Metal

The cutting of sheet metal by shearing and blanking was introduced in *Manufacturing Technology: volume 1*. This section examines some further processes for cutting sheet metal.

Piercing

The piercing process is used to produce holes in previously blanked-out components. The clearance required between the punch and the die, and the force required to cut out the hole, is calculated in the same manner as for blanking. Note that, when piercing, the punch is made the required hole size while, when blanking, the die orifice is made the required blank size. Figure 5.1 shows the principles of piercing and a typical small piercing tool.

'Follow-on' tools

To speed production, piercing and blanking operations are often combined together in one tool. Figure 5.2 shows a simple 'follow-on' tool for making washers. To ensure concentricity of the hole, the blanking punch is provided with a pilot to locate in the previously-pierced hole. This type of tool gets its name from the sequential arrangement of the piercing and blanking stations.

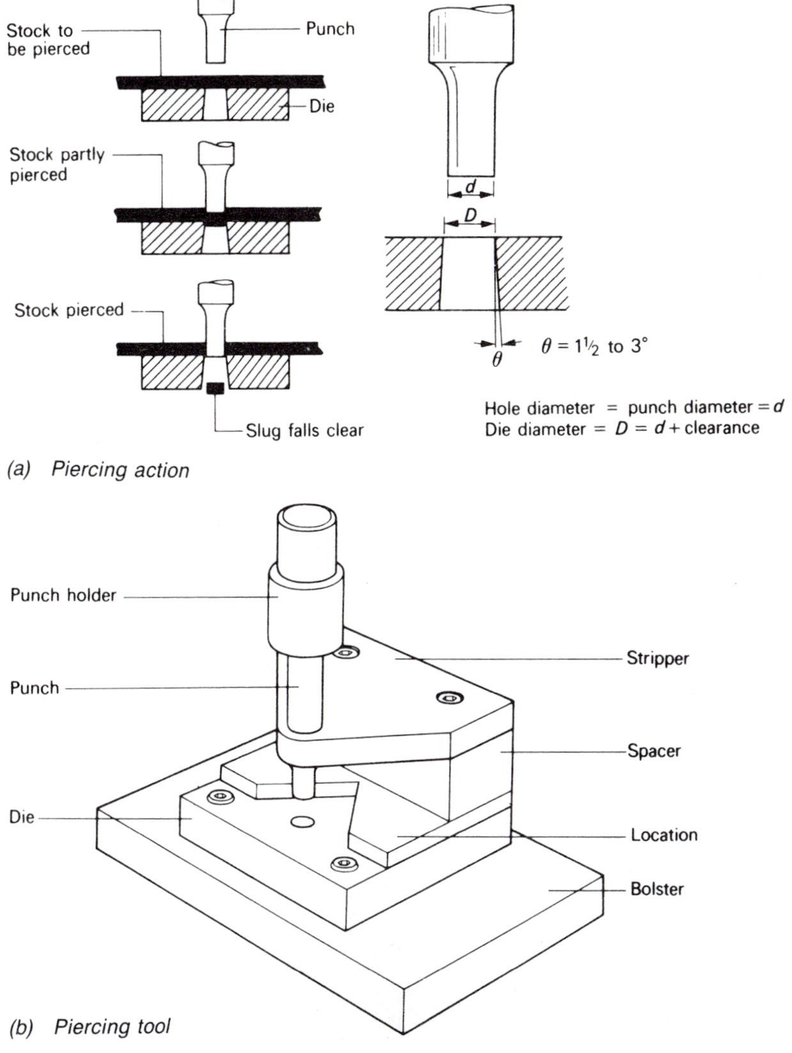

(a) Piercing action

(b) Piercing tool

Fig. 5.1 Piercing

Transfer tools

These develop the 'follow-on' design a stage further and incorporate forming operations. The blank and the partially-formed and pierced component is moved from stage to stage of the tooling by a transfer feed mechanism which is actuated by, and synchronised with, the crank or eccentric drive of the press. During each up-stroke of the ram, the partially-formed components are moved to the next stage.

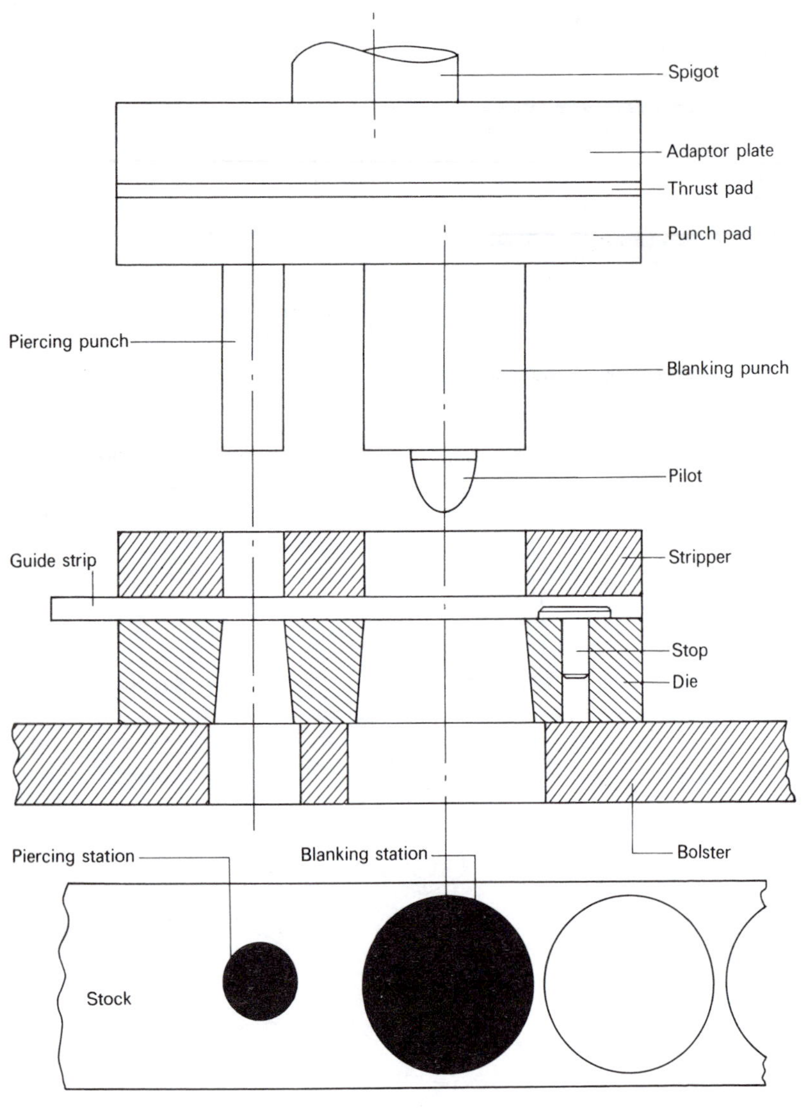

Fig. 5.2 Simple 'follow-on' tool

Combination tools

Like 'follow-on' tools, combination tools also combine operations. In these tools, various combinations of piercing, blanking operations and forming operations are combined together and are performed at the same time and not sequentially as in a 'follow-on' tool. Figure 5.3 shows a

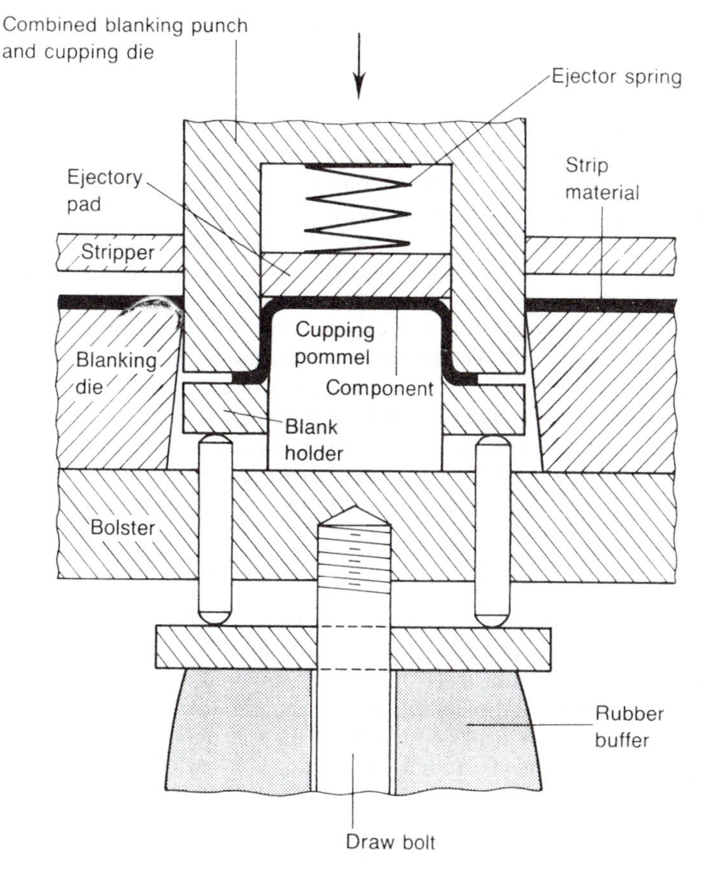

Fig. 5.3 Combination 'cut and cup' tool

simple 'cut and cup' combination tool in which the two operations take place at the same stroke of the press. If required, one or more holes could be pierced in the base of the cup in the same operation. Combination tools and 'follow-on' tools increase the rate of production of pressed components and reduce handling and inter-operation storage. However, they are more costly than simple tools and only become economical if the number of components warrants this extra cost. Also they are less flexible, i.e. only one combination of blank and hole piercing pattern is available, whereas with separate tools the same blank can be used with a variety of piercing tools to provide a range of components. The cutting force is also greater when the piercing and blanking operations are combined and care must be taken to ensure that the press is not over-loaded. The decision whether or not to use combination tools may be governed by the availability of suitable press capacity rather than optimum production rates.

5.2 Process selection (introduction)

The principles of metal machining were also introduced in *volume 1*, which included an introduction to the action of the cutting wedge, chip formation, and the application of these principles to single-point tools, and multi-point tools such as milling cutters and twist drills. The use of lubricants and coolants and the calculation of tool life was also discussed. Some further aspects of metal machining will now be considered.

When selecting a machining process, a compromise often has to be accepted between the ideal process to achieve the specified quality with maximum productivity and minimum unit cost, and the processes that are available within the limitations of the existing plant. It is only when very large quantities of a specific component are required — and that the requirement will be repeated — can the purchase of a particular item of plant be justified. If this outlay cannot be justified, it may be necessary to sub-contract some processes out to specialist companies. The selection process can be broken down into a number of stages, and some of the more important factors that have to be taken into account will now be considered.

Material

The ease with which a metal can be cut (its *machineability*) will affect the choice of process and machine. Some very hard alloys can only be cast and ground so that minimum metal removal is required. To machine components made from high duty alloys will require a heavier duty machine with a more powerful drive motor than that required to machine the same type of components from a free-cutting low-carbon steel.

Process

Modern machine and cutting tools can remove metal at such a high rate that the overall economics of the process can be easily overlooked. The conversion of stock material into swarf is wasteful, not only of the material, but also of energy to drive the machine. Further, high rates of metal removal and high volumes of metal removal lead to increased tool refurbishment and replacement costs. Minimum metal removal should be aimed for and machining should be looked upon, wherever possible, as a finishing operation after preforming by casting, forging or powder particulate processes for example.

Labour

The process should be selected so as to keep labour costs to a minimum. The time spent on any operation should be kept to a minimum. This not only includes the actual operation of the equipment but also the loading and unloading of the machine. Physical effort should be kept to a minimum by the use of mechanical aids so that the operator does not tire and work appreciably less effectively as the shift progresses. The skill of the operator must also be taken into account since highly-skilled workers invariably command higher wages than workers with lower skills. Some more specific factors affecting process selection will now be considered.

5.3 Process selection (surface geometry and process)

Figure 5.4 shows a component which consists of a number of surfaces that are basic geometrical shapes. The surfaces can be produced by selecting the appropriate machining process. For example:

(*a*) the circular plane surfaces on the end of the stem could be produced on a lathe at the same setting as the conical and cylindrical surfaces (*b*) and (*c*);
(*b*) the cylindrical surface could be produced on a lathe;
(*c*) the conical surface could also be produced on a lathe;
(*d*) the rectangular plane surface perpendicular to the axis of the cylindrical surface could also be produced on the lathe at the same setting as (*b*) and (*c*);
(*e*) the plane surfaces which are mutually parallel or perpendicular could be produced on a milling machine or on a shaping machine;
(*f*) the inclined plane surfaces could also be produced on a milling machine or on a shaping machine;
(*g*) the hole could be produced on a drilling machine.

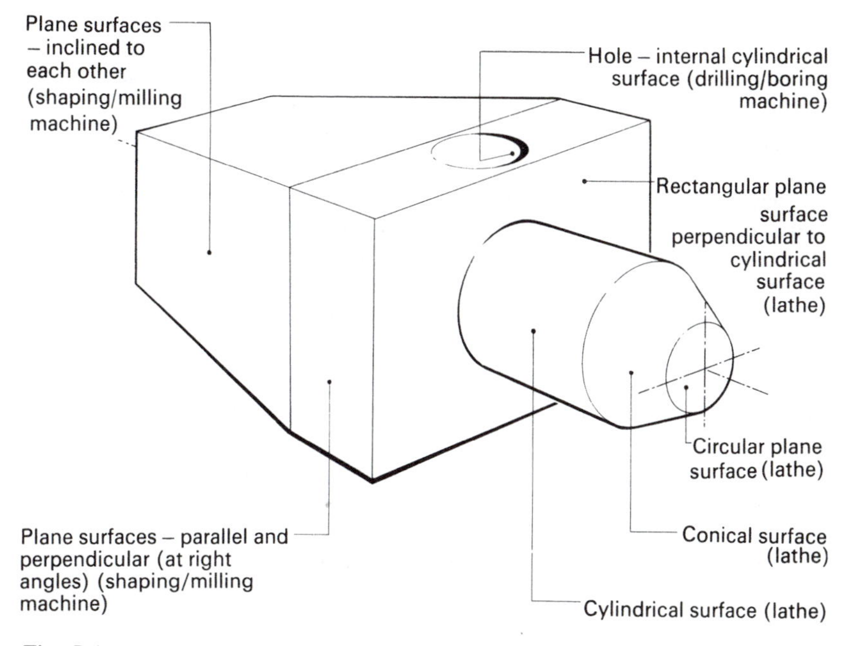

Fig. 5.4 Basic geometric shapes

Fig. 5.5 Copy turning

Should a high level of finish be required, then the turned surfaces could be finished on a cylindrical grinding machine, and the plane surfaces could be finished on a surface grinding machine. Most engineering components are designed as combinations of geometrical shapes with specific machining processes in mind. Where more complex surfaces are required, copy machining techniques, as shown in Fig. 5.5, or computer-controlled (CNC) machining techniques (see Chapter 6) can be employed. The generation and forming of geometrical surfaces was considered in *Manufacturing Technology: volume 1*. The machining processes available for producing various geometrical surfaces will now be listed.

- Plane surfaces: shaping, planing, milling, surface broaching, surface grinding.
- Cylindrical and conical surfaces (external): turning, cylindrical grinding.
- Cylindrical and conical surfaces (internal): drilling, parallel and taper reaming, boring, internal grinding, honing, lapping.
- Screw threads (sections 5.7 to 5.10): tapping, turning, grinding, use of die heads.
- Gear teeth (sections 5.11 and 5.12): milling, shaping, planing, hobbing, shaving, grinding.

- Contoured surfaces: copy milling, internal and external broaching, CNC machining, electric discharge machining (EDM), electrochemical machining (ECM).

5.4 Process selection (accuracy and surface finish)

From the above list it can be seen that there are alternative processes for producing any of the surfaces required. The alternative processes offer varying degrees of accuracy, varying size and mass of the component being machined, and varying process costs. The accuracy and surface finish of various manufacturing processes will now be considered.

It must be remembered that the closer the limits (smaller the tolerance), the more difficult and expensive it is to manufacture a component. It is no use choosing a process just because it offers low-cost production if it cannot achieve the tolerance specified. Table 5.1 is based upon BS 4500 and shows the standard tolerances from which tables of limits and fits can be derived (see *Manufacturing Technology volume 1*). It also shows how the International Tolerance (IT) number is related to standard tolerances for various ranges of linear dimension. The figures given are in microns (micrometres) (0.001 mm), so a tolerance grade of IT9 for a dimension of 12 mm would have a standard tolerance of 43 microns, that is 0.043 mm. It can be seen from the table that, as the International Tolerance (IT) number gets larger, the tolerance increases. The recommended relationship between process and standard tolerance is as follows:

IT16 sand casting, flame cutting
IT15 stamping; hot-rolling
IT14 die-casting, plastic moulding
IT13 press-work, extrusion
IT12 light press-work, extrusion
IT11 drilling, rough-turning, boring
IT10 milling, slotting, planing, cold-rolling
IT9 low-grade capstan, turret and automatic lathe work
IT8 centre lathe, capstan, turret and automatic lathe work
IT7 high-quality turning, broaching, honing
IT6 grinding, fine honing
IT5 machine lapping, fine grinding
IT4 gauge making, precision lapping
IT3 high-quality gap gauges
IT2 high-quality plug gauges
IT1 slip gauges, reference gauges

Surface finish quality is also related to process selection and accuracy. It is no use specifying close dimensional tolerances to a process whose inherent surface roughness lies outside that tolerance, as shown in Fig. 5.6(*a*). The dimensional tolerance and the surface finish of the manufacturing process must be matched, as shown in Fig. 5.6(*b*). Surface

Table 5.1 Standard tolerances

Tolerance unit 0.001 mm

Nominal sizes		Tolerance grades																	
Over	Up to and including	IT01	IT0	IT1	IT2	IT3	IT4	IT5	IT6†	IT7	IT8	IT9	IT10	IT11	IT12	IT13	IT14*	IT15*	IT16*
mm	mm																		
—	3	0·3	0·5	0·8	1·2	2	3	4	6	10	14	25	40	60	100	140	250	400	600
3	6	0·4	0·6	1	1·5	2·5	4	5	8	12	18	30	48	75	120	180	300	480	750
6	10	0·4	0·6	1	1·5	2·5	4	6	9	15	22	36	58	90	150	220	360	580	900
10	18	0·5	0·8	1·2	2	3	5	8	11	18	27	43	70	100	180	270	430	700	1100
18	30	0·6	1	1·5	2·5	4	6	9	13	21	33	52	84	130	210	330	520	840	1300
30	50	0·6	1	1·5	2·5	4	7	11	16	25	39	62	100	160	250	390	620	1000	1600
50	80	0·8	1·2	2	3	5	8	13	19	30	46	74	120	190	300	460	740	1200	1900
80	120	1	1·5	2·5	4	6	10	15	22	35	54	87	140	220	350	540	870	1400	2200
120	180	1·2	2	3·5	5	8	12	18	25	40	63	100	160	250	400	630	1000	1600	2500
180	250	2	3	4·5	7	10	14	20	29	46	72	115	185	290	460	720	1150	1850	2900
250	315	2·5	4	6	8	12	16	23	32	52	81	130	210	320	520	810	1300	2100	3200

Nominal sizes		Tolerance grades																	
Over	Up to and including	IT01	IT0	IT1	IT2	IT3	IT4	IT5	IT6†	IT7	IT8	IT9	IT10	IT11	IT12	IT13	IT14*	IT15*	IT16*
mm 315	mm 400	3	5	7	9	13	18	25	36	57	89	140	230	360	570	890	1400	2300	3600
400	500	4	6	8	10	15	20	27	40	63	97	155	250	400	630	970	1550	2500	4000
500	630	—	—	—	—	—	—	—	44	70	110	175	280	440	700	1100	1750	2800	4400
630	800	—	—	—	—	—	—	—	50	80	125	200	320	500	800	1250	2000	3200	5000
800	1000	—	—	—	—	—	—	—	56	90	140	230	360	560	900	1400	2300	3600	5600
1000	1250	—	—	—	—	—	—	—	66	105	165	260	420	660	1050	1650	2600	4200	6600
1250	1600	—	—	—	—	—	—	—	78	125	195	310	500	780	1250	1950	3100	5000	7800
1600	2000	—	—	—	—	—	—	—	92	150	230	370	600	920	1500	2300	3700	6000	9200
2000	2500	—	—	—	—	—	—	—	110	175	280	440	700	1100	1750	2800	4400	7000	11000
2500	3150	—	—	—	—	—	—	—	135	210	330	540	860	1350	2100	3300	5400	8600	13500

*Not applicable to sizes below 1 mm.
†Not recommended for fits in sizes above 500 mm.

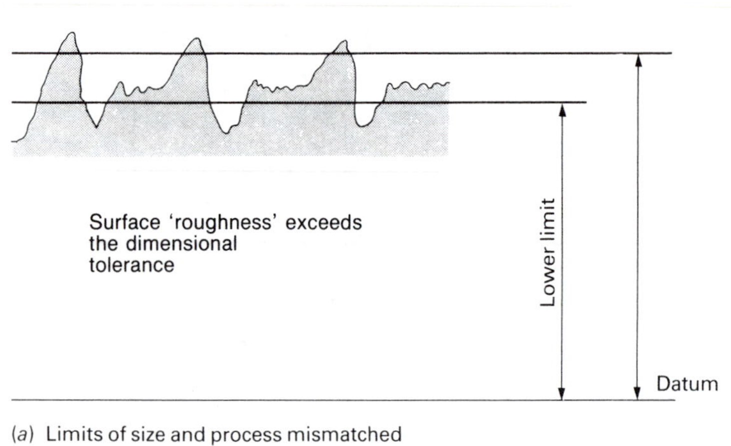

(*a*) Limits of size and process mismatched

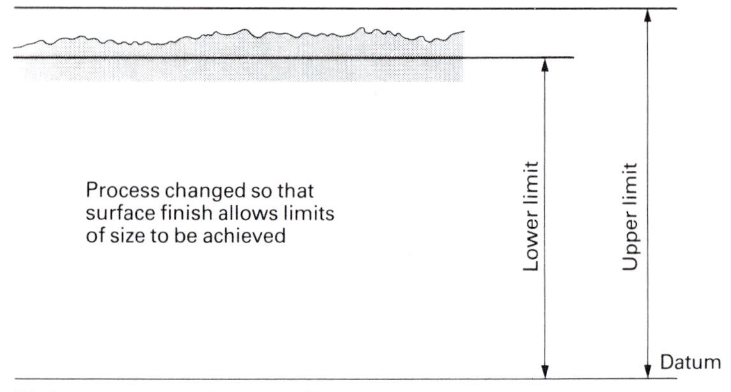

(*b*) Process suitable for limits of size

Fig. 5.6 Relationship between surface texture and dimensional tolerance

finish and its assessment is discussed in *Manufacturing Technology: volume 1* in some detail. Table 5.2 relates the roughness grade (N) number to the R_a value in microns and to various manufacturing processes. The R_a number is a measure of surface finish and its significance is explained in *volume 1*.

5.5 Process selection (tooling and cutting costs)

The calculation of cutting speeds and feeds, the power required for a given rate of metal removal and the calculation of tool life are all introduced in *Manufacturing Technology: volume 1*. These will now be considered further, in order to determine the overall tooling costs for a particular process. At first sight it would appear that since increasing the cutting speed and the feed reduces the process time, this should also reduce the process cost. However, the increase in cutting speed and the

Table 5.2 Relationship between surface texture and process

R_a value (mm)	Roughness grade number
50	*N*12
25	*N*11
12.5	*N*10
6.3	*N*9
3.2	*N*8
1.6	*N*7
0.8	*N*6
0.4	*N*5
0.2	*N*4
0.1	*N*3
0.05	*N*2
0.025	*N*1
0.0125	—

Process	*N*-value
Casting, forging, hot-rolling	*N*11–*N*12
Rough turning	*N*9
Shaping and planing	*N*7
Milling (HSS cutters)	*N*6
Drilling	*N*6–*N*10
Finish turning	*N*5–*N*8
Reaming	*N*5–*N*8
Commercial grinding	*N*5–*N*8
Finish grinding (tool room)	*N*2–*N*4
Honing and lapping	*N*1–*N*6
Diamond turning	*N*3–*N*6

feed also increases the tool wear so that 'down-time' for tool changing will be more frequent, with a corresponding loss of production, and the cost of tool refurbishment and replacement will be increased. Further, there is a greater chance of tool failure and the production of scrap components. Thus an optimum set of operating conditions has to be arrived at which attempts to achieve a balance between these various conflicting influences.

The experimental work of Takeyama and Murata showed that the relationship between tool temperature and tool life for any given set of cutting conditions was logarithmic. Other experimental work showed that tool temperature is related to cutting speed and feed by an equation of the form:

$$\text{Temperature } (K) = \text{some function of (cutting speed} \times \text{feed)}$$

Increasing the cutting speed increases the temperature more rapidly than increasing the feed, but increasing the feed not only increases the forces acting on the tool but also produces a rougher surface texture.

Since tool life is related to tool temperature and tool temperature is, in turn, related to the cutting speed and feed, it is reasonable to expect that tool life is related to cutting speed and feed. Taylor showed that an empirical relationship does exist between these variables and his tool life equation $Vt^n = C$ has already been introduced in *Manufacturing technology: volume 1*. A typical curve relating V and t is shown in Fig. 5.7(*a*). Cost is also related to tool life since, the longer the tool life, the lower will be the tool refurbishment and resetting costs. Thus tool cost is

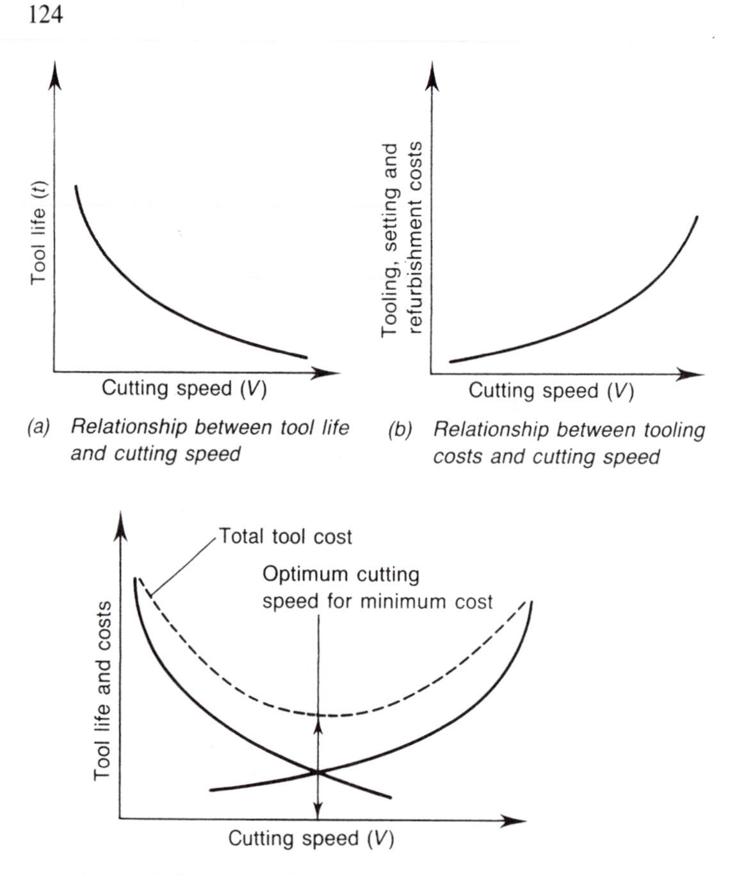

(a) Relationship between tool life and cutting speed

(b) Relationship between tooling costs and cutting speed

(c) Variation in machining costs with cutting speed

Fig. 5.7 Effect of cutting speed on tooling costs

proportional to $1/t$ and this relationship is shown in Fig. 5.7(*b*). By combining these figures, as shown in Fig. 5.7(*c*), it is possible to arrive at the optimum process conditions for lowest cutting cost and most economical cutting speed for a given set of conditions.

Example 5.1

The following data refer to the turning of a batch of 1000 components.

Cutting speed	40 m/min
Value of n for the tool/work combination	0.2
Value of C	80
Time to change tool tip and reset	3 min
Turning time per component	4 min
Machine cost	£0.15/min

$Vt^n = C$ $\qquad V = 40$ m/min

$t = \left(\frac{C}{V}\right)^{1/n}$ $\qquad n = 0.2$, $C = 80$

$= \left(\frac{80}{40}\right)^{1/0.2}$

= 32 min between tool changes

Number of components per tool change $= \frac{32}{4} = 8$

Tool change cost 3 min × £0.15 = £0.45

Tool change cost per component $= \frac{£0.45}{8} = £0.056$

Cost of turning a component = 4 min × £0.15 = £0.60
∴ Total cost per component = £0.60 + £0.056 = £0.656
∴ Total cost for a batch of 1000 components is £656
(*Note*: cost of tool changing per batch is £56.)

5.6 Process selection (cutting tool materials)

It has already been stated (section 5.5) that tool life is related to temperature and therefore to cutting speed. Figure 5.8 shows how the hardness of tool materials is related to temperature. It also shows that hardness at ambient temperatures can be misleading. For example, a hardened and tempered high-carbon steel is much harder at room temperature than high-speed steel. However, the carbon steel soon loses its hardness (its temper is 'drawn') as the cutting temperature increases, while high-speed steel remains hard up to a dull red heat.

Carbide- and ceramic-tipped tools

Cemented carbide- and ceramic-tipped tools can withstand even higher temperatures than high-speed steel without softening and are very much more resistant to abrasion. However, such materials are relatively weak in tension and also more brittle. Their lower mechanical strength requires greater support of the cutting edge. Therefore negative rake tool geometry is generally adopted when using carbide and ceramic tool materials, as shown in Fig. 5.9. To benefit fully from negative rake geometry the power and rigidity of the machine tool needs to be increased so as to increase the work done on the workpiece material at the cutting zone. This, in turn, increases the temperature of the cutting zone, which increases the plasticity of the workpiece material and the chip at the point of cutting. Increasing the plasticity of the workpiece and chip at the cutting zone reduces the mechanical forces acting on the cutting tool even when high-duty alloys are being machined.

Brittle cutting tool materials, such as carbides and ceramics, tend to be easily chipped. Therefore care is required in the fast approach of the tool

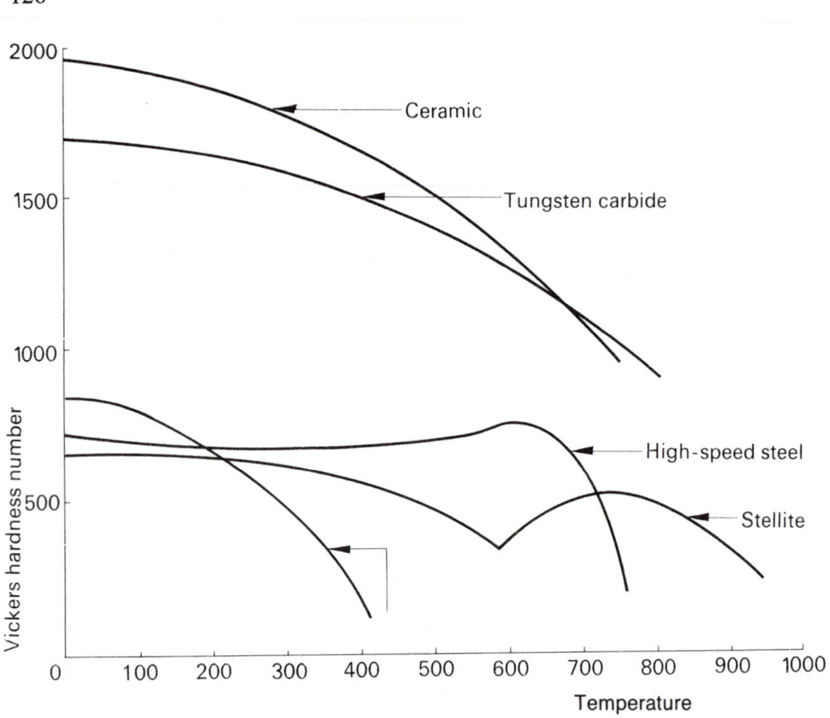

Fig. 5.8 Hardness–temperature curves for cutting-tool materials

to the workpiece. For this reason the more accurately and consistently controlled cutting conditions of automatic and computer-controlled machine tools result in improved tool life and fewer breakages when using carbide- and ceramic-tipped tools.

When using such tool materials the cost of replacement and refurbishment must be considered more carefully. To achieve the maximum performance from such materials the initial tool geometry must be strictly maintained, yet these materials are more difficult to re-grind than steel tools and usually need to be honed after grinding. For these reasons, modern practice favours the use of 'throw-away' disposable tool tips clamped into special holders. Such disposable tips are mass produced relatively cheaply to a high degree of accuracy so that consistency of performance is guaranteed. To extend the life of the tips, they usually have more than one cutting edge and they can be indexed round in the tool holder as each cutting edge becomes dulled, in order to present a fresh cutting edge. The cost of replacement is usually much less than attempting 'in-house' refurbishment.

Coated tools

For automatic and computer-controlled machines, *coated carbide* tipped tools are now widely used. The coating is usually *titanium nitride* (TiN)

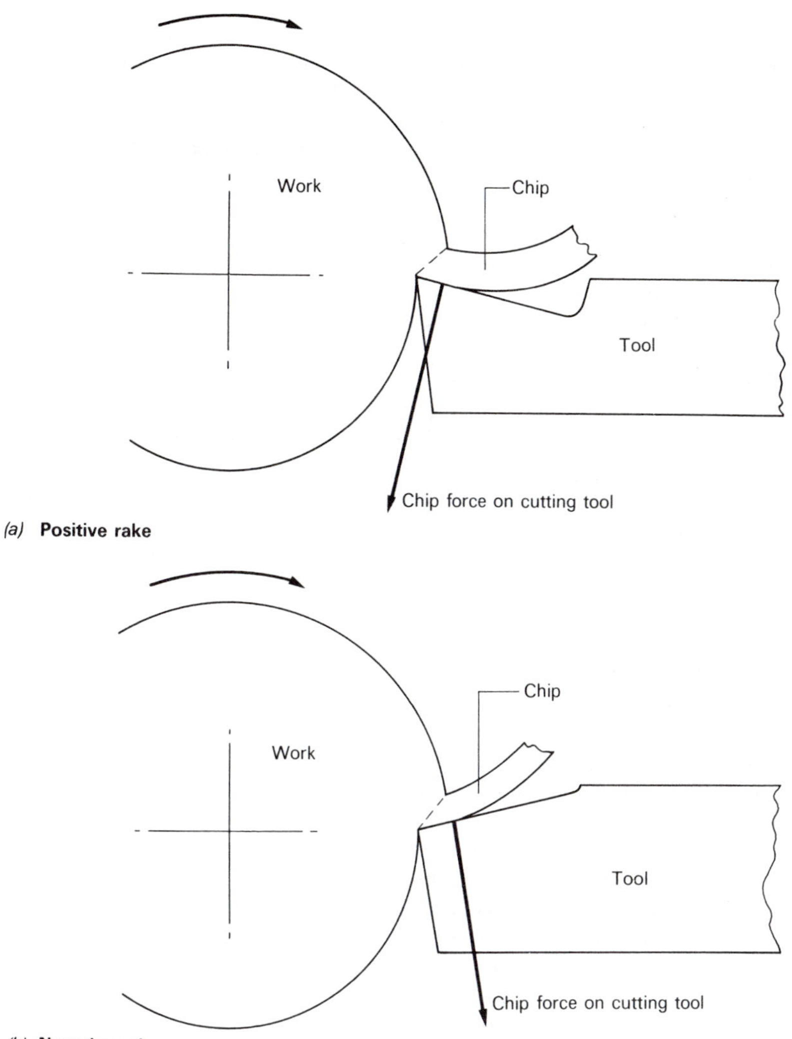

Fig. 5.9 Negative rake cutting

and the coating technique is either physical vapour deposition (PVD) or chemical vapour deposition (CVD). Coating thicknesses are typically between 1 micron and 6 microns, and the hardness of the coating can be as high as 3000 DPHN (Diamond Pyramid Hardness Number) (Vickers). The coefficient of friction for the coated surface is very much lower than for uncoated surfaces. This not only allows the chip to flow more freely across the rake face of the tool, but it also prevents chip welding and the formation of a built-up edge. Both these factors lead to a better workpiece surface finish. It is this lower coefficient of friction that permits coated tools to be operated at significantly higher speeds and feed without any increase in temperature and without any loss of tool life of the substrate. Typically, cutting speeds can be increased by up to 50 per cent and the tool life can be extended by up to eight times that of a similar uncoated tool. Tests on both drills and turning tool inserts have shown that the coating reduces both flank and crater wear. Such tests have shown that the cutting forces on coated tools remain constant over long periods, whereas the forces on uncoated tools increase progressively due to tool wear and the formation of a built-up edge. The coating can also be applied to high-speed steels.

Where tools are re-ground, as in the case of twist drills, the performance of the drill is not significantly affected providing the coating is still present in the flutes which provide the rake face of the tool and where the antifriction properties of TiN enhance chip clearance from the hole. Generally, holes drilled with coated drills have a smoother finish and improved dimensional and geometrical accuracy, often eliminating the need for reaming. Coating is expensive and coated tools can cost up to two and a half times the cost of uncoated tools. Thus, such tools can only be justified where their full potential can be exploited by the plant available.

Sialon ceramics

Sialons are a family of silicon nitrides containing differing levels of aluminium and oxygen substitution (Si—Al—O—N). These materials are particularly successful when cutting cast irons and nickel steels. They combine the high temperature (high cutting speed) performance of conventional solid ceramic tips (Al_2O_3) with the impact resistance of normal coated carbides and are thus suitable for intermittent cutting. Unlike conventional ceramic tool materials, sialons are also resistant to thermal shock and can be used with coolants. This improves the tool life even further and also improves the surface finish of the machined surface. The position of sialon ceramics in the cutting tool material hierarchy is shown in Fig. 5.10. From the same figure it can be seen that boron nitride heads the list for maximum cutting speed but only at very fine feeds.

Boron nitride

Polycrystalline cubic boron nitride (PCBN) has a hardness comparable with diamond, the hardest known substance. Tool life can be as high as 20 times that expected from coated carbide tooling providing the same cutting parameters are maintained. By reducing the 'down-time' for tool changing, the productivity of a machine tool using PCBN-tipped tools can be considerably (up to 20 per cent) improved. As for all very hard cutting tool materials, PCBN tips are easily chipped and care must be taken in tool approach and choice of feed rate. Again it must be remembered that sophisticated cutting tool materials such as the sialons

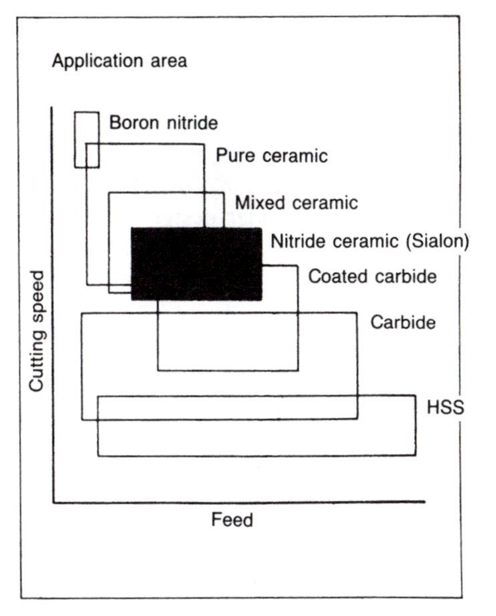

Fig. 5.10 Position of sialon ceramics in the cutting tool material hierarchy

and cubic boron nitride are substantially more expensive than more conventional materials such as coated and uncoated carbides. Therefore they should only be used where their properties can be fully exploited.

Diamond

Industrial diamonds are used for finish machining operations. Although very hard, diamonds have a low shear strength and can only be used with very fine feeds so that the forces acting on the tool are minimised. However, because of the very high measure of hardness possessed by diamonds, which is sustained at high temperatures, very high cutting speeds can be used. Diamond turning tools are used for finish turning such components as aluminium alloy pistons for internal combustion engines, aluminium alloy lens components for cameras, and copper rolls for printing colour illustrations. None of these components lend themselves to finishing by grinding because of the material from which they are made, but the finish attainable from diamond turning can be equally as good.

Finally, in selecting a suitable cutting tool material it must be remembered that a compromise has to be achieved between all the parameters discussed previously in this section. For example, despite the fact that the harder cutting tool materials are more brittle than conventional materials and have to be used with a finer feed, their productivity is greater because of the higher cutting speeds that can be used and the fact that tool changing is less frequent. On the other hand, they cost more and this additional outlay can only be recovered if the machines in which they are to be used have the power and rigidity to exploit the full potential of such materials. Again, the component itself and the material from which it is made must be compatible with the tooling selected. It would be not be cost effective to use a very sophisticated cutting tool on a low-powered and worn machine. Similarly, it would not be cost effective to use a cutting tool capable of very high rates of material removal if the component is of slender design which limits the clamping forces that can be applied and which, itself, cannot sustain the forces exerted upon it by such a cutter.

5.7 Screw-thread production (lathes)

So far, this chapter has only considered the production of plane, cylindrical and conical surfaces and combinations of these surfaces. It is now time to consider more complex geometrical surfaces. A screw thread is a practical application of a helix, and a helix can be defined as: *the locus (path) of a point travelling around an imaginary cylinder so that its axial and circumferential velocities maintain a constant ratio*. This is most easily understood by considering the configuration for cutting a screw thread on a centre lathe. It can be seen from Fig. 5.11 that the spindle of the lathe provides the rotational movement and that the lead screw provides the axial movement necessary to generate a helix. To maintain a constant velocity ratio between the two movements the spindle and the lead screw are connected by a gear train. By altering the gear ratio, the lead of the helix can be altered. The terms *lead* and *pitch* are often confused and their true meaning is shown in Fig. 5.12. Only single start threads, where lead and pitch are the same, will be considered in this chapter.

There are two main considerations when screw cutting:

- the lead and pitch of the thread;
- the profile of the thread (see section 8.8).

The tool is ground to the correct profile, and is set perpendicularly to the axis of the workpiece. It is fed more deeply into the rotating workpiece with each successive pass until the correct depth of thread is obtained. In practice the process is rather more complex and considerable skill is required on the part of the operator. Although widely used for

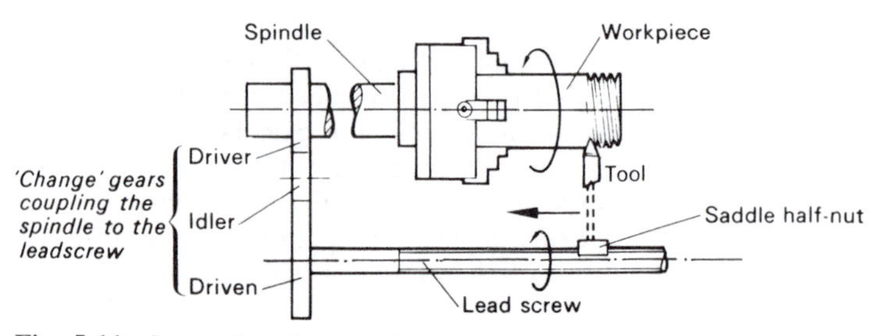

Fig. 5.11 Screw-thread generation

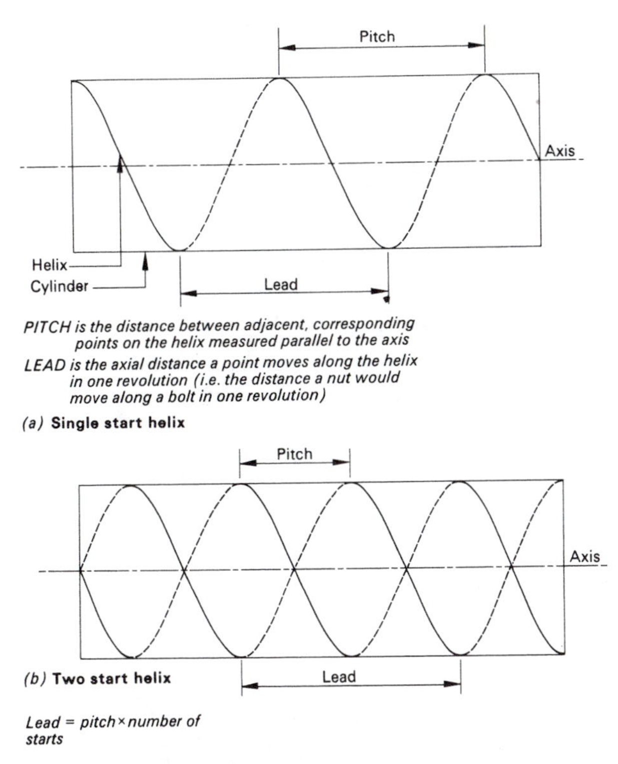

Fig. 5.12 Pitch and lead

proptotype development, screw cutting on the centre lathe is too slow and demanding in craft skill to be used for medium and high volume production. However, screw cutting with single point tools (similar to the centre lathe process) is used on CNC turning centres and is considered in *CNC in Manufacture and Computer Aided Machining* by R. Duffill and R.L. Timings (Longman, forthcoming). The flow forming of screw threads by rolling was considered earlier in section 2.9. Some alternative techniques used for quantity production will now be considered.

Capstan lathe

The capstan lathe was developed for medium volume production and is now being steadily superseded by CNC turning centres. However, they are still in use in industry and they are still being manufactured to a limited extent. Figure 5.13(*a*) shows an example of such a machine and Fig. 5.13(*b*) shows typical turret tooling. By using preset tooling which can be presented to the work sequentially and fed to preset stops, the skill of the centre lathe operator is not required, while the time per part is greatly reduced. Further, the accuracy and repeatability of the process is improved.

Self-opening die heads are used on capstan lathes for cutting internal and external threads, and two types of these are shown in Fig. 5.14

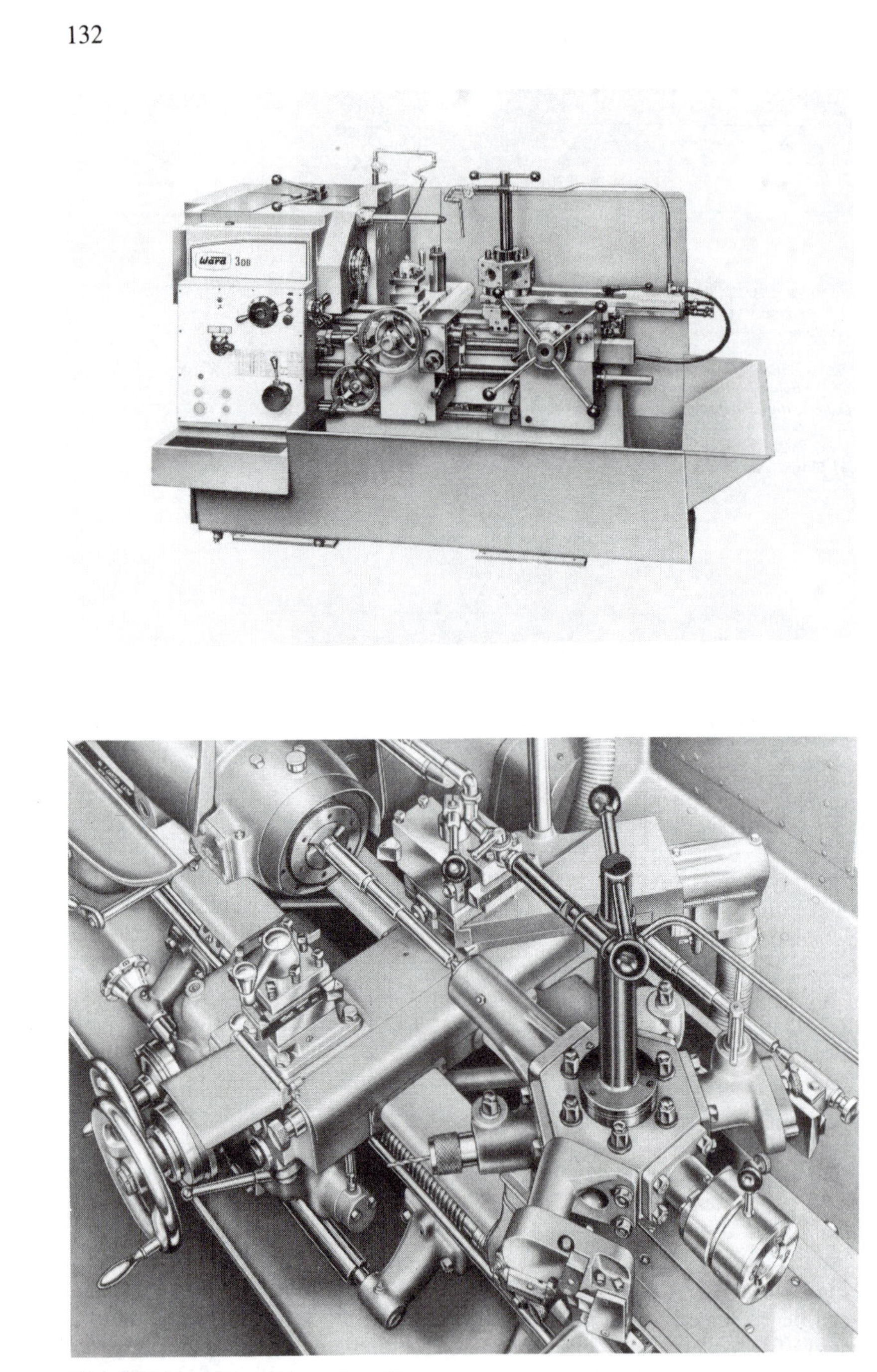

Fig. 5.13 Capstan lathe and tooling

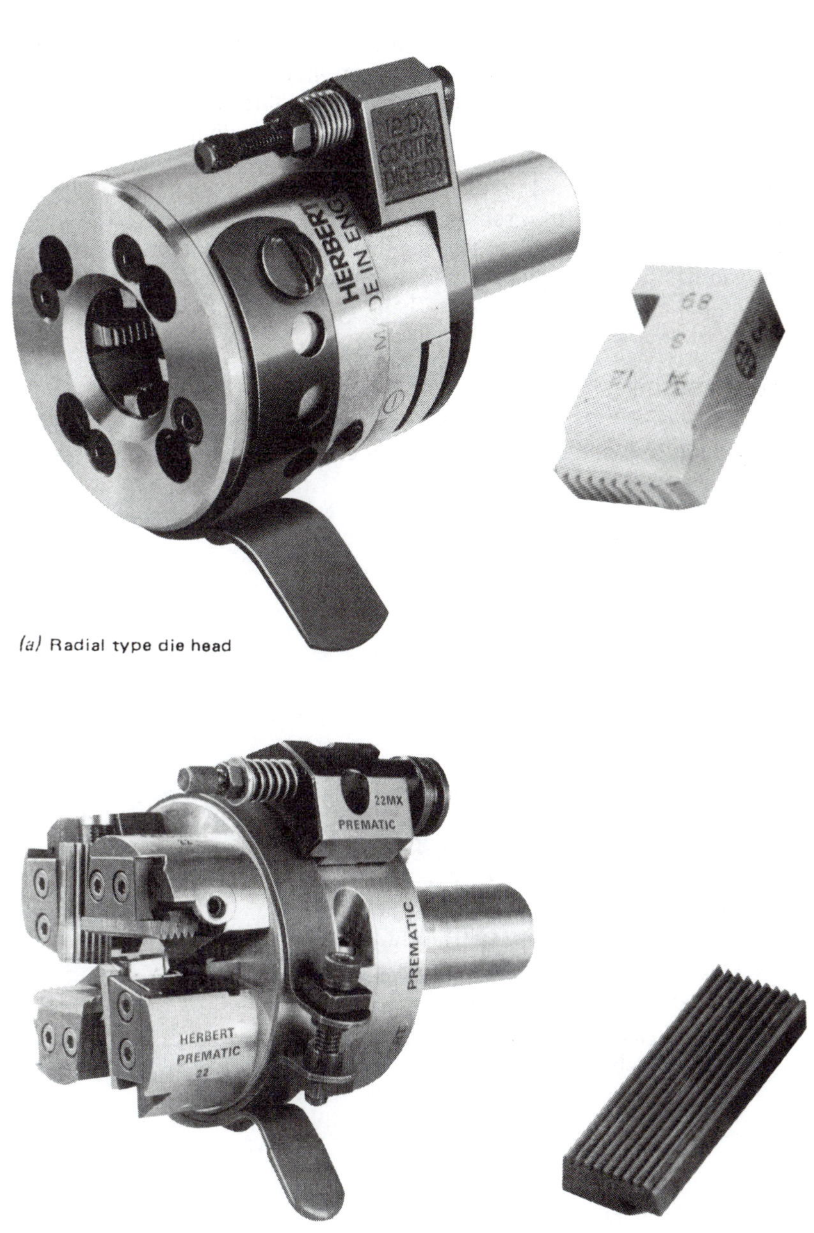

(a) Radial type die head

(b) Tangential type die head

Fig. 5.14 Die heads

together with their thread chasers. Both types of die head can be preset to size and are fitted with mechanisms whereby they can take a roughing cut followed by a finishing cut. When the end of the thread is reached, the die head opens automatically and can be withdrawn without having to stop or reverse the spindle.

Solid die heads are used on turret-type single-spindle automatic lathes as there is no operator to reset the die head between operations. The machine spindle is reversed so that the die head can be withdrawn when the thread has been cut. Sliding-head automatic lathes use single-point tools and the axial component of the helix is controlled by the traverse cam instead of a lead screw.

Thread chasing

Figure 5.15(*a*) and 5.15(*b*) show the difference between a thread cut with a single-point tool and a thread cut with a chaser. The nose radius of a

(a) Thread-form left by a single-point tool

(b) Thread-form left by a chaser

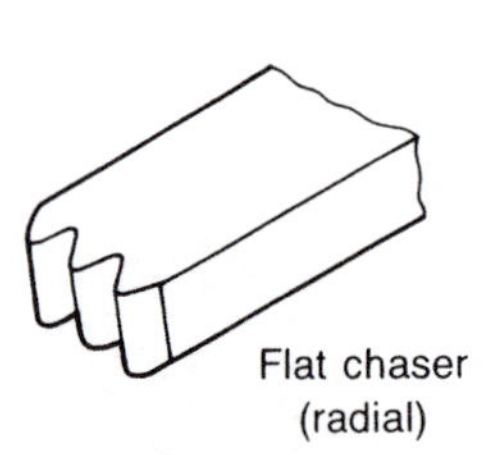

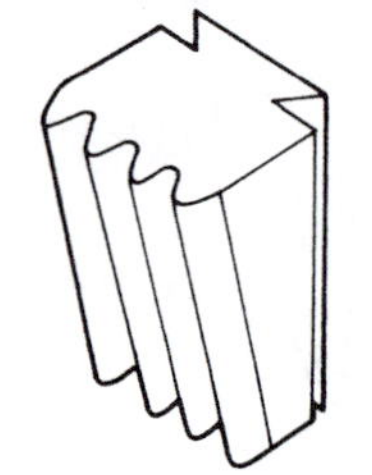

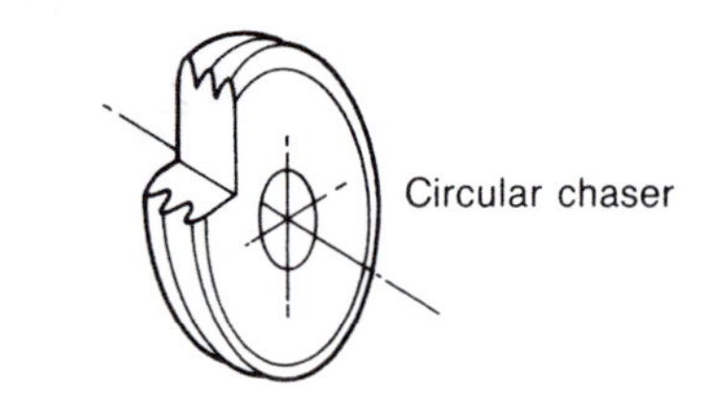

(c) Thread chasers

Fig. 5.15 Thread chasing

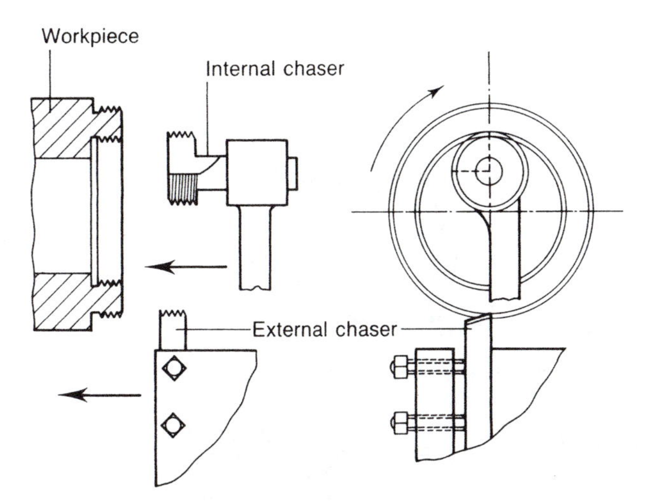

Fig. 5.16 Thread chasing: internal and external threads

single-point tool forms the root radius, but the crest of the thread is left flat and therefore it does not have a true form. The crest radius can only be formed by using a multi-tooth cutting tool called a *chaser*. Some typical chasers are also shown in Fig. 5.15(*c*). Thread chasing using die heads has already been described but single chasers can also be set in the tool posts mounted on the cross slide of a capatan lathe fitted with a short lead screw, as shown in Fig. 5.16. Internal and external thread chasing is used where the diameter of the screw thread is too large for a die head or where the screw thread is obstructed by a shoulder.

5.8 Thread milling

Thread milling is a rapid production process for cutting screw threads that are too large to be produced using die heads. Internal threads can also be cut.

Hobbing

This process uses a multi-tooth cutter and a full thread form is cut with radiused roots and crests. The principle of thread hobbing is shown in Fig. 5.17. It can be seen that the threads on the hob are *not* helical but are a series of discrete, annular, parallel threads. Some interference with the threads being cut does occur but this is minimal in large-diameter fine-pitch threads. The hob is gashed axially to produce cutting edges, and each tooth is form relieved to provide clearance. The hob is a few threads longer than the thread being cut and this limits the length of the workpiece thread. External hobs are set on a stub arbor, while hobs for internal threads have a shank for fitting into the machine spindle.

The workpiece and hob are carried on parallel spindles. The hob is

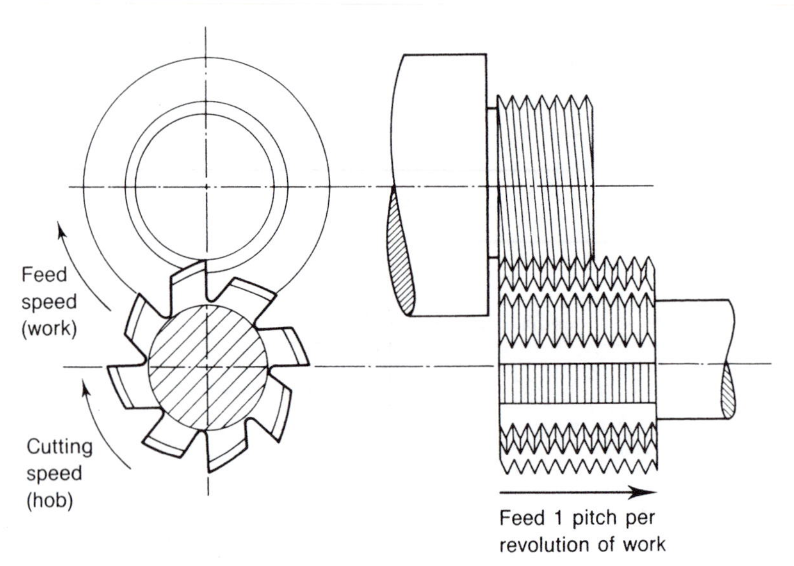

Fig. 5.17 Thread hobbing

rotated at the optimum cutting speed for the workpiece material, and the workpiece is rotated at a speed calculated to give the correct rate of feed. The cut is commenced by feeding the rotating hob into a stationary workpiece blank until the correct depth of thread has been achieved. The feed is engaged and the workpiece rotates at the correct feed speed. At the same time, the hob moves axially at the rate of one pitch per revolution of the workpiece. The relative directions of rotation and axial movement shown in Fig. 5.17 would produce a right-handed thread. By reversing the axial feed of the hob a left-hand thread would be cut. This process allows for threading close up to a shoulder.

Milling

The milling process is shown in Fig. 5.18. It is used for producing components such as worm gears which are, essentially, large-diameter coarse-pitch screw threads. A single form-relieved cutter is used and the axis of the cutter arbor and cutter is inclined to the axis of the workpiece spindle. This aligns the cutter with the angle of the helix and prevents interference with the flanks of the thread. Unlike thread hobbing, there is no limit to the length of the thread that can be cut. Unfortunately the thread cannot be cut up to a shoulder. The workpiece is rotated at one revolution per pitch for single-start threads and one revolution per lead for multi-start threads.

5.9 Thread grinding

Thread grinding is similar to thread milling except that a grinding wheel is used instead of a milling cutter or a hob. The thread form can either

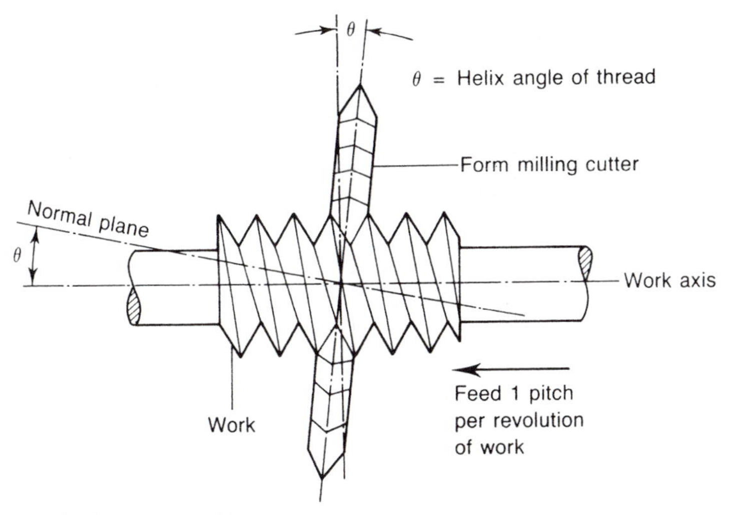

Fig. 5.18 Thread milling

be crushed or cut into the periphery of the grinding wheel. The former technique is used for fine-pitch threads. The wheel rotates slowly during crushing so that there is no slip between the abrasive wheel and the formed roller so that the roller is not abraded away. Alternatively, the profile can be cut into the periphery of the wheel using a diamond cutter controlled by a template and pantograph system, as shown in Fig. 5.19. This technique is used for coarse-pitch threads and worm gears.

Traverse grinding

The principle of traverse grinding is shown in Fig. 5.20. For fine threads a multi-ribbed wheel is used in which a number of annular thread forms are arranged side by side as in a thread milling hob. For coarse threads a single ribbed wheel is used and the wheel axis is inclined to the helix angle to prevent interference with the thread flanks. Fine threads are often ground from the solid from pre-toughened blanks but coarse threads are pre-machined, heat-treated, and thread-ground merely as a finishing and sizing operation.

Plunge cut grinding

The principle of plunge cut grinding is shown in Fig. 5.21. The wheel is plunged into the work to full thread depth with the work stationary. The work then makes one revolution while the grinding wheel traverses one pitch for a single-start thread or one lead for a multi-start thread. The thread can be cut up to a shoulder when plunge cut grinding.

5.10 Production tapping

Tapping machines and tapping attachments are available for use with conventional taps. The machines and attachments provide automatic

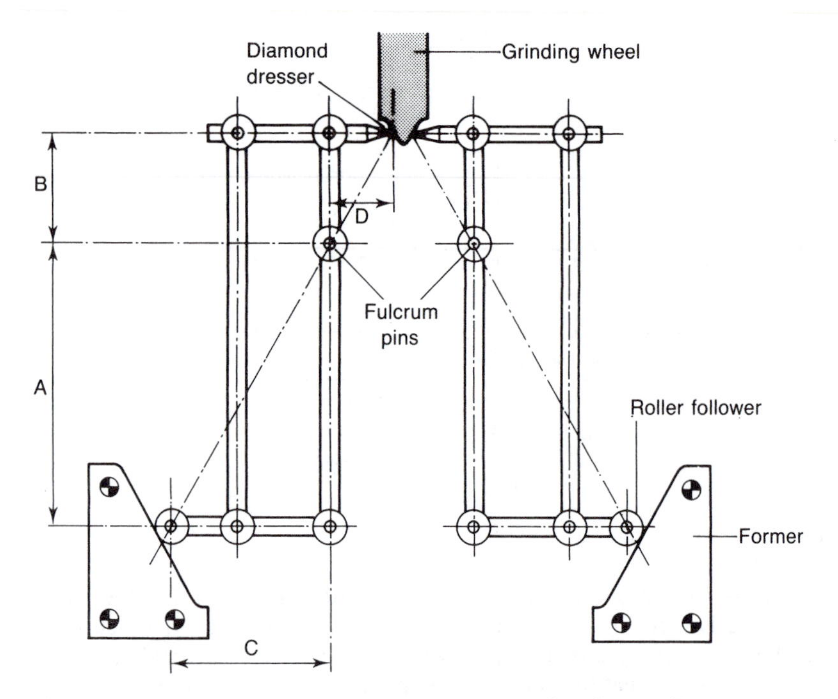

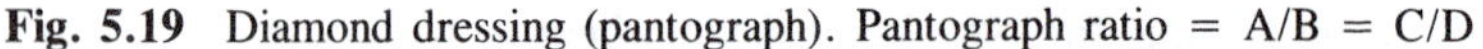
Fig. 5.19 Diamond dressing (pantograph). Pantograph ratio = A/B = C/D

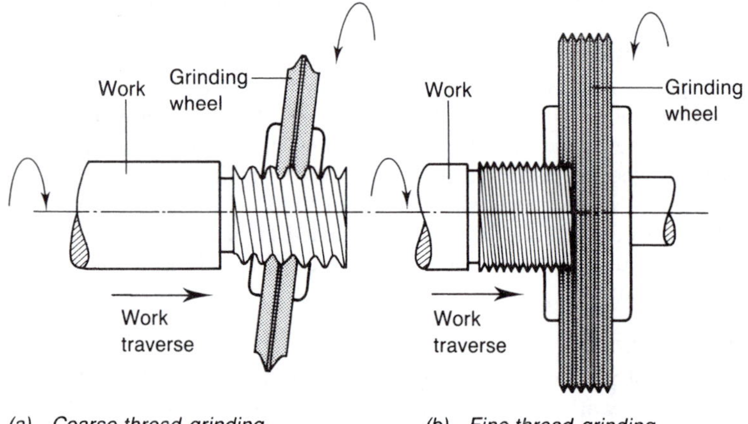

(a) Coarse-thread grinding

(b) Fine-thread grinding

Fig. 5.20 Traverse grinding

forward and reverse rotation of the tap, depending upon the direction of feed. A slight downward pressure on the tap engages the forward drive, while a slight upward pressure engages reverse drive and allows the tap to unscrew itself out of the threaded hole it has just cut. Another type of tapping attachment uses a spring-loaded clutch so that when the tap

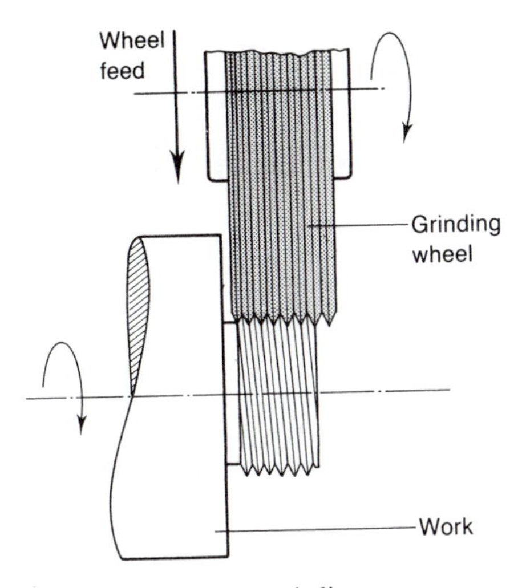

Fig. 5.21 Plunge cut grinding

reaches the bottom of the hole, or should the tap jam in the hole, the clutch slips and the machine spindle can be stopped and reversed. This latter type of tapping attachment is frequently used with large radial-arm type drilling machines.

Having to reverse the tap out of the hole is a waste of time. For nut tapping, where very large quantities are involved, *hook taps* are used, as shown in Fig. 5.22. The nuts run up the shank of the tap and spin off through a hole in the side of the tap carrier to be collected in an annular manifold. They fall away from the manifold by gravity into the work pallet.

5.11 Gear tooth production (forming)

A second complex surface that has to be considered is the *involute*. Figure 5.23 shows that an involute curve is generated by an imaginary taught cord BTC being unwound from a base circle centre O, commencing at the point A. The curve AC, so generated, has an *involute form*. The involute is the form of modern gear teeth and has the advantage that correctly meshing gear teeth of this form roll together with no sliding or scuffing to cause wear.

Milling is a form-cutting operation confined to single gears for prototype purposes, or very small batches of gears. Compared with the gear generation process to be described in section 5.12, gear forming is a very slow and uneconomical method of production. Figure 5.24(*a*) shows a typical form-milling gear cutter, and Fig. 5.24(*b*) shows the set-up for producing the gear on a horizontal milling machine. The cutter is the profile of the space between the gears, and it passes through the blank at a depth of cut dependent upon the strength of the cutter, the rigidity of the set-up and the machineability of the workpiece material. After each

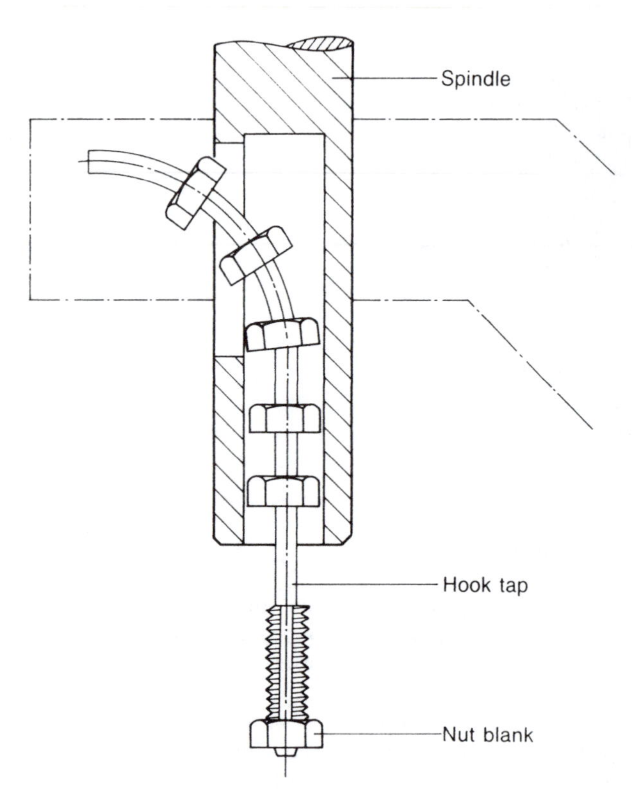

Fig. 5.22 Production tapping

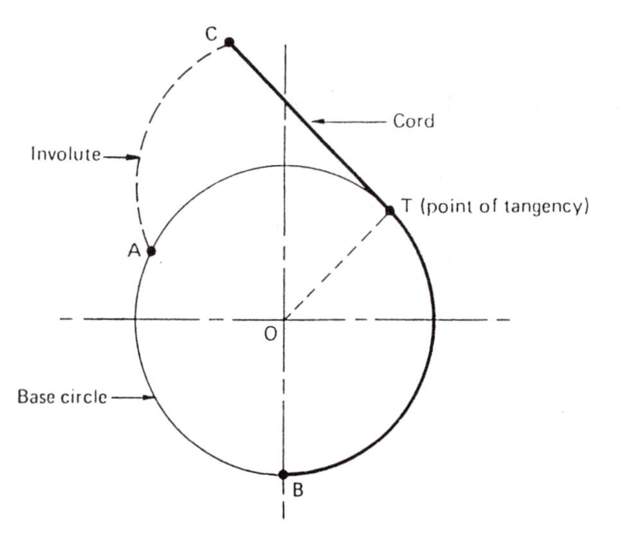

Fig. 5.23 The involute curve

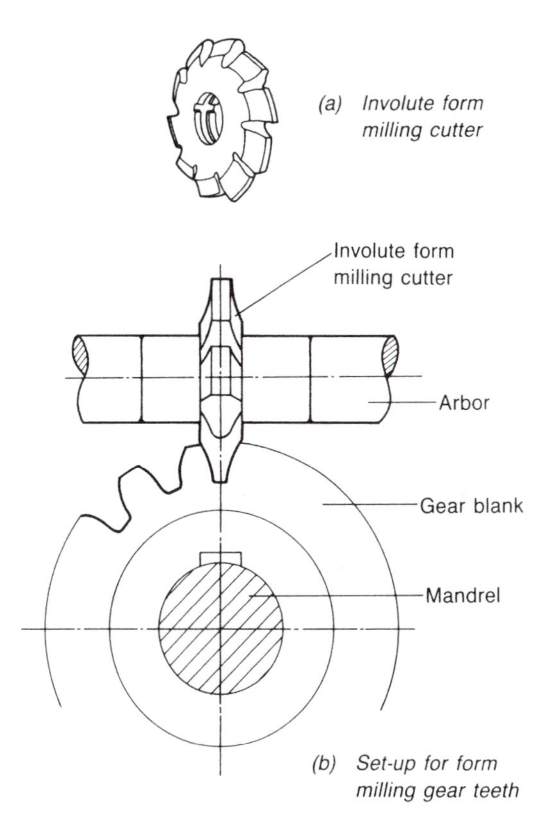

Fig. 5.24 Form-milling gear teeth

pass the gear blank is rotated (indexed) to the next position by a device called a dividing head upon which the gear blank is mounted. When all the spaces have been cut the cutter is set deeper into the blank and the process is repeated until the required tooth height is attained.

5.12 Gear tooth production (generation)

For the medium and high volume production of gears, generating processes are used. Generation depends upon the fact that a *rack*, with easily-machined straight-sided teeth, is an involute gear of infinite radius and will mesh correctly with a corresponding involute gear as shown in Fig. 5.25. All gear-generation processes are superior in respect of accuracy, finish, and rate of production compared with gear forming on a milling machine.

Planing

Figure 5.26 shows a gear blank, devoid of teeth, mounted on a mandrel which is geared to a slide carrying the rack cutter. The slide carrying the

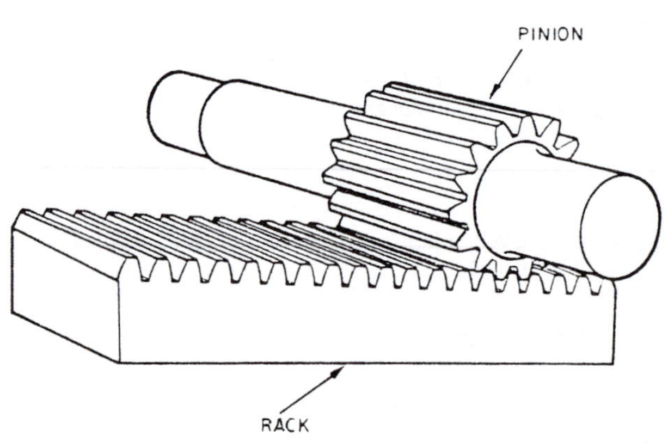

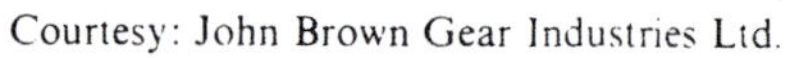
Courtesy: John Brown Gear Industries Ltd.

Fig. 5.25 Rack and pinion

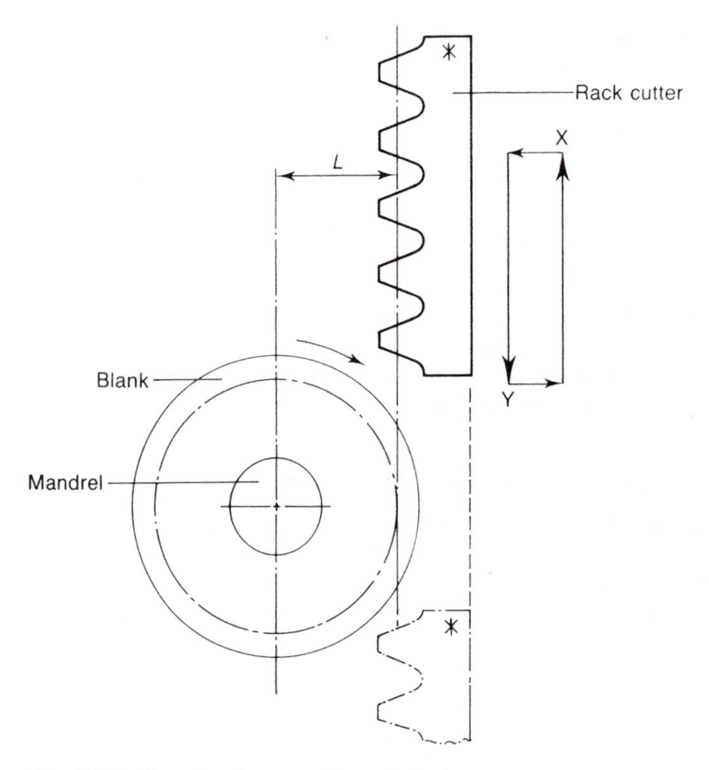

Fig. 5.26 Use of rack cutter (gear planing)

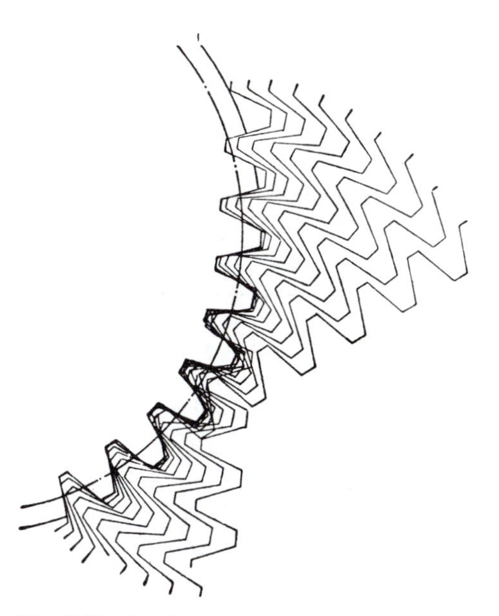

Fig. 5.27 Involute generation

cutter can move in the direction XY and is placed so that the distance L between the pitch line of the cutter teeth and the axis of the blank will give the correct depth of cut. The gearing between the cutter slide and the mandrel carrying the blank is arranged so that as the blank rotates, the cutter slide moves along the path XY exactly as if the teeth being cut engaged the teeth of the rack cutter without slip.

In order to cut the teeth, the rack edges are made into cutting edges by providing them with rake and clearance angles. The cutter slide and cutter are arranged so that they reciprocate in a plane parallel to the blank axis. Compared with this reciprocating motion, the rotation of the blank and the movement of the cutter along XY is relatively slow since this movement provides the feed of the cutter. The feed takes place intermittently during the return (non-cutting) stroke of the cutter.

Figure 5.27 shows the generating action of a rack cutter. For the sake of clarity, the blank is shown fixed and the rack is shown rolling around it. This provides the same relative motion between the blank and the rack. When the rack reaches the end of its XY travel it is automatically withdrawn and returns to its starting position.

Gear planing is a comparatively simple process using relatively cheap and easily-made cutters. These cutters can be made to a high degree of accuracy and are easy to regrind. It is the only process for cutting large diameter gears and can be used for the production of single- and double-helical gears.

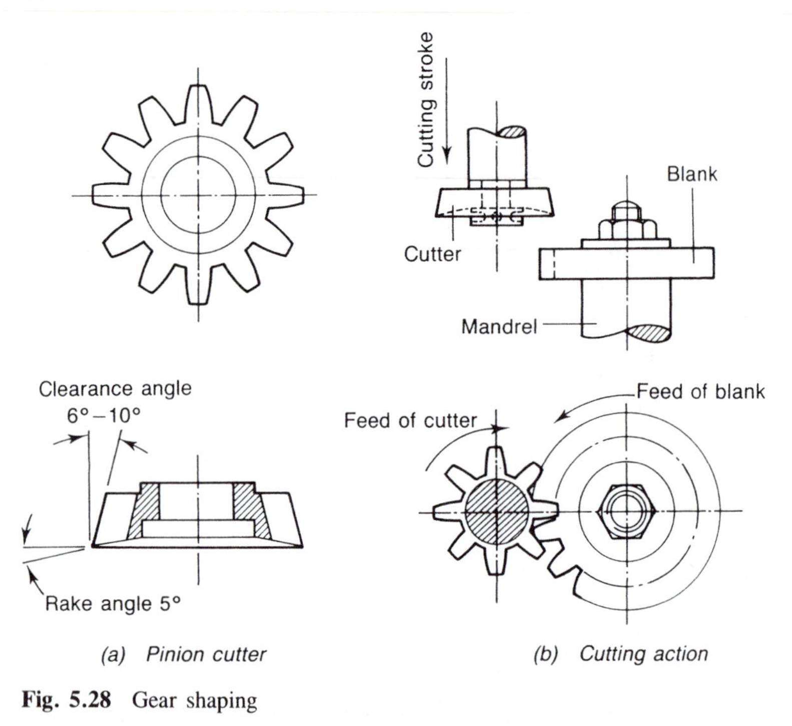

Fig. 5.28 Gear shaping

Shaping

Shaping is similar to gear planing except that the workpiece blank is set with its axis vertical and the cutter also reciprocates in a vertical plane. Further, a circular pinion-shaped cutter is used in place of a rack cutter. Figure 5.28(*a*) shows a diagram of a pinion cutter and indicates the rake and clearance angles. Figure 5.28(*b*) shows the configuration of the set-up for cutting involute teeth in the blank. While cutting takes place, the cutter and blank rotate together slowly as though they are two gears in mesh. Gear shaping is quicker than gear planing because the cutting process is continuous and the cutter does not have to keep being stepped back. Gear shaping is the most versatile of all gear cutting processes and is capable of producing internal and external gears. It can also be used for manufacturing close-coupled cluster gears and splines. The pinion cutter itself is produced by planing since the rack cutter is the fundamental tooth form.

Hobbing

Hobbing is the most productive of all gear-cutting processes since cutting takes place continuously. Compare this continuous cutting action with planing and shaping where cutting only takes place on the forward stroke of a reciprocating cutter which is retracted clear of the work on the return stroke. Hobbing cutters for gear cutting have a tooth form that is

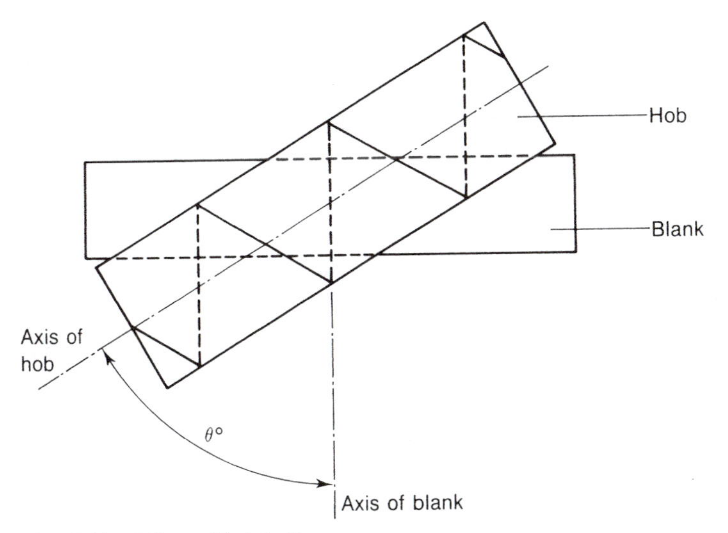

Fig. 5.29 Effect of hob helix

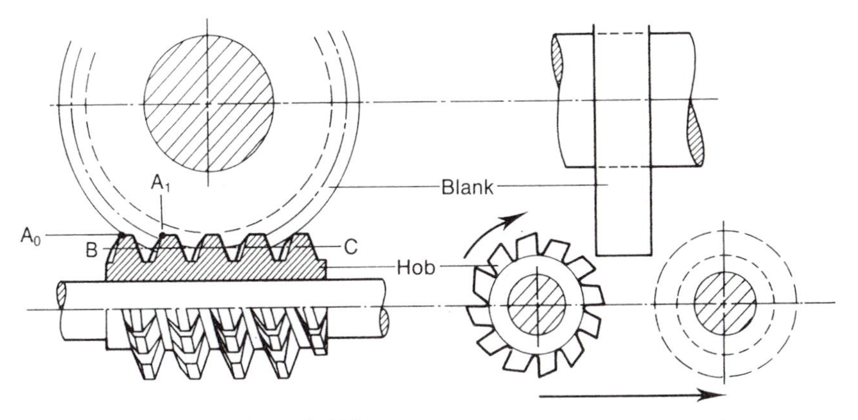

Fig. 5.30 Principle of gear hobbing

cut helically, unlike threading hobs where the tooth form is annular. For this reason the hob and its spindle are set at an angle to the work. This angle is equivalent to the helix angle of the cutter as shown in Fig. 5.29. The set-up for generating a gear using a hobbing cutter is shown in Fig. 5.30. Hobbing can only be used for producing spur gears and also worm wheels. It cannot be used to cut internal gears and it cannot work up to a shoulder.

Finishing

Gears can be finished by shaving or by grinding where high accuracy and surface finish are required to reduce noise in high-speed applications.

Gears are usually deburred using electrochemical machining as described in section 5.18.

Before moving on to non-conventional machining processes, it is necessary to consider two further machining processes associated with medium and high volume production.

5.13 Broaching

The broaching process uses a cutting tool called a *broach*, and was originally developed as a means of machining non-circular holes. More recently the broaching process has been adapted for the machining of external surfaces. Some typical broached surfaces are shown in Fig. 5.31.

The principle of broaching is shown in Fig. 5.32. This shows the simple broaching operation of cutting a keyway. The hole is first bored to the correct diameter and the broach is then pushed or pulled through it. Since the teeth of the broach stand progressively further out from the broach body, until the end is reached, each tooth successively deepens the keyway until the finished depth is reached. The broaching process can produce accurate internal and external surfaces and, for external surfaces, it competes with milling for large volume production where the quantity warrants the higher cost of the cutter. Figure 5.33 shows the elements of a typical broach and its teeth. There are three basic broaching techniques.

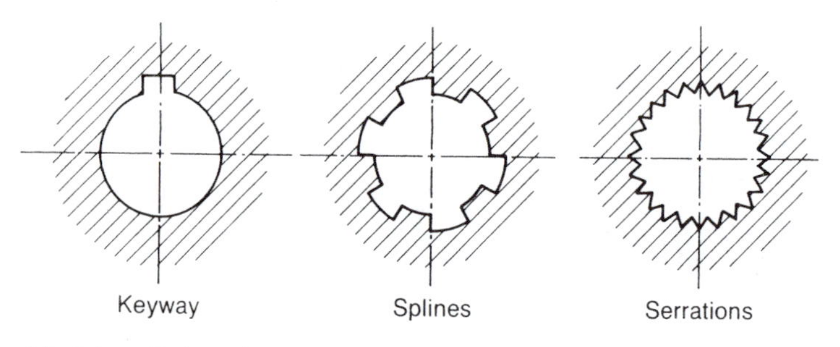

(a) Internally-broached surfaces

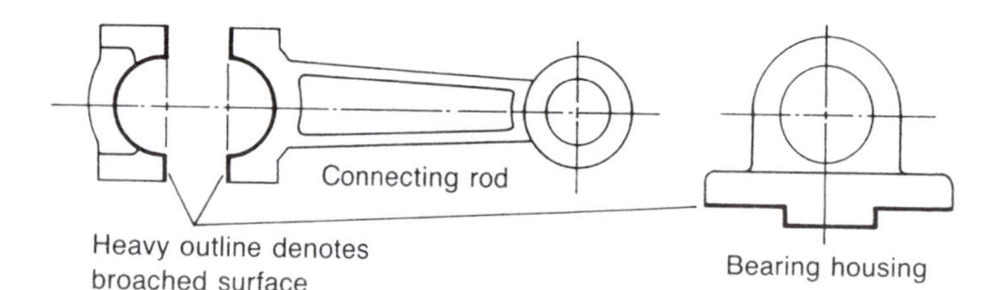

(b) Extermally-broached surfaces

Fig. 5.31 Typical broached surfaces

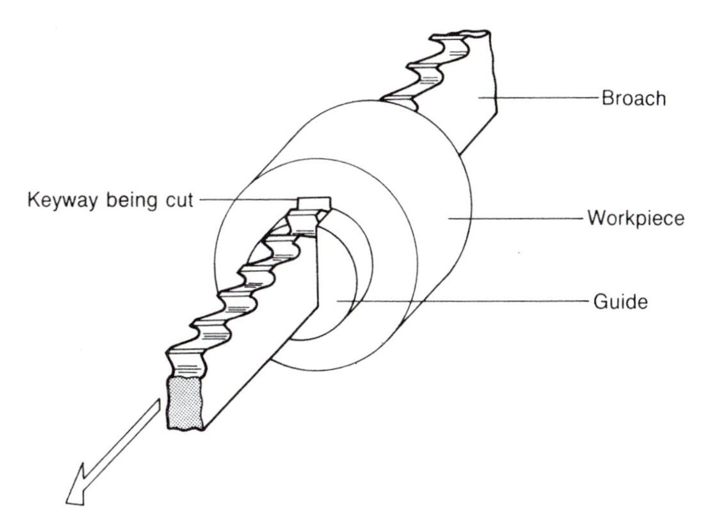

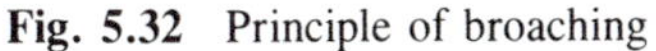
Fig. 5.32 Principle of broaching

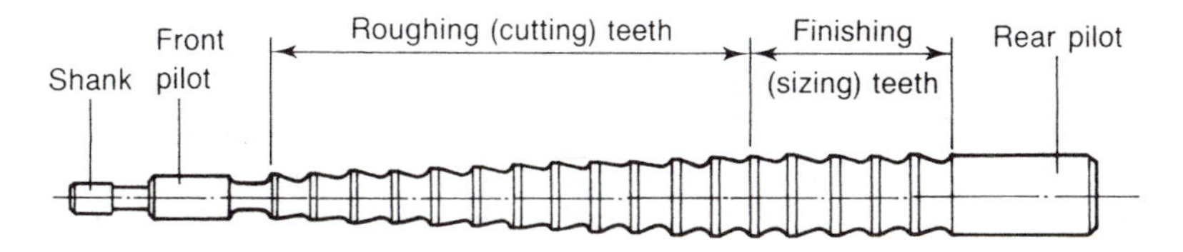

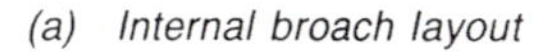
(a) Internal broach layout

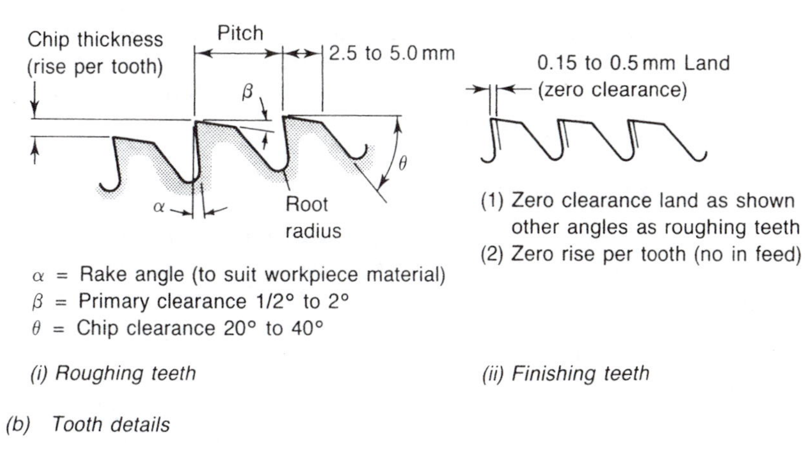

(b) Tooth details

Fig. 5.33 Broach details

Pull broaching

As the name implies, the broaching machine pulls the broach through the workpiece which must always be pierced with a pilot hole equal to the root diameter of the finished profile. Because of the inherent stability of drawing the broach through the workpiece, long broaches can be used and the work finished in one pass. Since the broach needs to be threaded through the component it must be fitted with a quick release shank for attachment to the machine.

Push broaching

Push broaching uses short broaches to prevent deflection and buckling. However, a short broach cannot remove as much metal in one pass as a long broach. Where large amounts of metal have to be removed, a sequence of separate broaches are pushed through the same component. Push broaches are easier and cheaper to make than the long pull broaches and they are also less likely to distort or crack during hardening. Because of the shorter stroke required, the press used with push broaching is simpler and less costly. Since push broaching is usually done in the vertical plane, work holding is easier and machine loading is quicker as the broach does not have to be coupled to the machine ram.

Surface broaching

The push and pull broaching techniques discussed so far have been concerned only with internal broaching. However, external or *surface broaching* techniques have been developed as an alternative to the milling process. Surface broaches are very expensive and are usually built up in short sections, not only for ease of manufacture, but so that in the event of uneven wear or accidental damage only part of the broach has to be repaired or replaced. Compared with milling, broaching is more expensive in tooling costs, but gives greater accuracy and improved surface finish to the work, coupled with higher rates of production. This is because the broaching teeth which take the roughing cuts never have to take the finishing cuts, and the slide which carries the broach is the only moving part of the machine during the cutting cycle so that the machine as a whole can be made more rigid.

5.14 Centreless grinding

Cylindrical grinding was introduced in *Manufacturing Technology: volume 1*. However, for many applications, the cost of mounting components in chucks or between centres is prohibitive. In other instances, the components may not lend themselves to such conventional methods of workholding.

Centreless grinding is an alternative manufacturing process which lends itself to the accuracy, production rates and cost control demanded by the high volume industries. Centreless grinding can provide metal removal rates of 0.25 mm per pass when roughing out, yet maintain a dimensional tolerance of 0.005 mm on finishing cuts. The process readily adapts to automatic sizing techniques and also to automatic component feed systems.

As the name suggests, the work is not held between centres but is supported on a work-rest blade and held up to the face of the grinding wheel by a second abrasive wheel called the control or regulating wheel. This control wheel not only rotates the workpiece but also provides the through feed in the case of long parallel work. Figure 5.34(*a*) shows the

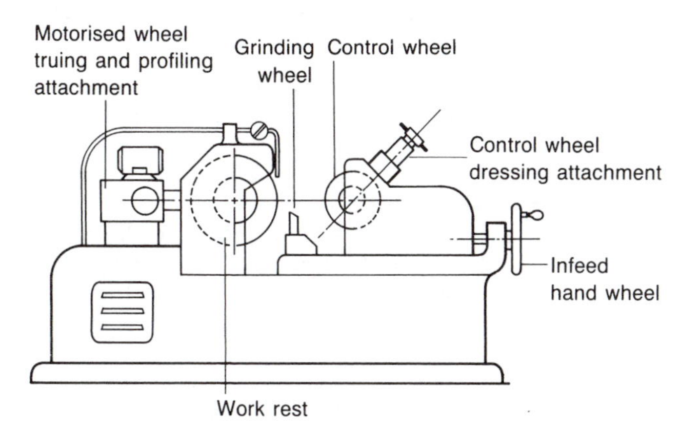

(a) Main features of a typical centreless grinding machine

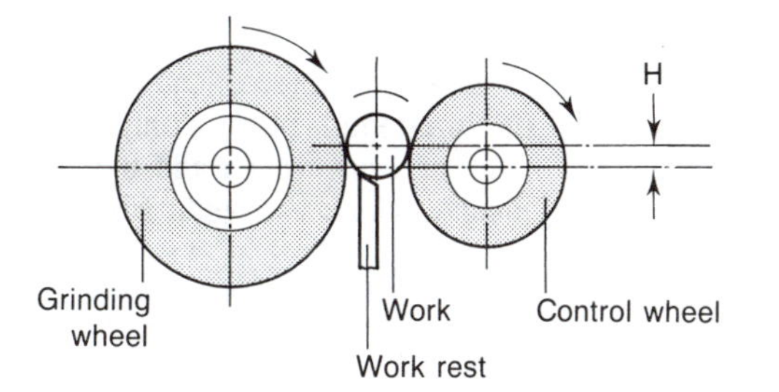

(b) Relationship between work and grinding/control wheels

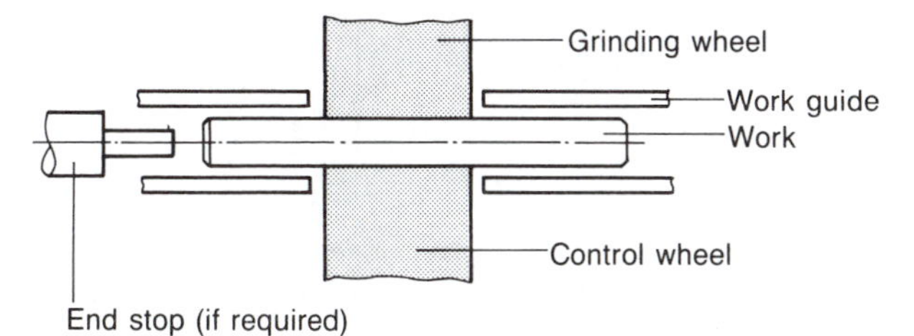

(c) Work guides and end stop

Fig. 5.34 External centreless grinding

layout of a typical centreless grinding machine, while Fig. 5.34(*b*) shows the relationship between the grinding wheel, the control wheel and the workpiece. Figure 5.34(*c*) shows the work guide and end stop in plan view. The end stop is not always required. The angularity of the work-rest blade is necessary to:

- assist in 'rounding-up' the workpiece and preventing lobing;
- keep the workpiece from being drawn into the grinding wheel and to keep the workpiece in contact with the control wheel;
- provide the control to ensure that the workpiece is brought gradually into contact with the grinding wheel when plunge grinding.

The work rest is set so that the axis of the workpiece is slightly below the centre line of the grinding wheel and the control wheel to help prevent lobing.

Through-feed (traverse) grinding

Through-feed grinding is used for parallel work of any length such as rods of silver-steel or the rollers of roller bearings. These components are of constant diameter and have no shoulders to prevent them from passing between the grinding wheel and the control wheel. In order to provide axial movement to the workpiece, the control wheel is inclined as shown in Fig. 5.35. This inclination produces an axial component velocity and, as is evident from the velocity diagram, the greater the inclination, the greater will be the feed rate (F).

Plunge grinding

Plunge grinding is used for short, shouldered workpieces, multi-diameter work, and form work. It is essential that the length of the component being ground is less than the width of the grinding wheel and the control wheel. Figure 5.36(*a*) shows some typical workpieces, while Fig. 5.36(*b*)

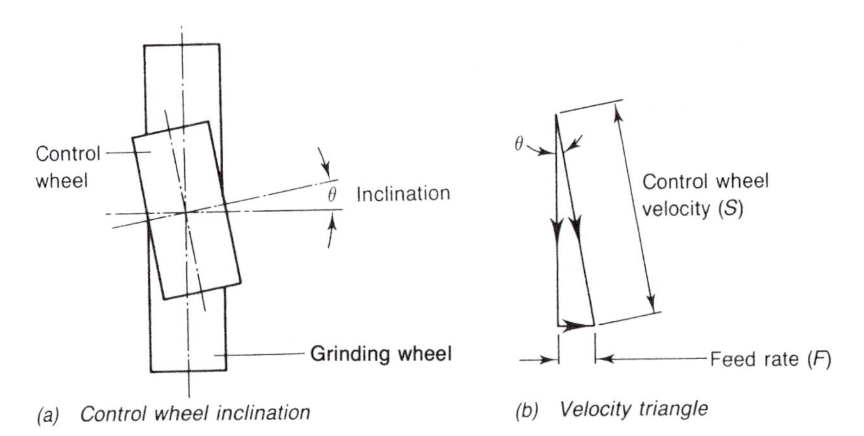

(a) *Control wheel inclination* (b) *Velocity triangle*

Fig. 5.35 Control wheel inclination for 'through-feed' grinding

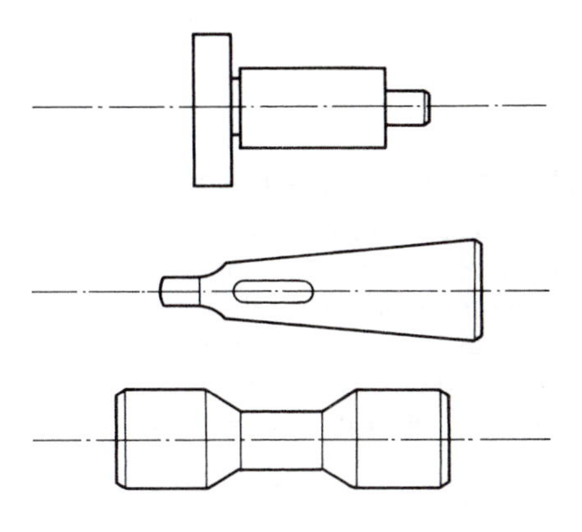

(a) *Typical plunge ground components*

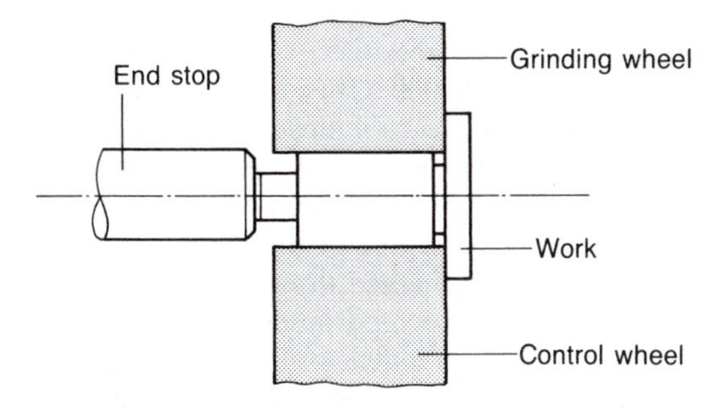

(b) *Set-up for plunge grinding*

Fig. 5.36 Plunge grinding

shows the principle of the technique. An end stop is used to position the work axially and the control wheel is given a slight inclination not exceeding 0.5° to keep the work fed up to the stop.

After the workpiece has been positioned against the stop, the control wheel is fed forward and advances the rotating workpiece up to the grinding wheel. When the control wheel slide is arrested by a positive stop, the control wheel is allowed to dwell while the grinding wheel 'sparks-out' to leave the workpiece the correct diameter. The slide and the control wheel are withdrawn and the work is automatically ejected by the end stop. As previously mentioned, the angularity of the work-rest blade ensures that the workpiece falls back against the control wheel and prevents it from being drawn into the grinding wheel.

End-feed grinding

This is a hybrid technique embodying the principles of both through-feeding and plunge-grinding techniques. It is used for work that is too long for plunge grinding but which cannot be fed through the machine because of a shoulder or other obstruction.

5.15 Electrical discharge machining (EDM) — principles

In addition to the conventional metal-cutting processes previously described there are several other processes which are increasingly being used. The first of these to be considered is electrical discharge machining (EDM) which is also called 'spark erosion' since it uses the thermal energy of electric sparks to remove workpiece material.

The earliest practical system was developed by the Lazarenko brothers in Russia during the 1940s, using the circuit shown in Fig. 5.37. This is included since its simplicity is useful in describing the principle of the process. The capacitor is charged from a direct current source through the resistor. Charging continues until the potential difference (voltage) across the capacitor exceeds the breakdown potential of the spark gap between the tool electrode and the workpiece. The spark then jumps across the shortest distance between the workpiece and the electrode. The extremely high spark temperature (~ 20 000°C) causes local melting and vaporisation of both the workpiece and the tool electrode. However, by connecting the workpiece to the positive side of the supply and the tool electrode to the negative side of the supply, the rate of erosion of the workpiece is made very much greater than that of the tool electrode. After each discharge, the capacitor is recharged through the resistor and the process is repeated. The resistor has two purposes: firstly, to control the time-constant of the circuit, i.e. the frequency of the discharges; secondly, to limit the current flow from the supply to a safe value during the charge and discharge cycle.

Both the tool electrode and the workpiece are immersed in a dielectric fluid, such as paraffin, or de-ionised water. The functions of the dielectric fluid are to cool the tool electrode and to flush away the debris caused by the discharge. Although the dielectric is normally an insulator, the very high potential across the spark gap immediately preceding the discharge causes local ionisation of the dielectric. This allows the spark to jump across the gap.

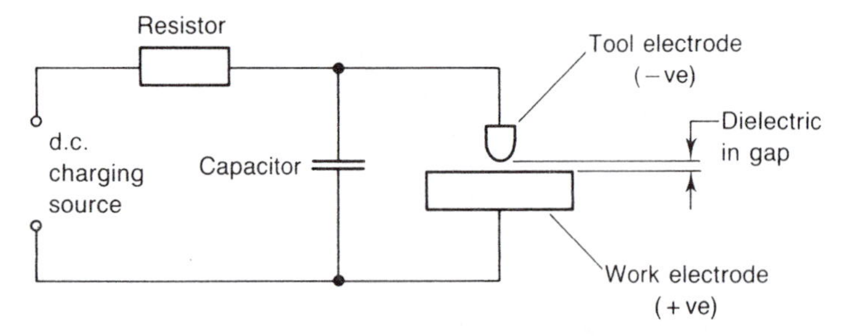

Fig. 5.37 Lazarenko R.C. (Resistance — Capacitance) circuit

Controlled pulse generation

This simple circuit is not satisfactory for actual production requirements as too much time is spent in charging the capacitor and too little time in effective machining. This problem has been overcome by the development of controlled-pulse generators which allow precise control of spark frequency and duration, thus allowing much higher metal removal rates.

Control of gap width

To attain maximum material removal, an optimum gap width has to be used and maintained at a constant value. This is achieved by means of a servo system which controls the movement of the tool electrode slide. The electronic control system constantly monitors the potential across the spark gap and compares it with a reference potential. If the potential across the spark gap rises relative to the reference potential, the servo closes the gap slightly. If the potential across the gap falls, the servo widens the gap slightly. A typical gap lies between 0.02 mm and 0.08 mm.

Dielectric flushing

Continuous flushing of the spark gap by the dielectric fluid is necessary in order to remove the debris of the erosion process. If the debris is allowed to build up in the spark gap a short circuit will occur. It is essential that the dielectric is filtered before it is recirculated. Further, continual flushing of the machining area improves the cooling of the tool electrode. Figure 5.38(*a*) shows a conventional flushing arrangement. Other techniques that are used are to:

- pump the dielectric through a hole or holes in the electrode, as shown in Fig. 5.38(*b*);
- draw the dielectric through the component by suction, if it already has holes drilled in it, as shown in Fig. 5.38(*c*).

Electrode design

Electrodes range from simple rods to complex three-dimensional shapes for producing cavities in dies and moulds. A typical machining sequence for a complex cavity would require several replica electrodes. All but one of the electrodes would be required for roughing out, using relatively high current densities resulting in rapid metal removal. The final electrode would be used to obtain the finished size and required surface finish at a lower current density and lower metal removal rate.

Electrode materials must be electrically conductive (as must the workpiece material) and, preferably, must have a low-erosion wear rate. The basic cost of the electrode material, the cost of shaping the material to the required size and shape, and the working life of the material must all be taken into account when selecting an electrode material. Typical electrode materials include: copper alloys, zinc-based alloys, aluminium alloys, tungsten carbide, and graphite. Graphite is the most widely used

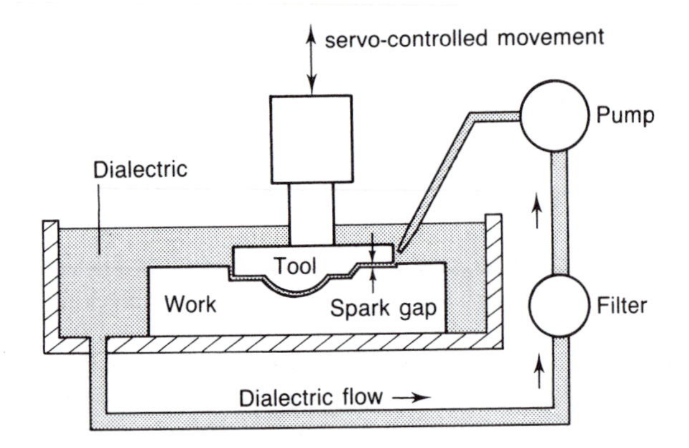

(a) Conventional dielectric flushing

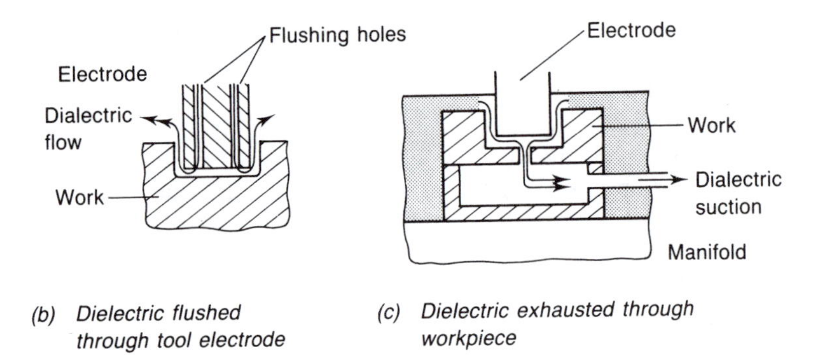

(b) Dielectric flushed through tool electrode

(c) Dielectric exhausted through workpiece

Fig. 5.38 Dielectric flushing

material as it has the best all-round capability. Its main drawbacks are its susceptibity to damage while being handled and set in the machine, and the fact that the dust produced while machining it to shape represents a severe health hazard if inhaled. Comprehensive dust-extraction facilities must be provided.

Application of CNC to EDM

Conventional CNC control systems can be applied to the x and y axes of the EDM machine table and also to the z motion of the electrode slide. This greatly increases the flexibility of the process and allows complex contoured shapes to be machined by the use of very simple circular or rectangular rod electrodes.

Wire-cutting

EDM wire-cutting is an important variant of the process. The erosion principles remain the same as those previously described, but the electrode is a continuously-fed wire, as shown in Fig. 5.39. The process

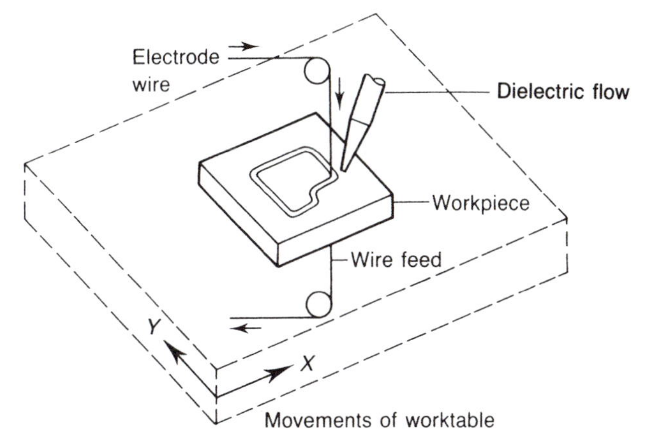

Fig. 5.39 EDM wire cutting

is particularly suitable for cutting complex profiles in pre-hardened press-tool die blanks using CNC profiling control. The use of pre-hardened blanks avoids distortion and cracking problems which can occur when heat treatment is carried out after conventional machining. Although a relatively slow process, it does not require attention and can continue 'lights out' if required. Further, the final pass of the electrode not only provides an accuracy and surface finish that is sufficiently good to avoid the necessity for further machining, but can also provide the taper or 'draught' necessary in blanking dies.

5.16 Electrical discharge machining — applications

The EDM process can be used to cut any electrically-conducting material regardless of hardness. Most of its applications are associated with the 'one-off' production of tools and dies. One advantage of the use of this process for producing tools and dies is the fact that the blanks can be hardened and tempered before machining. This prevents the cracking and distortion that can occur when heat treatment is carried out after machining due to variations in cross-section.

There are some instances where EDM is used for low-volume production, particularly in the aerospace industry, for machining exotic materials that are not easily cut by other means. The process can also be used to cut very fine holes (0.05 mm to 1.0 mm diameter), particularly fine holes with a high depth/diameter ratio, without the usual drill breakage problems. The advantages and limitations of the process can be summarised as follows:

Advantages

- Any electrically-conductive material can be cut, regardless of hardness and without causing distortion.
- The workpiece does not have to resist any cutting forces.

- The process is readily adaptable to automatic operation.
- The process is compatible with CNC control systems.
- EDM may be the only process available for machining some of the more 'difficult' aerospace alloys.
- Produces a non-directional surface finish. The surface consists of tiny craters with no definite lay.
- The surface texture can be varied as required, usually in the range 1.5 μm to 5.0 μm R_a. However, mirror finishes of 0.05 μm R_a are possible, as are deliberately-created coarse patterned surfaces.

Disadvantages

- The metal removal rate is slow compared with conventional machining.
- EDM can only be used with electrically-conducting workpiece materials.
- The repeated vaporisation and melting forms an 'as cast' surface layer. This causes later problems of surface cracking leading to fatigue failure.
- There are potential health hazards from contact with the dielectric fluid and the inhalation of spark-induced fumes from the dielectric fluid.
- Electrode wear can vary from almost zero with some workpiece material/electrode combinations, to situations where the tool electrode wears faster than the workpiece material, necessitating several electrode changes.

5.17 Electrochemical machining (ECM) — principles

This process can be used with any electrically-conducting workpiece material and dissolves away the material by electrolysis. The principles involved are the same as for electroplating, except that electroplating takes place at the cathode (negative electrode) of the cell, whereas electrochemical machining (ECM) takes place at the anode (positive electrode) of the cell. Figure 5.40 shows the essentials of the process. The tool electrode is fed into the workpiece at a controlled rate. Note that the shank of the tool is insulated except for a small land at the end. This is to confine the reaction to the end of the tool only, otherwise it would continue along the full length of the tool, causing a tapered hole to be machined in the workpiece. In the following description of the reactions which take place, it is assumed that a *steel* workpiece is being electrochemically machined.

When a potential difference is applied across two electrodes, the reaction which takes place at the anode is called *anodic dissolution* of the workpiece. That is, metallic ions are released from the workpiece surface as shown in reaction (1).

$$Fe \rightarrow Fe^{2+} + 2e \qquad (1)$$

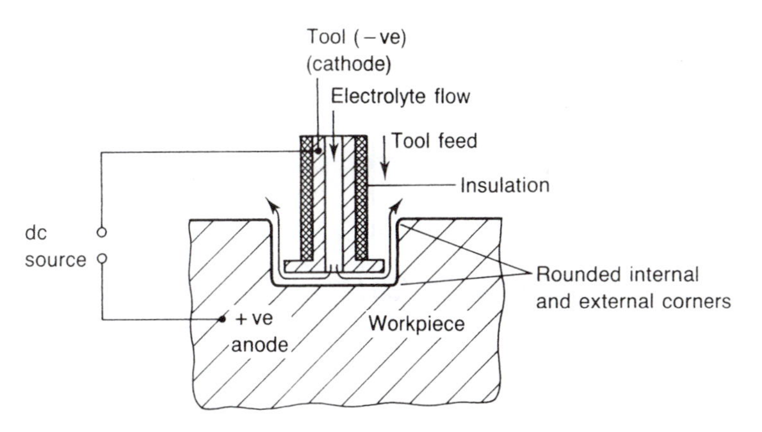

Fig. 5.40 Electrochemical machining (ECM)

Note that the metallic ion has lost two electrons and, therefore, carries a positive charge. Also two free electrons have been released.

At the cathode, the reaction is the generation of hydrogen gas — which bubbles off — and the production of negative hydroxyl ions. The electrolyte is a water-based solution (e.g. sodium chloride in water). Reaction (2) shows what happens at the cathode.

$$2H_2O + 2e \rightarrow H_2 + 2(OH^-) \qquad (2)$$

The positively-charged metallic ions released at the anode combine with the negative hydroxyl ions released at the cathode to form ferrous hydroxide which precipitates out and is flushed away by the electrolyte flow. Thus the overall reaction is:

$$Fe + 2H_2O \rightarrow Fe(OH)_2 + H_2 \qquad (3)$$

The ferrous hydroxide [$Fe(OH)_2$] precipitate initially forms a dark green sludge, but this later oxidises in air to form the typical reddish-brown sludge of ferric oxide [$Fe(OH)_3$].

5.18 ECM machine tools and applications

ECM machines appear in a variety of configurations but, typically, consist of a table for the workpiece and a vertical ram with controlled feed on which can be mounted the tool electrode. ECM machining poses many problems including electrical insulation, electrolyte containment and handling, and fume extraction since hydrogen is an explosive gas. A typical ECM configuration is shown in Fig. 5.41.

Control of the process

Compared with EDM, there is no significant electrode wear during ECM, so there is no need for a closed-loop servo system to maintain a

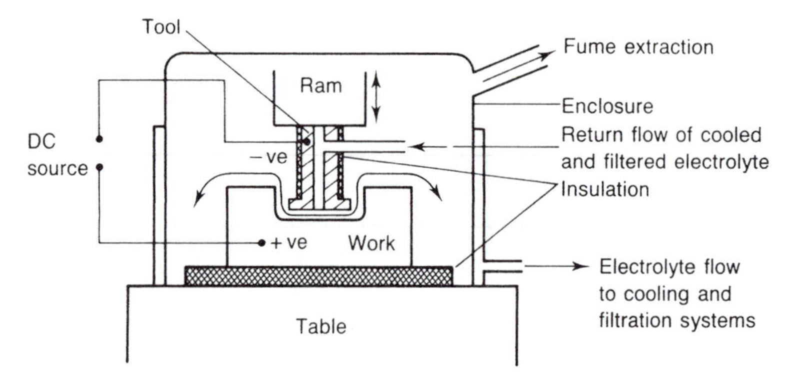

Fig. 5.41 Typical ECM configuration

controlled gap width. It is only necessary to feed the tool electrode into the work at a constant speed dependent upon the metal removal rate. The tool cuts a mirror image of itself into the workpiece. The rate of metal removal is proportional to the electrical current flow between the electrodes. Too high a current flow can cause overheating and boiling of the electrolyte, excess generation of gases, and a poor surface finish. Further, the increased potential difference between the electrodes, required to cause such a heavy current flow, could result in arcing between the electrodes.

Electrolytes

The electrolyte completes the circuit between the anode and the cathode and allows the electrochemical reaction to occur. It also transports away the machined particles and the heat of the reaction. The electrolyte has to be filtered and cooled before being recirculated. Water-based solutions of sodium chloride and sodium nitrate are the most commonly used electrolytes.

ECM applications

As an alternative to EDM, electrochemical machining can be used to cut any electrically-conducting material regardless of hardness. The main uses for this process are as follows: the 'one-off' production of complex internal cavity shapes in dies and moulds; the low-volume production of components made from materials that are difficult to machine by conventional methods or where distortion-free machining needs to be carried out on components that have already been hardened; and also for deburring after conventional machining.

A variation on the process is electrolytic grinding. Special grinding wheels are used which are porous and electrically conductive. These can be similar to conventional wheels but contain graphite powder, or diamond-impregnated copper wheels for the finest work. The wheel is kept lightly in contact with the work and the abrasive acts mainly as an insulator controlling the distance between the electrically-conducting elements of the wheel and the workpiece. The surface finish and rate of material removal depends largely upon how hard the wheel presses onto the workpiece. At high pressures some EDM as well as ECM takes place so that metal removal rates are relatively high at the expense of the surface finish. At low pressures only ECM takes place and the equivalent of precision lapped surfaces can be produced.

Advantages

- ECM can cut any electrically-conducting materials, regardless of hardness and without distortion.
- There are no cutting forces acting on the workpiece.
- ECM may be the only economically-viable process for materials that are difficult to machine or that have been pre-hardened.
- The surface produced is a very close copy of the electrode surface so, for example, a highly-polished electrode will produce a highly-polished workpiece surface.
- A high order of accuracy is obtainable due to the small gap between work and tool electrodes and the absence of wear on the tool electrode (typical gap 0.25 mm).
- The process is readily adaptable to automatic deburring. If a tool electrode is placed close to an area needing deburring, the current densities are highest at the peaks of the surface irregularities (the burrs) and these are rapidly removed.

Disadvantages

- Metal removal rates are slow compared with conventional machining.
- ECM can only be applied to electrically-conducting workpiece materials.
- There are difficulties with handling and containing the electrolyte.
- There are difficulties in safely removing and disposing of the explosive hydrogen gas generated during the process.
- The workpiece needs to be cleaned and oiled immediately after machining because of the corrosive effects of the electrolyte residue.
- The process cannot produce sharp internal or external corners and allowance also has to be made for overcutting, as shown in Fig. 5.42. For deep holes the tool electrode is often fitted with an insulating sleeve so that cutting only takes place at the end of the electrode.
- The pumping of high pressure electrolyte into the narrow gap between the workpiece and the tool electrode can give rise to large forces acting on the work and on the electrode.

5.19 Chemical machining

The term covers a variety of processes which use chemical action to remove material. The machining action is localised by protecting areas,

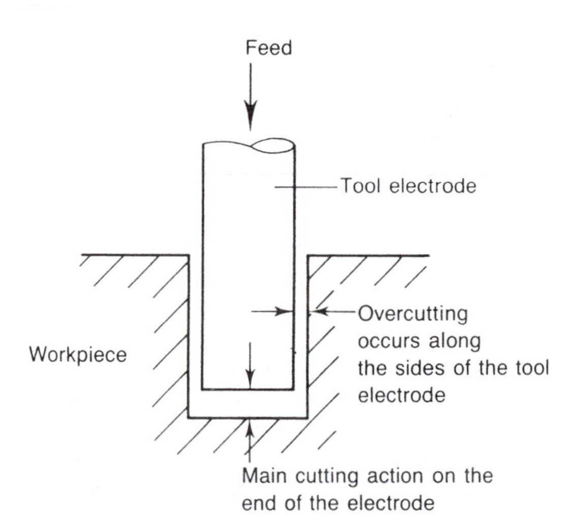

Fig. 5.42 Overcutting (ECM)

which do not need machining, with a chemically-resistant film called a '*resist*'. Usually the resist is applied all over the component and then removed in selected areas, either by mechanical cutting and peeling or by photographic processing. Alternatively, the resist can be applied only to the required areas by a screen-printing technique. The chemical solutions used to remove material are called '*etchants*' and can be either acid or alkaline depending upon the material being machined. When the workpiece is to be chemically machined only to a certain depth the process is called *chemical milling* or *chemical etching* but when the material is cut right through leaving a profile behind, the process is called *chemical blanking* or *chemical piercing*. In the former case the piece of metal removed from the sheet is the required component, while in the latter case it is the hole that is required. Figure 5.43(*a*) illustrates the general principles of chemical milling (etching) while Fig. 5.43(*b*) illustrates the general principles of chemical blanking.

Chemical milling

In this process the resist (also called a maskant) is applied to the component material surface by dipping or spraying. When the resist has dried, the required pattern is cut using a hand-knife guided by a template and the unwanted areas of maskant are peeled off. Alternatively, where greater accuracy is required, a CNC-controlled low-power laser may be used to cut the required shape in the maskant. A photographic technique similar to that described in the next (chemical blanking) section may also be used.

The workpiece material is then immersed in the etchant for a predetermined time period to obtain the depth of machining required. The etchant only acts on the unmasked areas, although some overcut occurs in

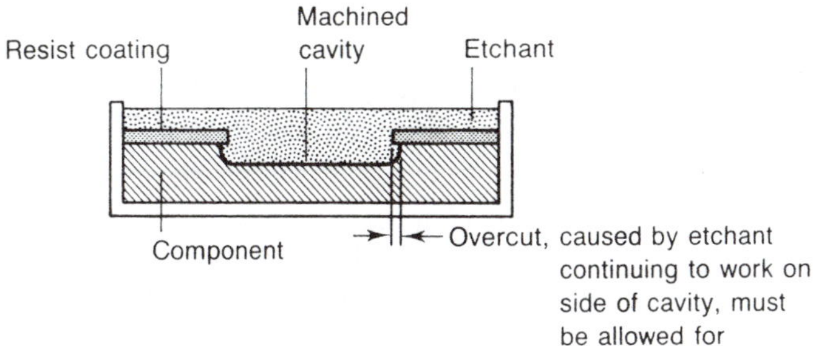

(a) *Chemical milling*

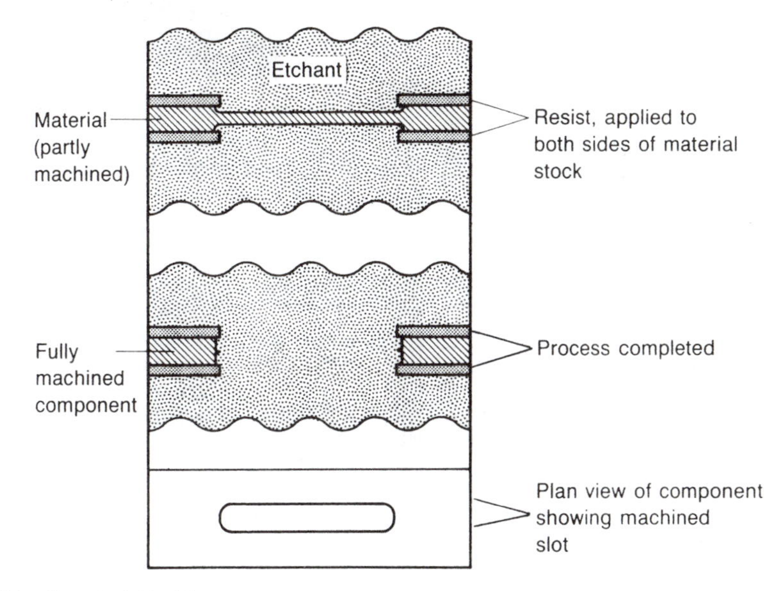

(b) *Chemical blanking*

Fig. 5.43 Chemical machining

the sides of the machined cavities as shown in Fig. 5.43(*a*), and this must be allowed for. Finally the component is rinsed to remove the etchant, after which the remaining maskant is either peeled off or dissolved by immersion in a suitable solvent.

A typical application for the process is for the production of aerospace components where the objective is to remove metal selectively to give a lighter component but to leave behind a lattice of integral strengthening ribs. The related process of chemical engraving uses exactly the same principles but the workpiece is only given a very brief exposure to the etchant to produce decorative artwork or labelling only lightly etched into the workpiece surface.

Chemical blanking

Chemical blanking almost always uses a photographic resist. This is a material which polymerises after exposure to ultraviolet light and, thereafter, becomes resistant to the action of the etchant. The process has the following stages:

(1) The required product shape is drawn out by hand or CAD at a known magnification onto a plastic sheet, as shown in Fig. 5.44(*a*). The magnification is necessary to capture the fine detail of the

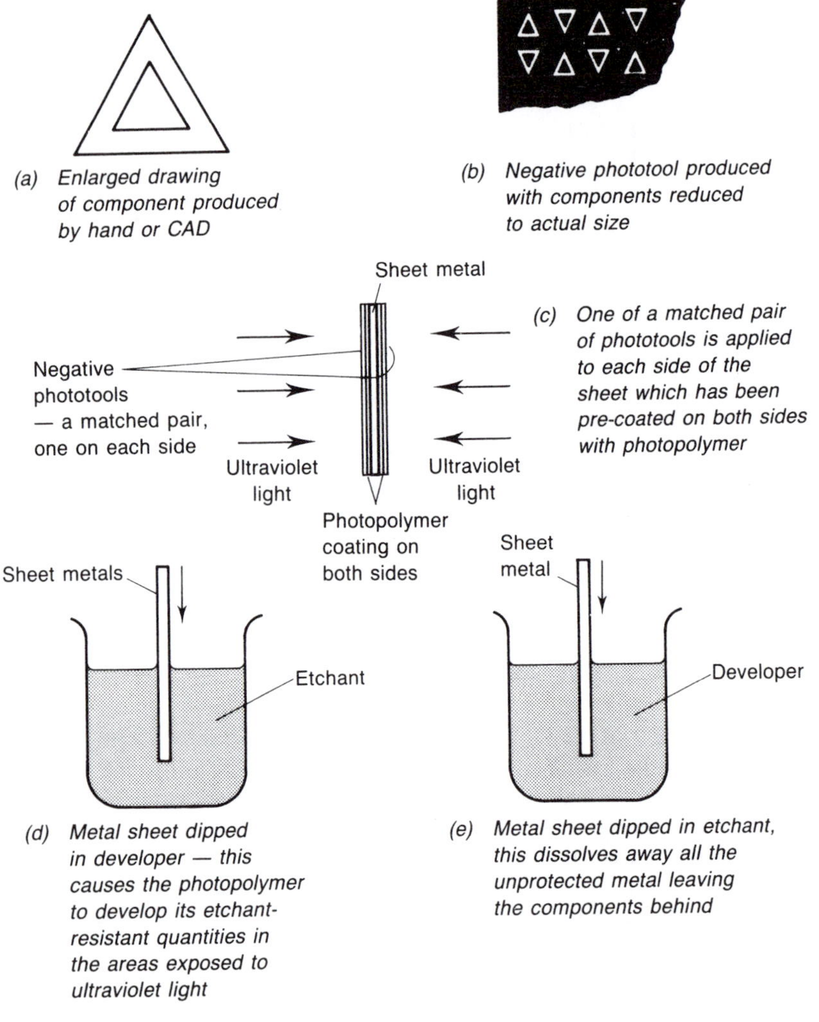

Fig. 5.44 Stages in chemical blanking

components since this will, typically, be small in size and complex in shape.

(2) This artwork is then photographically reduced to actual size to produce a photographic negative of the component. This is called a *phototool*. Where large quantities of a small component are required, a single phototool might have hundreds of economically-spaced components reproduced over its entire area, as shown in Fig. 5.44(*b*).

(3) The sheet of material to be processed is cleaned and coated on both sides with the photopolymer. Each side is then covered with a matched pair of phototool negatives.

(4) Both sides of the workpiece sheet are then exposed to ultraviolet light as shown in Fig. 5.44(*c*). The exposed sheet is then dipped into a developer, as shown in Fig. 5.44(*d*), which causes the coating to polymerise in the exposed areas and become resistant to the etchant.

(5) The unexposed, unwanted, photoresist is then removed by means of a solvent so as to expose the surface of unwanted areas of the metal sheet.

(6) The sheet of material and exposed resist is dipped into the etchant, as shown in Fig. 5.44(e), for the time required to dissolve away all the material not protected by the photoresist. All that remains are the required component shapes which are then removed and rinsed to halt the reaction.

Applications

For blanked components from thin foils and sheets, the process is an alternative to fine blanking and laser cutting. When the process is used to cut a cavity or form into the workpiece, it is an alternative to conventional milling, EDM or ECM. The main advantage of chemical machining is that it does not require extensive investment in expensive capital equipment and tooling where 'one-off' and low-volume production is required. Further, lead times when using chemical machining are also short. Where high-volume production is required, automated plant is available and this can be very costly depending upon the degree of automation and accuracy required. Chemical machining is widely used in the electronics industry where it is used to produce printed circuit boards and solid state devices.

Advantages

- Chemical machining can be used with almost any metal or alloy.
- It can be used for hardened materials.
- No cutting force is exerted on the workpiece.
- Chemical machining does not leave any residual stress or surface phenomena.
- When the complexity requirements are too great for fine blanking or the accuracy requirements are too great for laser cutting, chemical

machining may be the most economical process even for high-volume production.

- The components are free from burrs or fraze.
- For low-volume production only simple equipment is required.

Disadvantages

- Production rates are very slow compared with conventional machining or blanking.
- The process is limited to metals.
- Safety hazards: as with all corrosive chemicals care must be taken in the handling and disposal of the etchants. Fumes produced during etching must also be disposed of safely.
- Only rounded internal and external corners can be produced.

5.20 Laser cutting (principles)

- 'LASER' is an acronym for Light Amplification by Stimulated Emission of Radiation.
- Laser light is *monochromatic*, i.e. of one wavelength.
- Laser light is *coherent*, i.e. all the light waves vibrate at the same frequency, in the same direction, and in the same phase with each other so that their peaks and troughs match, as shown in Fig. 5.45(*a*).

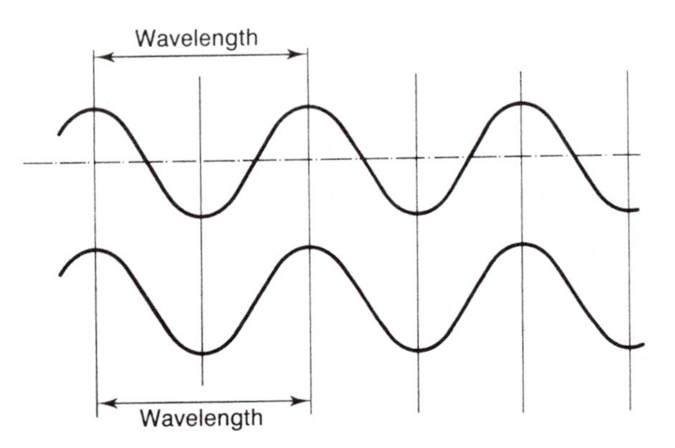

(a) Monochromatic, coherent laser light

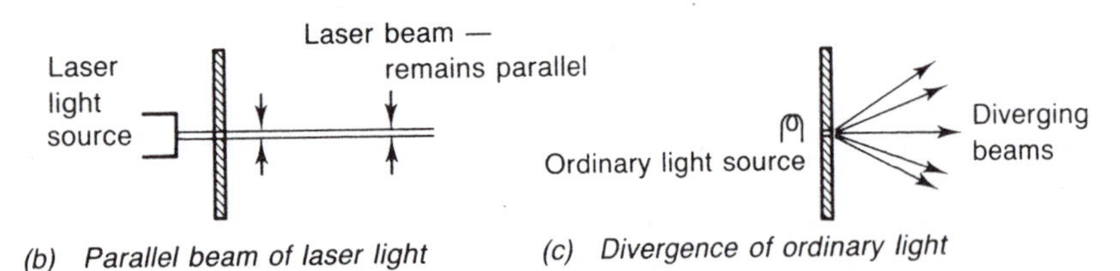

(b) Parallel beam of laser light *(c) Divergence of ordinary light*

Fig. 5.45 Characteristics of laser light

These features mean that laser light can be transmitted in an almost perfectly parallel beam, as shown in Fig. 5.45(*b*), unlike a point source of ordinary light which is highly divergent, as shown in Fig. 5.45(*c*). Laser light can also be focused onto a small area so that its energy is concentrated. With some types of laser this gives rise to localised high temperatures causing melting and vaporisation of the material.

Principles of laser operation

The types of laser suitable for metal cutting use either a transparent solid material or a clear gas mixture as the lasing material. An example of the solid state type is Nd–YAG which consists of a small amount of neodymium in a matrix of yttrium, aluminium and garnet. A typical metal-cutting 'CO_2' gas laser uses a mixture of carbon dioxide, nitrogen and helium in the ratio 5:55:40 as the lasing material.

Regardless of whether the lasing material is a solid or a gas, the basic laser configuration is as shown in Fig. 5.46. An intermittent external energy source is needed to initiate and maintain the lasing action. For solid state lasers, the energy is derived from light flash sources similar to those used for flash photography and stroboscopes. The light energy passes through the transparent walls of the lasing material. This is called *optical pumping*.

In a gas laser, the lasing medium is contained within an electrical discharge tube and pulses of energy are passed into the gas via electrical discharges between the anode and the cathode. This is called *discharge pumping*.

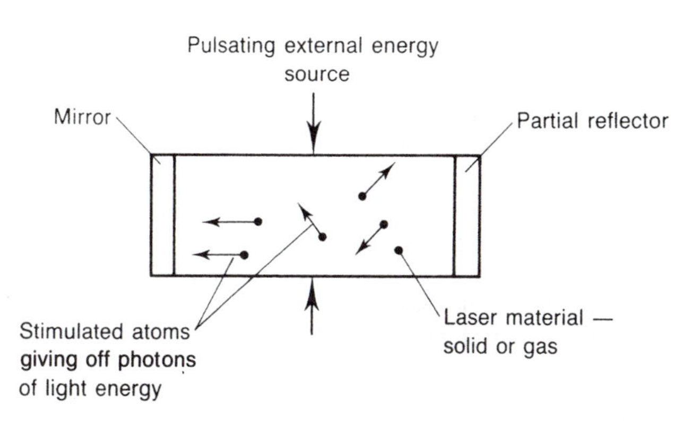

An intermittent external energy source is needed to initiate and maintain the lasing action. For solid state lasers, the energy is derived from flash lamps. The light energy passes through the transparent walls of the lasing material. This is called optical pumping

Fig. 5.46 Basic laser configuration

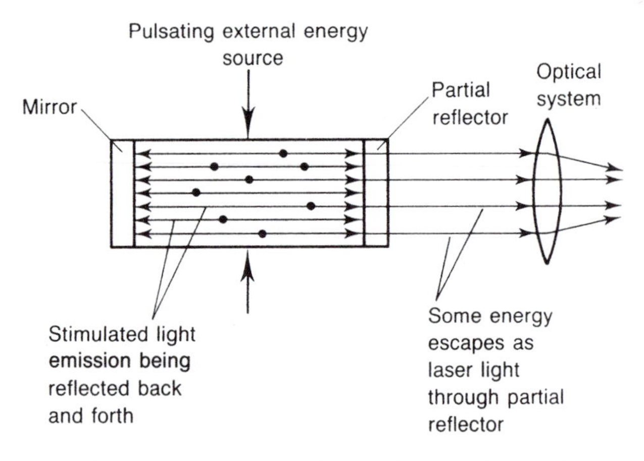

Fig. 5.47 Configuration of a laser with external optical system

No matter which system is used, atoms within the lasing material are stimulated by the external energy source into giving off photons of light energy in all directions. Some of this stimulated light emission occurs along the axis of the laser and is reflected back and forth between the mirror and the partial reflector as shown in Fig. 5.47. This causes a 'cascade effect' where more and more atoms are stimulated into giving up their photon energy so that they are all in phase with the pulsing of the light source. Some of the energy is emitted as a continuous stream of laser light through the partial reflector and can then be focused by an optical system. Only optically-pumped solid state lasers and discharged-pumped CO_2 gas lasers have sufficient power and reliability for cutting operations.

Optically-pumped solid state laser

The Nd–YAG solid state laser-cutting configuration is shown in Fig. 5.48. It is relatively compact, robust and reliable. It is also easy to maintain in a harsh workshop environment. The main limitation is the difficulty in keeping the solid lasing material cool enough to prevent degradation. This limits the power to a maximum of 1 kW, with 0.5 kW being a more common rating. Such a laser is ideal for drilling and profile cutting in sheet metal up to 5mm thick. It can also be used for spot and seam welding.

Discharge-pumped CO_2 gas laser

The CO_2 laser is not so compact as the Nd–YAG type, but it is much more powerful with up to 15 kW available as output. Figure 5.49 shows the configuration of a gas discharge laser. The greater output power is obtainable because of the inherently greater efficiency of the lasing medium and the relative ease of cooling. The gas itself can be passed at high velocity through a heat exchanger, giving very effective heat

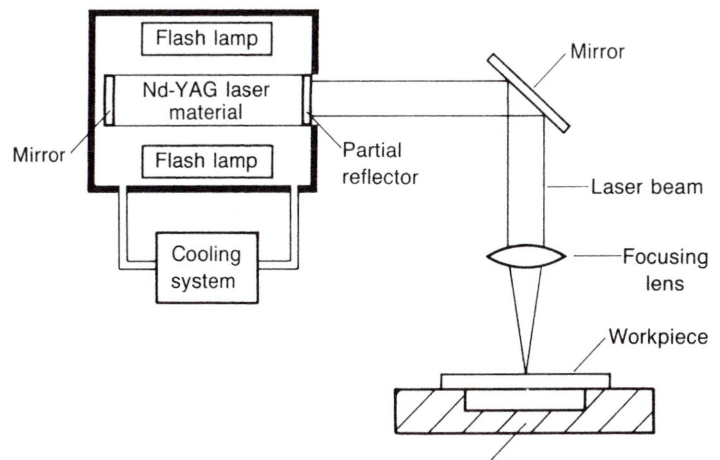

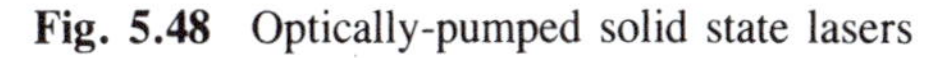
Fig. 5.48 Optically-pumped solid state lasers

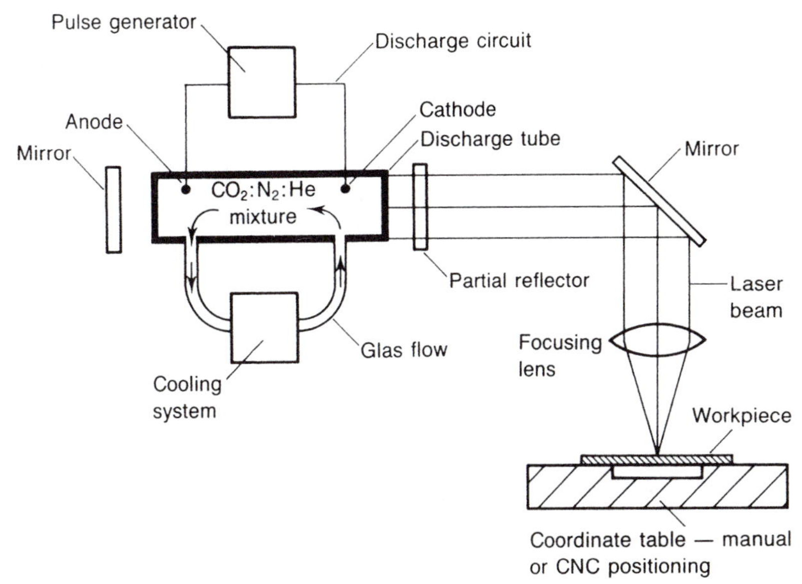

Fig. 5.49 Discharge-pumped CO_2 gas laser

control. The applications are as for the solid state laser except that the greater power allows profile cutting of metal plate up to 20 mm thickness.

Laser cutting applications

For cutting operations the laser light is focused on a small spot. The concentrated energy causes local melting, evaporation and ablation. Most

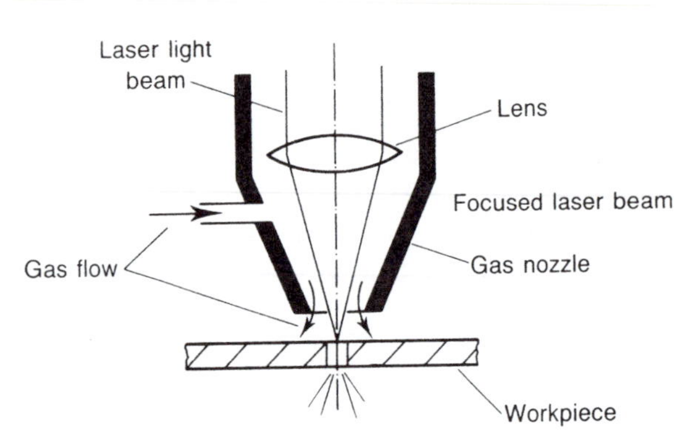

Fig. 5.50 Gas-assisted laser cutting

practical cutting applications of lasers use an assisting gas to increase the efficiency of the process. The functions of the assisting gas are cooling and cleaning the cut area of molten and evaporated material. With materials such as steel a reactive gas (oxygen or air) is used to cause an exothermic reaction. The additional heat energy generated increases the cutting speed. With highly reactive materials such as aluminium and its alloys, a protective gas atmosphere (nitrogen or argon) is used to prevent excessive oxidation. The assisting gas is usually delivered to the cutting zone through a co-axial nozzle as shown in Fig. 5.50.

The process can be used to mark, drill and cut a wide range of materials. These include metals such as the nimonic alloys which are difficult to cut with conventional tools, and non-metals such as ceramics, plastics, glass, wood, leather and cloth.

Advantages

- High cutting speed.
- Small heat affected zone.
- Good profiling accuracy due to small cutting area.
- Smooth finish along the cut edge.
- Compatible with CNC systems.
- Readily adaptable to robots.
- The workpiece is not subjected to any cutting forces.
- The cutting beam can be directed and focused over long distances and into inaccessible positions.

Disadvantages

- The relatively high capital cost of the equipment and the relatively high operating cost of the consumable gases.
- Hazards resulting from the laser beam itself and from the vaporised workpiece materials.

Assignments

1. With the aid of sketches explain the difference between *combination* and *follow-on* press tools.
2. Justify the cutting process selection for a component of your choice in terms of cost, accuracy, repeatability, and finish.
3. Compare and contrast the advantages and limitations of the following cutting tool materials:
   (a) high speed steel (HSS);
   (b) sintered carbide;
   (c) 'Sialon'.
4. Compare the advantages and limitations of 'coated' carbide tool tips with conventional carbide tool tips.
5. Compare and contrast the production of screw threads by turning, milling, hobbing, and grinding, in terms of cost and quality.
6. Describe, with the aid of sketches, the difference between gear tooth production by forming and by generation.
7. (a) Describe the processes of:
      (i) gear planing;
      (ii) gear shaping;
      (iii) gear hobbing.
   (b) Compare the advantages and limitations of each of the above processes in terms of their applications, accuracy, finish, and process cost.
8. With the aid of sketches, describe the basic principles of the broaching process.
9. (a) With the aid of sketches describe the principles of the centreless grinding process.
   (b) Explain how 'out of roundness' is avoided when centreless grinding.
   (c) Compare and contrast the advantages and limitations of centreless grinding with plain cylindrical grinding between centres.
10. With the aid of sketches decribe the basic principles of:
    (a) electric discharge machining (EDM);
    (b) electro-chemical machining (ECM);
    (c) chemical machining.
11. For a component of your choice, justify its manufacture by an appropriate process from those listed in question 10, giving reasons for your choice of process.
12. (a) With the aid of sketches, describe the principle of the laser cutting process.
    (b) Compare and contrast laser cutting processes with other thermal cutting processes (e.g. oxy-fuel gas cutting).

6 Advanced manufacturing technology

The computer numerial control of machine tools (CNC) was introduced in *Manufacturing Technology: volume 1*, Chapter 5 and, at the end of that chapter, two simple part programs were included as an example of manual programming. This chapter will develop the techniques of manual part programming still further and will also examine some methods of computer-aided part programming.

6.1 Subroutines

When a program contains fixed sequences of frequently-repeated patterns, these sequences can be stored in memory as a sub-program or *subroutine* to simplify the task of programming. The principles of constructing a subroutine are essentially the same as for conventional computer programming since a subroutine is simply a set of instructions that can be called up and inserted repeatedly into the main body of a program by entering an identifying code. Figure 6.1 shows the general principle where it can be seen that the subroutines themselves can call up other subroutines if required. Figure 6.2 shows a typical component with a repeated pattern of milled slots which are to be cut 5 mm deep using a 20 mm diameter slot drill.

The following example shows how the main program and the subroutine are constructed to suit a Fanuc controller using the codes M98 (subroutine call) and M99 (subroutine end) together with the letter

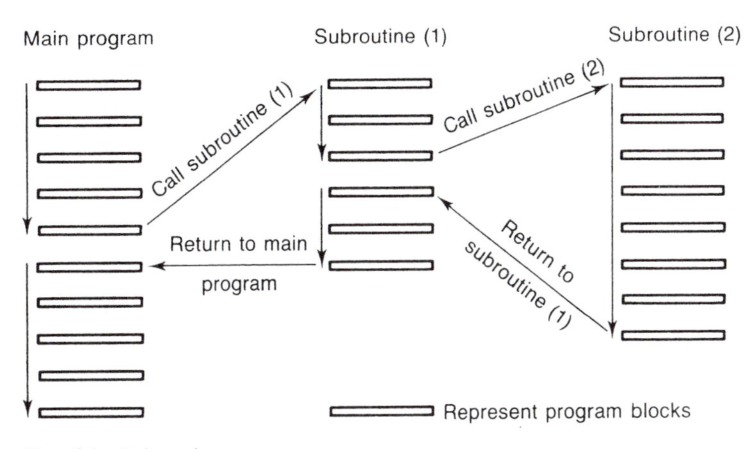

Fig. 6.1 Subroutines

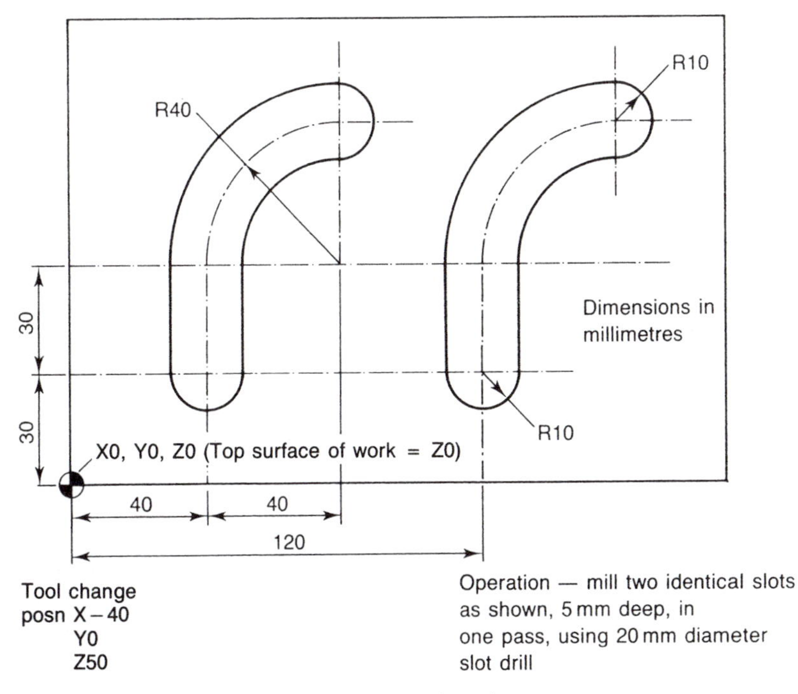

Fig. 6.2 Typical component requiring a subroutine

address P to identify the subroutine. The following program segment assumes that earlier program blocks have established correct feeds and speeds and tool offset. The program uses tool centreline programming only and assumes that radius compensation is not needed. Absolute mode is active at the start of the main program.

Subroutine

%	
:1020	Colon followed by identification number specifies the start of subroutine number 1020.
N1030 G91	Set to incremental programming.
N1040 G0 Z-40	Rapid descent to 10 mm above work surface.
N1050 G01 Z-15	Feed-rate descent to cutting depth.
N1060 Y30	Feed-rate move along slot to start of curve.
N1070 G02 X40 Y40 I40 J0	Clockwise circular move to end of slot.
N1080 G01 Z15	Feed-rate withdrawal of cutter from work to 10 mm above work surface.
N1090 G0 Z40	Rapid move to safe Z height to clear clamps, etc.
N2000 M99	End of subroutine.

Main program segment

o	
N300 G0 Z50	Rapid positioning in absolute to safe Z height.
N310 X40 Y30	Rapid positioning over start of first slot.
N320 M98 P1020	Call subroutine number P1020.
N330 G90	On return from subroutine, reset to absolute.
N340 G0 X120 Y30	Rapid positioning over start of second slot.
N350 M98 P1020	Call subroutine number P1020.
N360 G90	On return from subroutine, reset to absolute.
N370 G0 X-40 Y0 M02	Return to tool change position. End of program.

6.2 Subroutines with a loop

It is sometimes useful for a subroutine to be repeated a set number of times. As before, the letter address P is used to identify the subroutine number. In addition, when using a Fanuc or similar controller, the letter address L is used to specify the number of repeats. For example, N400 M98 P2030 L2 means go to subroutine number 2030 and repeat it twice before returning to the main program.

The technique can be illustrated using a turning program to produce the component shown in Fig. 6.3. The operation is to turn down the 28 mm diameter to 20 mm diameter in two equal passes of the tool for a length of 50 mm. Note that because of the simplicity of this example, it could also be performed by a canned cycle on most machines.

The following program segment assumes that earlier program blocks have set appropriate feeds and speeds and have established the radius programming mode, absolute positioning mode, and set the tool offsets.

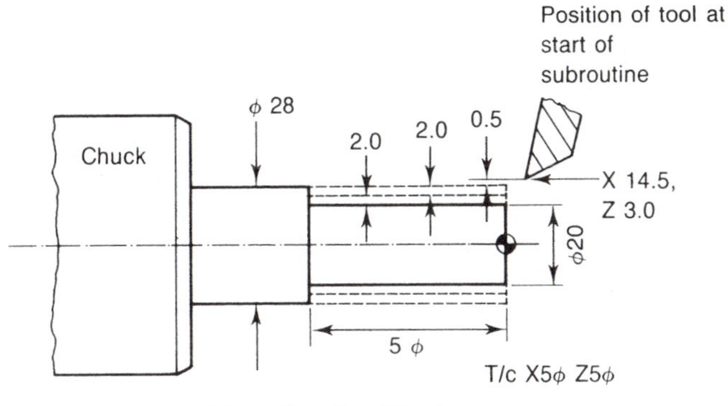

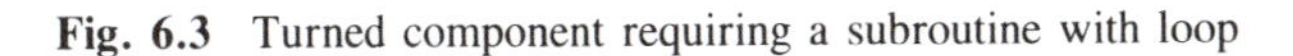

Fig. 6.3 Turned component requiring a subroutine with loop

Subroutine

:3030	Specifies the start of subroutine number 3030.
N3040 G91	Incremental programming.
N3050 G01 X-2.5	Feed-rate move to cutting depth clear of the job.
N3060 Z-53	Feed-rate move to turn diameter.
N3070 X0.5	Feed-rate move of tool 0.5 mm away from work at end of cut.
N3080 G0 Z53	Rapid return with tool 0.5 mm clear of previously turned diameter.
N3090 M99	End of subroutine.

- Since the main program will call for the subroutine to be repeated twice, the component will be turned to the finished size with 2 mm being taken off the radius (4 mm off the diameter) at each pass.
- The 2.5 mm infeed may cause some confusion. Remember that the tool point is 0.5 mm clear of the work at the start of each pass and that this must be added to the required depth of cut.

Main program

N90 G0 X14.5 Z3	Rapid positioning in absolute to start of subroutine.
N100 M98 P3030 L2	Call subroutine 3030 and do twice.
N110 G90	On return from subroutine, reset to absolute.
N120 G0 X50 Z50 M02	Rapid return to tool change position, end of program.

The main advantage gained from the use of subroutines is the time saved in programming, particularly if there is a need for the programmer to write out repetitive blocks of programming which involve the tool making a series of identical moves at different stages in the machining.

6.3 Macros

A macro is a special type of subroutine. The essential difference between a macro and a conventional subroutine is that the use of variables and calculated values is allowed. Canned cycles (see *Manufacturing Technology: volume 1*) are, in fact, macros provided by the control system manufacturer. However, the term *macro* is usually reserved for sections of programming provided to meet the needs of individual machine tool user companies. Macros may be provided by the machine tool supplier or they may be written 'in house' by user-company personnel.

For example, suppose a company has a frequent need to machine bolt holes on a pitch circle with equal hole spacing but that the radius of the pitch circle and the number of holes varies from job to job. If the machine controller does not have a pitch-circle canned cycle, then the company may find it very worthwhile to write a suitable macro and store it permanently in the machine memory.

Figure 6.4 shows a typical bolt-hole pitch circle with its parameters. A macro would enable calculation, within the machining program, of the coordinates of each hole pitch circle for any particular values of A, R

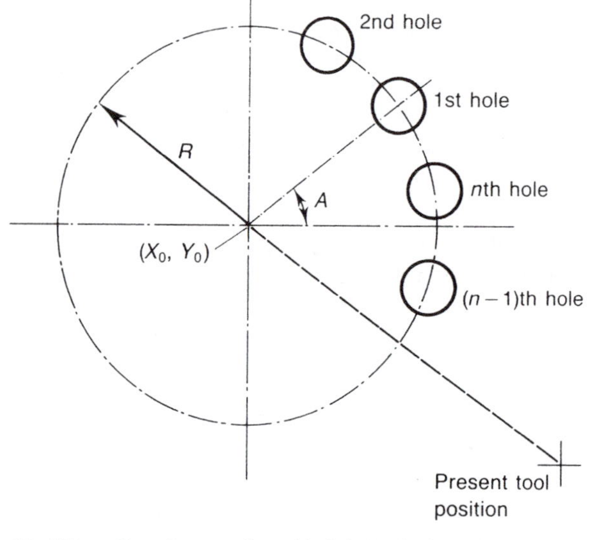

(X_0, Y_0) = Coordinate value of bolt-hole circle centre
R = Radius of bolt-hole circle
A = Start angle
H = Number of holes

Fig. 6.4 Typical bolt-hole circle

and H needed for a given job. The programmer would merely need to enter a single line statement containing the required values of A, R and H to cause the coordinate values to be calculated. A suitable canned drill cycle could then be used to perform the actual machine movements.

The macro call statement takes the following form:

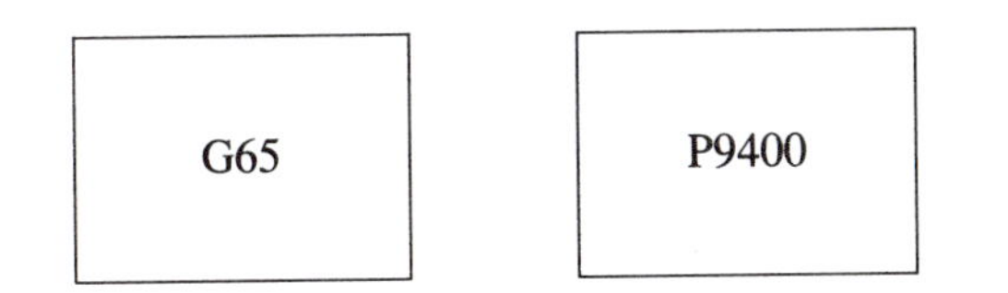

G code to call macro.	Macro number (same format as for subroutine).	Data for a given bolt-hole pitch circle (see Fig. 6.4).

The macro called P9400 might take the following form:

```
:9400
#30 = #101;                              Store reference point
#31 = #102;
#32 = 1;
While [#32 LE ABS(#11)] Do 1;            Repeat by number of holes
#33 = #1 + 360 * [#32-1] / #11;
#101 = #30 + #18 * cos[#33];             Hole position
#102 = #31 + #18 * sin[#33];
X#101 Y#102;
#100 = #100 + 1;                         Increase hole count by 1
#32 = #32 + 1;
End 1
#101 = #30                               Return to reference point
#102 = #31
M99                                      End of macro
```

Note that numbers with # in front are variables in the normal computing sense, where the variables have the following meanings:

#100	Hole number counter
#101	X coordinate reference point
#102	Y coordinate reference point
#18	Radius *R*
#1	Start angle *A*
#11	Number of holes *H*
#30	Storage of X coordinate reference point
#31	Storage of Y coordinate reference point
#32	Counter for the *n*th hole
#33	Angle of the *n*th hole

The macro example is given for the illustration of the general form only. Note that the syntax used is similar to that of conventional computer languages such as Pascal. To write macro subroutines for a given control system would require significant training and access to the software manual for the system.

6.4 Scaling

The scaling feature allows the X and Y coordinates in milling and drilling, or the X and Z coordinates in turning, to be increased or decreased by a scaling factor from their stated values in the program. For example, Fig. 6.5 shows a component requiring two slots to be machined with their widths equal to the cutter diameter. To simplify programming, a subroutine could be written to machine the inner slot and this could be called twice: once without scaling active to machine the inner slot and the second time with a scaling factor of 1.5 (150/100) active to machine the outer slot. Equally, the subroutine could be written for the outer slot and this could be called for a second time with a scaling factor of 2/3 active to machine the inner slot.

6.5 Mirror imaging

This is a useful feature, available with some control systems, which allows either a whole program or a subroutine to have the signs of its coordinate data selectively reversed. The facility is applicable to milling and drilling operations and either X coordinates or Y coordinates or both may be reversed.

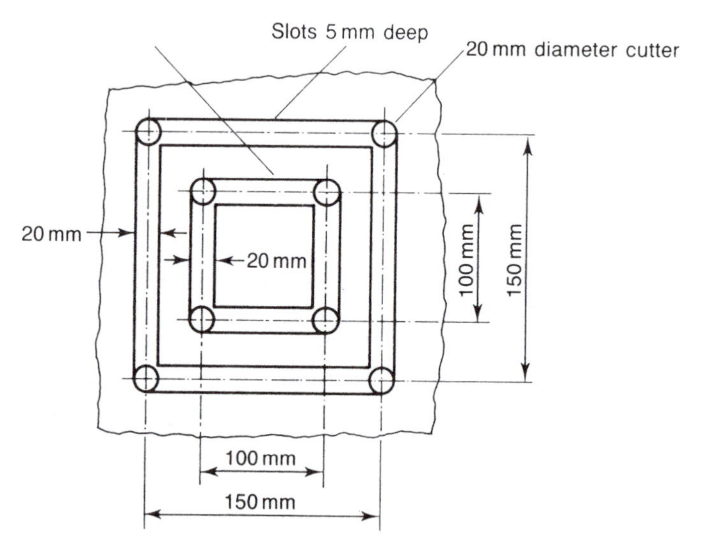

Fig. 6.5 Scaling

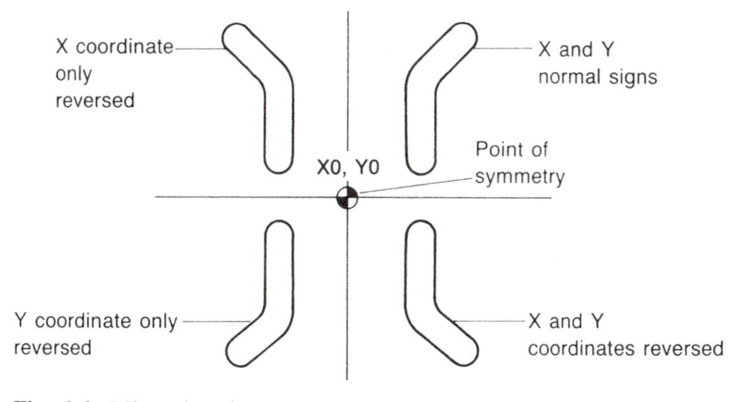

Fig. 6.6 Mirror imaging

The facility can be illustrated by referring to Fig. 6.6, which shows a cavity machined into a die block at four symmetrical positions. With the program datum (point of symmetry) defined as shown, the subroutine to produce the cavity would normally result in the top right-hand quadrant being machined. If the subroutine is then called with the X reversal facility active, then the top left-hand quadrant would be machined. If the subroutine is called with the X and Y reversal facilities active, then the bottom left-hand quadrant would be machined. Finally, if only the Y reversal facility is active, then the bottom right-hand quadrant would be machined. Clearly, a significant amount of programming time can be saved using the mirror-image facility and there is also less opportunity for programming errors to be introduced.

The ISO code for mirror imaging is G28 but control systems vary greatly in their adherence to standards, and other methods of activating the feature may be encountered. Assuming that the control system does use G28, the following program lines illustrate how mirror imaging may be called up. For example:

N110 G28 X — This reverses the sign of the X coordinates only for the subsequent subroutine.

____

____

N190 G28 Y — This reverses the sign of the Y coordinates only for the subsequent subroutine.

____

____

N290 G28 XY — This reverses the signs of both the X and the Y coordinates for the subsequent subroutine.

6.6 Rotation

Rotation is a useful feature available with some control systems which allows the whole coordinate system to be rotated by a stated angle. This

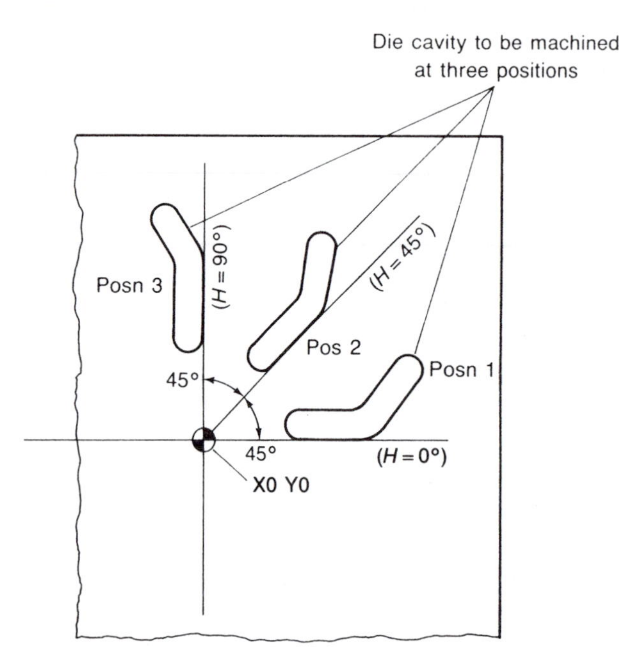

Fig. 6.7 Rotation

facility is applied to milling and drilling operations and is particularly useful when a machine feature is repeated at various angular positions around a common centre, as shown in Fig. 6.7.

The most efficient way to prepare a program to machine this component would be to write a subroutine to carry out the machining of the cavity and then to use the rotation feature to rotate the coordinate axis system each time the subroutine is called. The ISO word address code for rotation is G73, with the H address used to store the rotation angle. However, these standards are not universally applied and other methods of activation may be met. Assuming the ISO system to be in use, the general form of the program to machine the component shown in Fig. 6.7 would be as follows:

———

———

N200	call subroutine to machine cavity.	Causes cavity to be machined at position 1
N210	G73 H45	Rotates coordinate system by 45° from angle zero
N220	call subroutine to machine cavity.	Causes cavity to be machined at position 2
N230	G73 H90	Rotates coordinate system by 90° from angle zero

N240 call subroutine to machine cavity.	Causes cavity to be machined at position 3
N250 G73 H0	Sets coordinate system back to normal (zero°)

———
———

6.7 Zero shift

Zero shift is a facility commonly available on most control systems. It allows the program datum to be changed within the program. The component shown in Fig. 6.8 will be used to explain this feature. A program has to be written to machine the large hole A and then the smaller holes B which are arranged in a pattern about a centre point.

The most convenient datum for hole A is datum (1) and a convenient tool-change position is X-45, Y0, and Z20 with reference to datum (1). The first line of the program would therefore define the start position of the tool (the tool-change position) using a G92 code and the machine operator would position the tool at this point at the start of the program. For example:

N100 G92 X-45 Y0 Z20	Define tool change position relative to datum 1
———	
———	Program lines for machining hole A
———	
N1200 G00 X-45 Y0 Z20 M06	Return to tool change position and change tool

In order to minimise the calculations for the pattern of holes B, the adoption of datum (2) would be most convenient. Therefore the program would continue with a G92 code to tell the control system where the

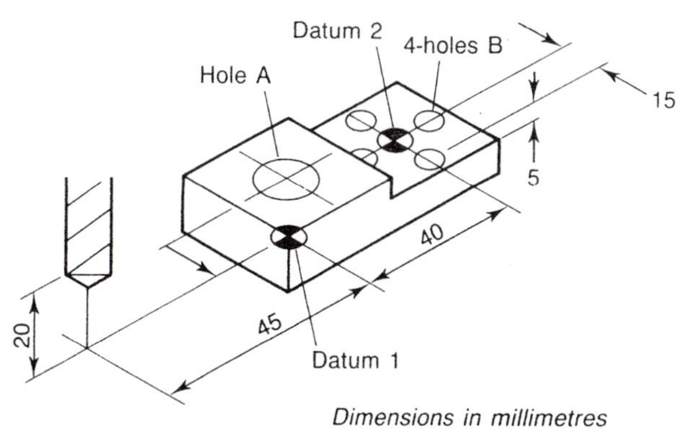

Fig. 6.8 Zero shift

existing tool-change position is relative to datum (2). This continuation of the program would read:

N1250 G92 X-85 Y-15 Z20	Define tool change position relative to datum (2).
______	
______	Program lines to machine the four smaller holes.
______	
N3000 G00 X-85 Y-15 Z20 M02	Return to tool change position at end of program.

6.8 Computer-aided part programming

So far, the manual procedures followed when part programming may be summarised as follows. The programmer:

(1) studies the drawings and decides on a basic plan of action including a manufacturing sequence, the tooling required, and suitable means of workholding and location. This stage may involve the production of a formal planning sheet;
(2) then considers the cutter paths needed in more detail and calculates the required coordinate data;
(3) selects the tooling and calculates the speeds and feeds;
(4) writes the program in word address format using G, M, and other codes specified for the control system to be used.

In general, the process of manual programming is slow and error prone, it may require complex calculations, and it will require an expert knowledge of different word address languages for each control system used. On the other hand, computer-aided part programming techniques use a computer to generate the part programmming code needed to machine the component. The advantages of this include:

- simpler part programming;
- quicker part programming, particularly for complex components;
- less likelihood of calculation errors;
- the part programmer needs only to learn one language, no matter how many different machine tools and control systems may be in use.

Computer-aided part programming languages differ in detail but all of them, essentially, have three parts:

- — Component shape definition by computer language statement. For example:
  P1 = X0 Y0 defines a point at a program datum;
  S1 = V100 defines a vertical line 100 mm from the program datum.

- — Component shape definition by interactive graphics.
- — Component shape definition by importing the component shape from a CAD system.

- Processing.
- Post-processing.

There is a wide choice of computer-aided part programming languages. Some are 'universal' languages applicable to a wide range of CNC processes and are independent of any particular machine control system. Other types are designed to be used for a particular range of CNC operations or are specific to a particular control system.

The original universal language was APT (automatically programmed tool) developed during the 1960s in the USA. Some modern APT-like languages that are used in the UK are PEPS (production engineering productivity system) and GNC (graphical numerical control). The PEPS language will be used to give an example, based upon the simple component shown in Fig. 6.9, of the steps involved in preparing a part program using a computer-aided part programming system. The profile shown is to be externally finish machined to a depth of 10 mm in one pass.

Stage 1: component shape definition

A typical PEPS installation would make use of a twin screen set-up with a mouse and graphics tablet. One screen would show program menus and the PEPS language statements as they were written. The other screen would show the interactive graphics. As mentioned earlier, one method of component shape definition is to import the required information directly from a CAD system. Alternatively, the component may be drawn on-screen using menu commands. This is done in a similar way to a

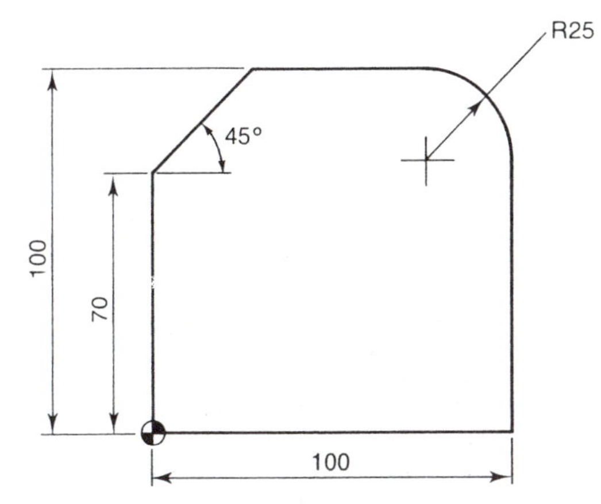

Fig. 6.9 Component to be programmed using 'PEPS'

draughtsperson working on a conventional drawing board: first, construction lines are created and then the required shape is defined using heavier lines.

Using PEPS, the equivalent of these construction lines are shown as dotted lines and circles, which appear on the graphics screen in response to the operator's commands. Simultaneously the graphical information is converted into PEPS language statements which appear on the other screen. Alternatively, a programmer conversant with the PEPS language can write the language statements direct causing the graphical information to be produced. Figure 6.10 shows the graphics screen after completion of the construction lines for the component and, below it, the equivalent PEPS language statement.

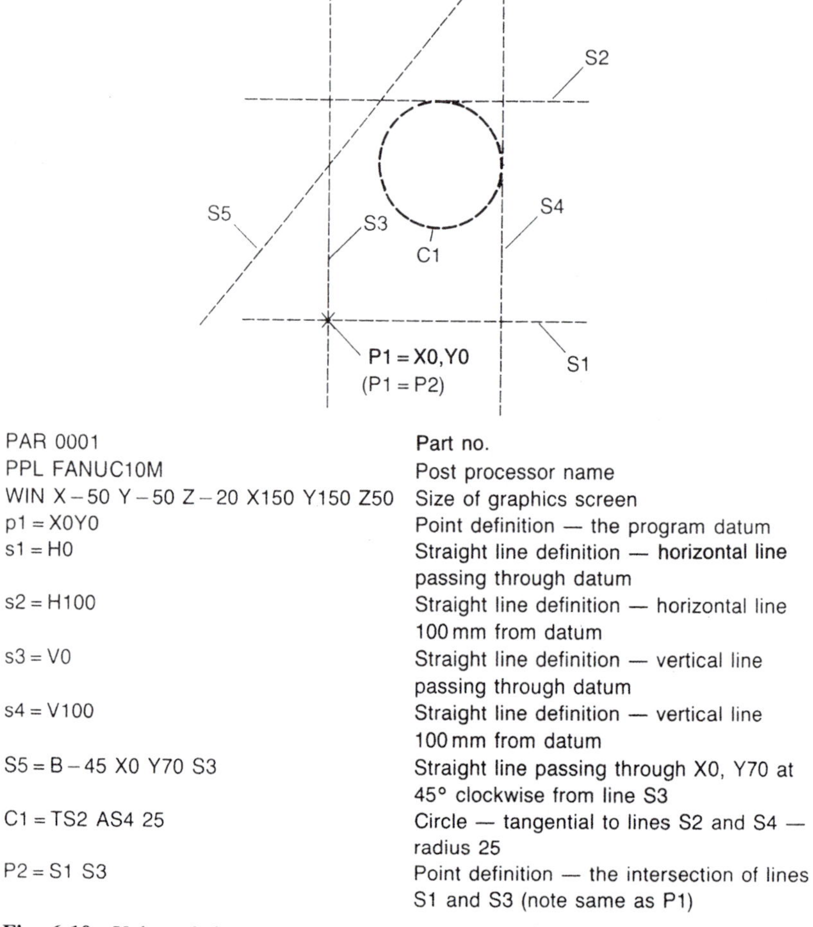

PAR 0001	Part no.
PPL FANUC10M	Post processor name
WIN X − 50 Y − 50 Z − 20 X150 Y150 Z50	Size of graphics screen
p1 = X0Y0	Point definition — the program datum
s1 = H0	Straight line definition — horizontal line passing through datum
s2 = H100	Straight line definition — horizontal line 100 mm from datum
s3 = V0	Straight line definition — vertical line passing through datum
s4 = V100	Straight line definition — vertical line 100 mm from datum
S5 = B − 45 X0 Y70 S3	Straight line passing through X0, Y70 at 45° clockwise from line S3
C1 = TS2 AS4 25	Circle — tangential to lines S2 and S4 — radius 25
P2 = S1 S3	Point definition — the intersection of lines S1 and S3 (note same as P1)

Fig. 6.10 Unbounded geometry and PEPS language statements for component shown in Fig. 6.9

Note that T and A in the definition of the circle C1 stand for 'tangential' and 'antitangential'. When a line touches a circle or when two circles touch, the tangency point may be resolved by considering the directions of the two touching features. When the two element directions are the same, the condition is described as tangential (T). When the two element directions are different, the condition is described as antitangential (A). In PEPS the direction of a line depends upon the way it is defined, but the direction of a circle is always considered to be clockwise. Figure 6.11 illustrates this by showing the directions of lines S2 and S4 and the circle C1, and shows how the A and T conditions arise.

When the stage shown in Fig. 6.10 has been reached, all the necessary construction geometry is present. This is sometimes referred to as the *unbounded geometry* since the actual boundary of the component has not yet been defined. The next stage in shape definition is to create the actual component shape or *bounded geometry*. This is done in PEPS using a *kurve*. A kurve is defined as a two-dimensional shape made up of straight and/or circular segments with a defined start and end point. Figure 6.12 shows the kurve for the sample component with its language statement equivalent. Remember that although these language statement equivalents look complicated, the computer is preparing them automatically without the assistance of the programmer. The easiest way to create a kurve is to select the appropriate menu option and then work round the component shape by picking off the various intersection points with the graphics cursor. The kurve appears as a heavy yellow line on the graphics screen. The language statement associated with Fig. 6.12 means:

> 'a kurve k1 starts at point P2, moves tangentially along S3, tangentially along S5, tangentially along S2, tangentially around C1, antitangetially along S4 and, finally, antitangentially along S1 back to the starting point. (EK means end of kurve.)'

The component shape definition is now complete.

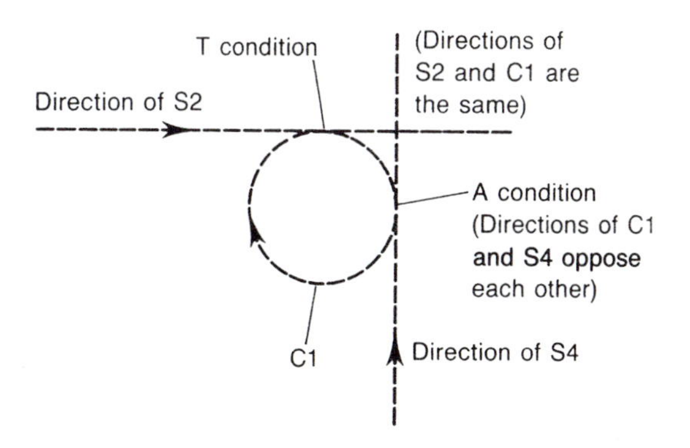

Fig. 6.11 Tangent definitions

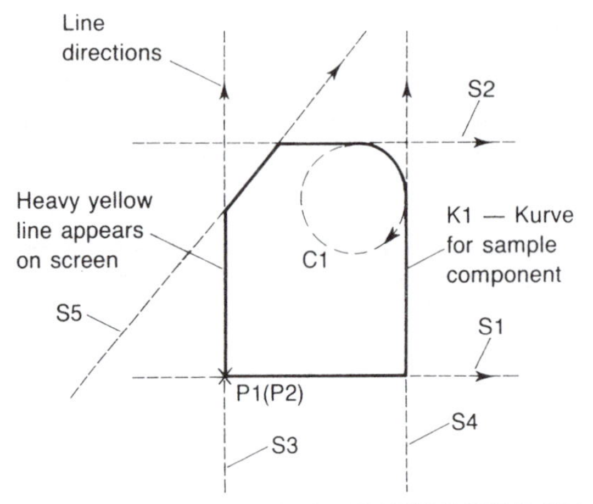

Equivalent language statement K1 P1 TS3 TS2 F TC1 F AS4 AS1 P2 EK

Fig. 6.12 Kurve for sample component

Stage 2: processing

In Stage 2, the cutter paths and various machine functions are determined. Again, working from the menu commands, the operator enters statements which control machine operations and movements. One of the great strengths of the system is that the machine movements and cutter paths are displayed interactively on the graphics screen so that full simulation of the machining process takes place. Figure 6.13 shows the full PEPS program including the milling technology statements and the equivalent graphics screen.

The sample program shows the ease with which a series of machining moves can be defined and the required coordinate data generated. The key to this simplicity is the profiling command which quickly and automatically generates the cutter path around a defined kurve. This command can be used to generate *roughing* and *finishing* cutter paths around a workpiece. This can be done by varying the offset distance. In the sample program, the line 'OFF L0' prior to the profiling command 'PRO T K1' tells the software to calculate the cutter centre-line path which is just the cutter radius distance from the finished workpiece profile. If, for example, the command 'OFF L2' had been used, the software would have calculated a cutter centre-line path that was offset by the cutter radius plus 2 mm from the workpiece, and this would have been a suitable allowance for finishing.

Stage 3: post-processing

During this stage, the general purpose machine control and cutter location data that has been produced during the processing stage is post-processed to obtain a machine-specific CNC word address program. The post-processor is a software program and a different post-processor is required

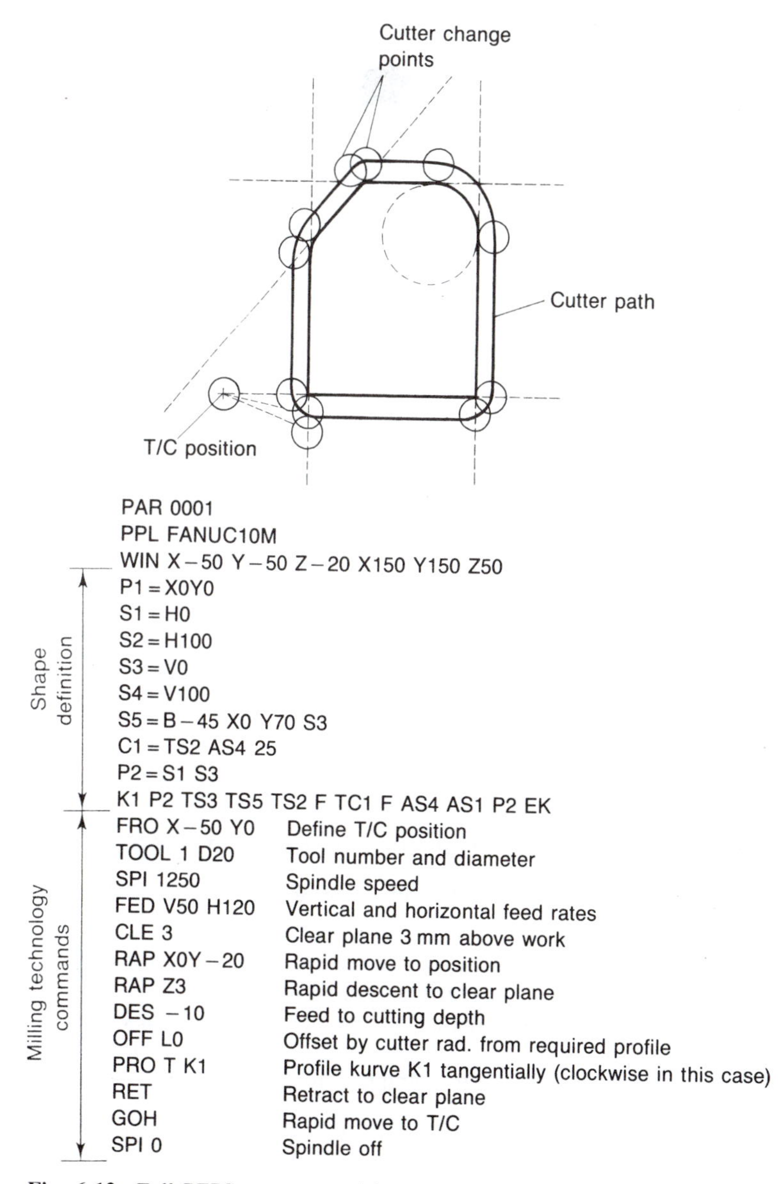

Fig. 6.13 Full PEPS program and language statement

for each machine control system used by a company since word address systems are not standardised.

Usually the supplier of the computer-aided part programming software can supply compatible post-processors for all the popular control systems 'off the shelf'. The post-processed output for the sample component (Fig. 6.9) written for a Fanuc 10M control system is as follows:

```
%
00001
N10G21G40G49G80G90
N20G55X-50.0Y0.0
N30T1M6
N40S1250M3
N50G0X0.0Y-20.0
N60G43Z3.0H1
N70G1Z-10.0F50
N80Y-10.0F120
N90G2X-10.0Y0.0I0.0J10.0
N100G1Y70.0
N110G2X-7.071Y77.071I10.0J0.0
N120G1X22.929Y107.071
N130G2X30.0Y110.0I7.071J-7.071
N140G1X75.0
N150G2X110.0Y75.0I0.0J-35.0
N160G1Y0.0
N170G2X100.0Y-10.0I-10J0.0
N180G1X0.0
N200G49G53Z0.0
N210G0X-50.0Y0.0
N220M5
N230M5
N240M30
%
```

6.9 Benefits of computer-aided part-programming

Compared with manual part-programming, the benefits of computer part-programming are as follows:

- Programming procedures are more structured and greatly simplified.
- More efficient use is made of the programmer's time.
- Cutter path coordinate calculations are performed more quickly and more accurately.
- Most systems include interactive graphics to simulate the programmed movements.
- Post-processing is performed automatically, reducing programmer transcription errors.
- The programmer does not need to know the word address format for each machine. He/she has only to learn one language using English type words. The post-programmer automatically converts the program into the correct G and M codes.
- Most systems have powerful functions for scaling, mirror imaging, rotation, translation, etc.
- Programs for families of parts can be written using variable parameters for the dimensions.

- Some systems are capable of producing programs for simultaneous contouring in multiple axes. This would be beyond the capability of a manual programmer.
- All systems can transmit the post-processed output either directly to a storage medium such as paper tape or magnetic tape. Alternatively, the output can be downloaded directly to the machine tool using a *direct numerical control* (DNC) link. This saves time and reduces the likelihood of conversion errors.

6.10 Trends in computer-aided part programming

The tendency has been for part programming to be based in specialist departments away from the shop floor. Although it can be claimed that this makes for economic use of expensive hardware and software, nevertheless it does mean that programmers can become divorced from shop-floor realities. Priorities can become blurred and program preparation lead times can increase.

There are often many advantages to be gained if programming can be done on the shop floor, provided that productive manufacturing time is not lost. For example, shop-floor personnel will know what cutting tools, clamps and fixtures are currently available without having to make enquiries. Further, the machine operators can generally be more responsive to changing priorities.

Shop-floor programming systems that allow setter/operators to generate component shapes and cutter paths interactively at the machine are increasingly available. These systems are very user friendly and allow the programming for one job to proceed while the machine cuts a job programmed earlier. An example is the Fanuc APT system (FAPT).

6.11 Tooling systems for CNC machine tools

One of the most useful features of CNC machine tools is their flexibility. However, a wide variety of cutting applications can only be carried out if a correspondingly wide range of cutting tools is available. Machine tool using companies will often have a range of machines of different ages and varying degrees of technical sophistication. The machines will have different tool-mounting methods, spindle nose tapers, draw bar arrangements, etc., and this can lead to each machine needing a unique set of tools and holders. This, in turn, can lead to an unacceptably high investment in tooling and also introduce storage and retrieval problems.

Fortunately, tooling manufacturers offer a variety of solutions to the problem of locating and clamping a range of cutting tools in a variety of machine tools. These 'tooling systems' consist of a range of standard cutting tools and tool holders together with appropriate adaptors for each type of machine used. This allows the inventory of standard tools and holders to be minimised since they can all be used on any machine with some general characteristics. Therefore using a well-designed tooling system offers the following advantages.

- Maximum tool interchangeability between different machines.
- Easier process planning and tool layout planning.
- Reduced overall tooling costs, since the tooling inventory is minimised.
- Downtime due to tooling shortages is minimised.
- Good repeatability of location accuracy between tool to adaptor and adaptor to machine.
- Standard holders and adaptors make pre-setting of the machine easier.
- Tooling systems are compatible with automatic tool changing devices.
- Tooling systems are compatible with touch-trigger probe systems.

Tooling systems may be either manual or automatic. Manual systems are quite satisfactory for jobbing and low-volume production but, the higher the volume of production, the more economical it becomes to consider investment in automatic tool-changing systems. Thus it is advisable to purchase a tooling system that can adapt to automatic tool changing even if it is slightly more costly in the first instance. Figure 6.14 shows a typical tooling system which is equally applicable to manual or automatic tool changing.

6.12 Tool changing

Tool changing can be performed manually or automatically. Automatic tool-changing facilities add considerably to the cost of a machine, so manual tool changing is still widely used for small-quantity batch work. When manual tool changing is used it is essential to minimise the changeover time. This can be achieved by using quick-change tool holders and preset tooling. The tools are kept in a 'crib' placed conveniently beside the machine. The 'crib' is a stand in which the tools are not only stored when not in use, but are located in the order in which they are to be used, and in such a position that they can be easily grasped and removed from their location.

Automatic tool-changing systems are classified according to the way in which the tools are stored.

Indexable turrets can be programmed to rotate (index) so as to present the tools mounted in the turret in the order in which they are required. Indexable turrets are widely used on turning centres and also on some milling and drilling machines. In the latter case, the drive to the cutting tool is also transmitted through the turret. Such a system lacks the rigidity of a conventional machine head and spindle assembly and is usually only used on comparatively light duty machines.

Tool magazines are indexable storage facilities and are used only on machining centres. Two systems are shown in Fig. 6.15. The tool magazine is indexed to the tool-changing position and an arm removes the current tool from the machine spindle and inserts it into the empty socket

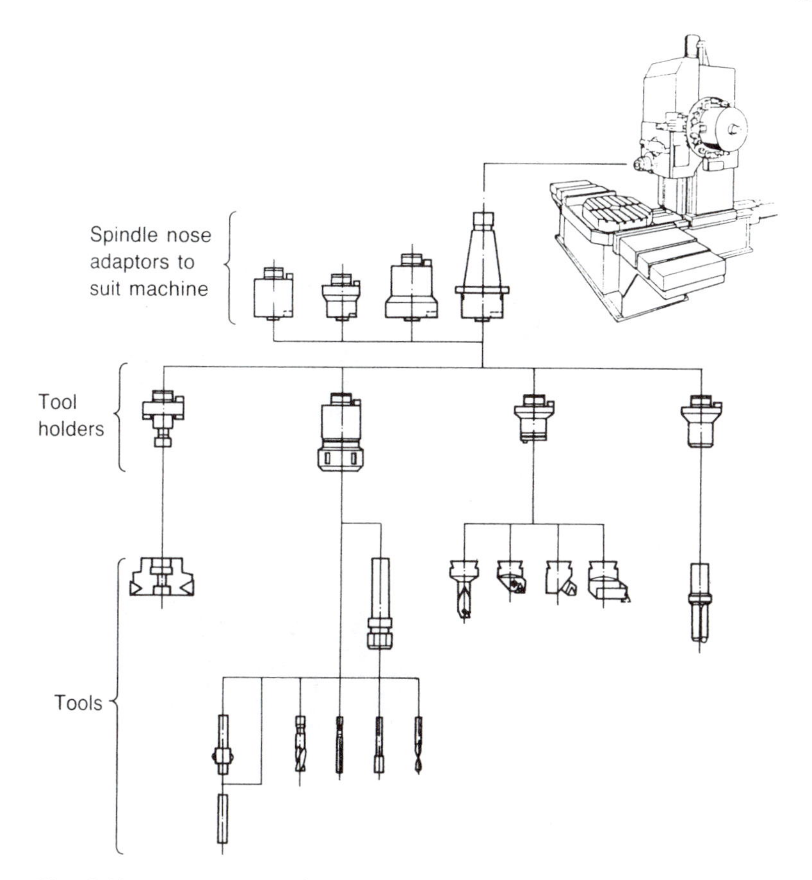

Fig. 6.14 Tooling system for machining centres

in the magazine. The magazine then indexes so that the next tool to be used is presented to the tool change position. The arm then extracts the tool holder and tool from the magazine and inserts it into the machine spindle ready for use.

6.13 Tool replacement

When tooling has to be replaced due to wear or breakage, one of the following situations will arise:

- A suitable 'non-qualified' tool will be fitted. That is, a tool will be used which is suitable for the job but which is not a direct replacement for the previous tool. The operator will, therefore, have to re-datum the tool and edit the tool-length offset file in the machine memory before machining can recommence. This is obviously time-consuming and expensive and unacceptable if 'just-in-time' production is being employed.

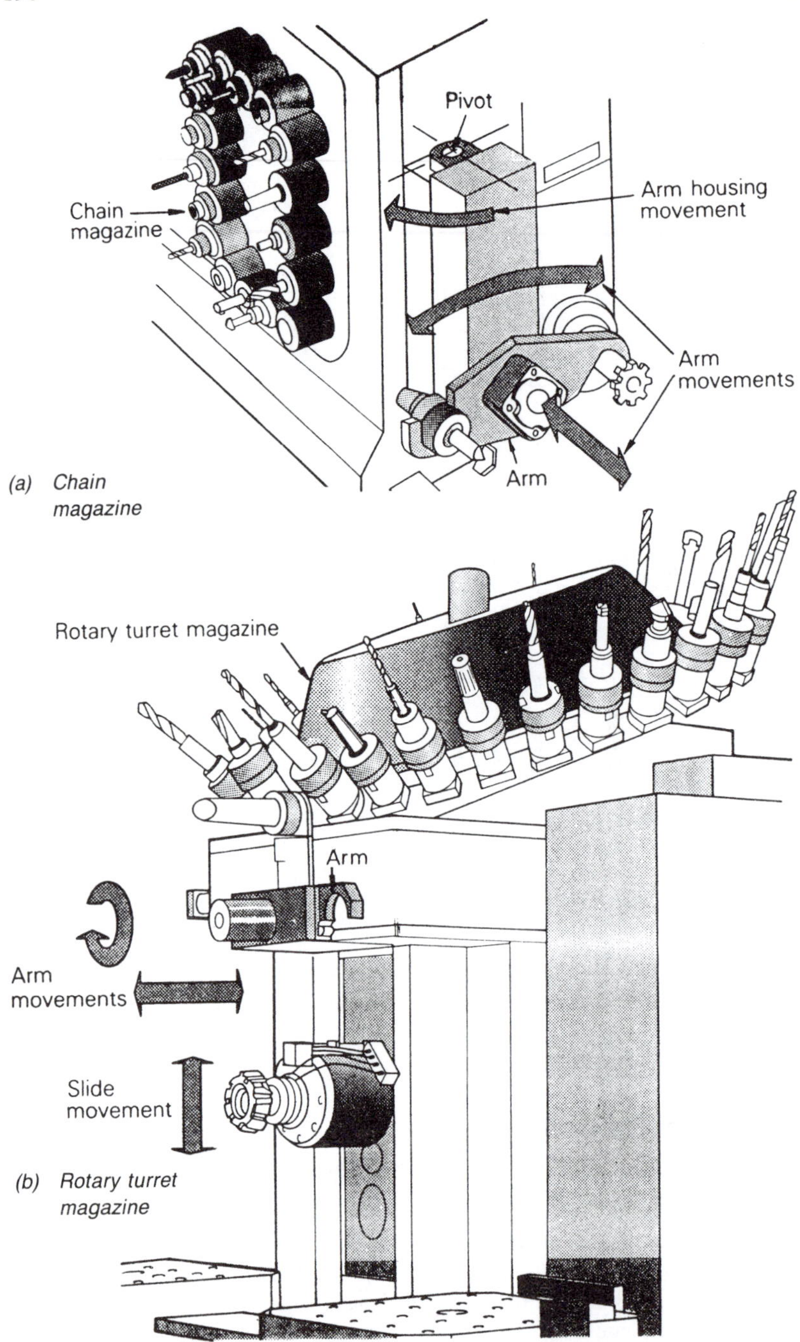

Fig. 6.15 Automatic tool changing (a) Chain magazine; (b) Rotary turret magazine

- A 'qualified' replacement tool will be substituted. This is a tool that has identical dimensions to the one it is replacing within known tolerances. This will allow machining to continue without having to re-datum the tool and with a minimum loss of time. Figure 6.16 (ISO 5608: 1989) shows how different dimensions of a tool can be qualified.
- A 'preset' replacement tool will be used. This is a tool that is not necessarily a direct replacement for the worn tool, but it is a tool that has been preset in a setting fixture to a known offset value. This is shown in Fig. 6.17. Although the operator/setter will need to edit the offset file, there will be no need to re-datum the tool before recommencing machining.
- The final possibility, only available on the more advanced machines, is the use of automatic tool datum and offset edit facilities, using touch-trigger probes interfaced to the machine control system. With this system, the tool is touched onto a probe and the appropriate offset values are entered automatically into the offset file. The use of probes for tool setting and in-process inspection are described in *CNC in Manufacturing and Computer Aided Machining* by R. Duffill and R.L.Timings (Longman, forthcoming).

It is no longer considered economical to re-grind worn cutting tools made from cemented carbides. Nor is it desirable in the case of coated carbides since the coating would be destroyed. Further, re-grinding alters the size of the tool and this is not acceptable when 'qualified' tooling is being used; the exception being twist drills of various types. Normal practice is to use disposable tip tools as described in section 5.6. Figure 6.18 shows a typical disposable tip system. The tips are designed so that they can be indexed round to a new cutting edge several times before finally being discarded.

6.14 Flexible manufacturing systems (FMS)

Traditionally, workshops adopted 'process layouts' where all the lathes were grouped together, all the milling machines were grouped together, all the drilling machines were grouped together and so on. Such workshops were used and, in many instances still are used, for jobbing and small-batch production. Such layouts caused many problems. Complex product routings, queuing, high levels of work in progress, poor quality, and lack of employee accountability are all characteristic of this type of layout. Nevertheless, despite its many disadvantages, this type of layout has always been considered the only way of coping with product variety inherent in batch production. A typical layout is shown in Fig. 6.19.

One alternative is to adopt a 'product layout', as shown in Fig. 6.20, where machines and processes are arranged in the required sequence so that the work can move quickly from machine to machine. Where the quantities of a particular product are very large so that continuous

Tolerances in millimetres

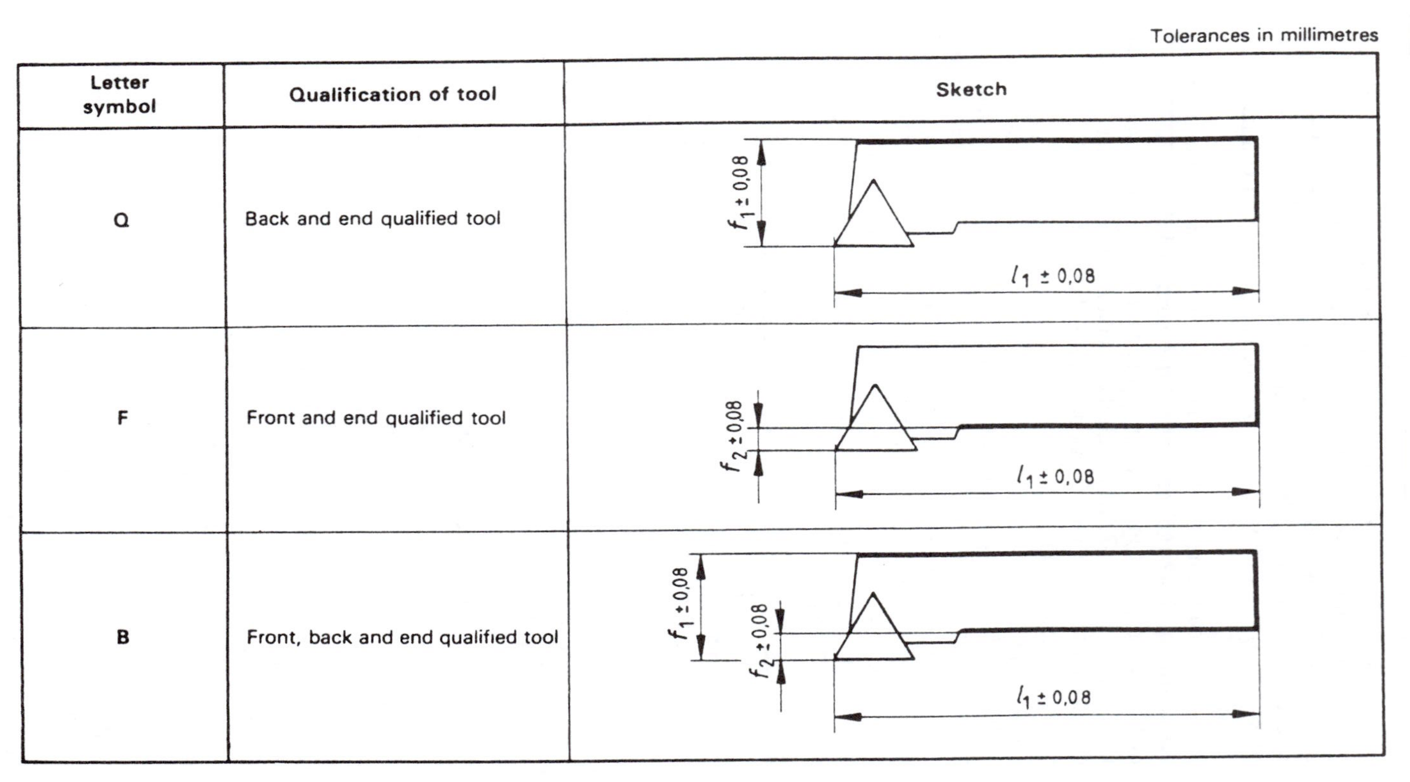

Letter symbol	Qualification of tool	Sketch
Q	Back and end qualified tool	$f_1 \pm 0{,}08$; $l_1 \pm 0{,}08$
F	Front and end qualified tool	$f_2 \pm 0{,}08$; $l_1 \pm 0{,}08$
B	Front, back and end qualified tool	$f_1 \pm 0{,}08$; $f_2 \pm 0{,}08$; $l_1 \pm 0{,}08$

Fig. 6.16 Qualified tooling

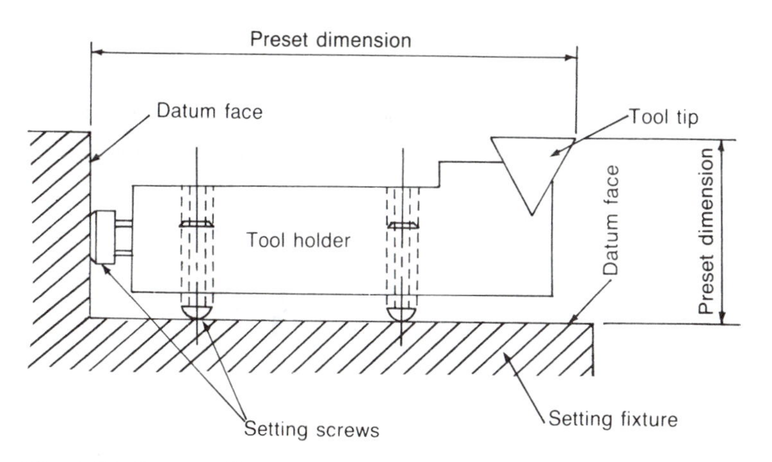

Fig. 6.17 Preset tooling

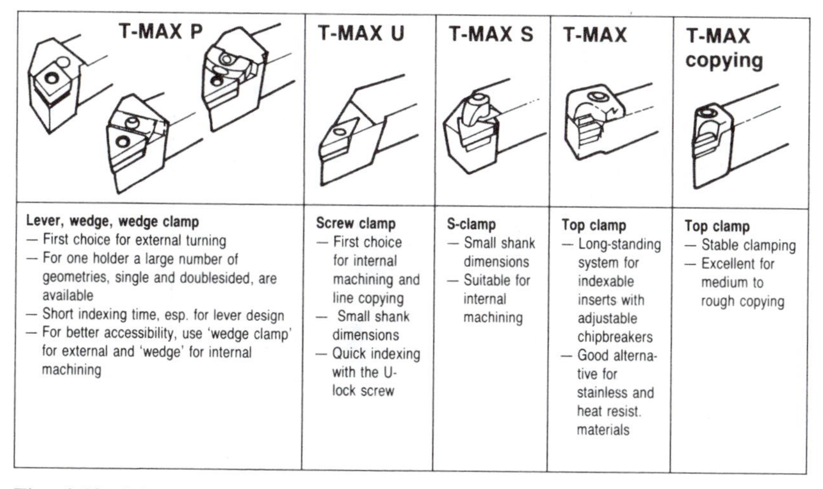

T-MAX P	T-MAX U	T-MAX S	T-MAX	T-MAX copying
Lever, wedge, wedge clamp — First choice for external turning — For one holder a large number of geometries, single and doublesided, are available — Short indexing time, esp. for lever design — For better accessibility, use 'wedge clamp' for external and 'wedge' for internal machining	**Screw clamp** — First choice for internal machining and line copying — Small shank dimensions — Quick indexing with the U-lock screw	**S-clamp** — Small shank dimensions — Suitable for internal machining	**Top clamp** — Long-standing system for indexable inserts with adjustable chipbreakers — Good alternative for stainless and heat resist. materials	**Top clamp** — Stable clamping — Excellent for medium to rough copying

Fig. 6.18 Disposable tip system

production techniques can be adopted, this type of layout offers many advantages over the 'process layout'. For example, the 'transfer batch' between processes can be low (ideally a transfer batch of one), resulting in low levels of work in progress and short lead times. The product flows smoothly and there is greatly reduced complexity of production management.

A dedicated flow line can be seen to be the most clear-cut example of a product-type layout. The equipment is dedicated to one product only and there is usually a high level of automation both in the work transfer arrangements and in the processes themselves. Traditional thinking

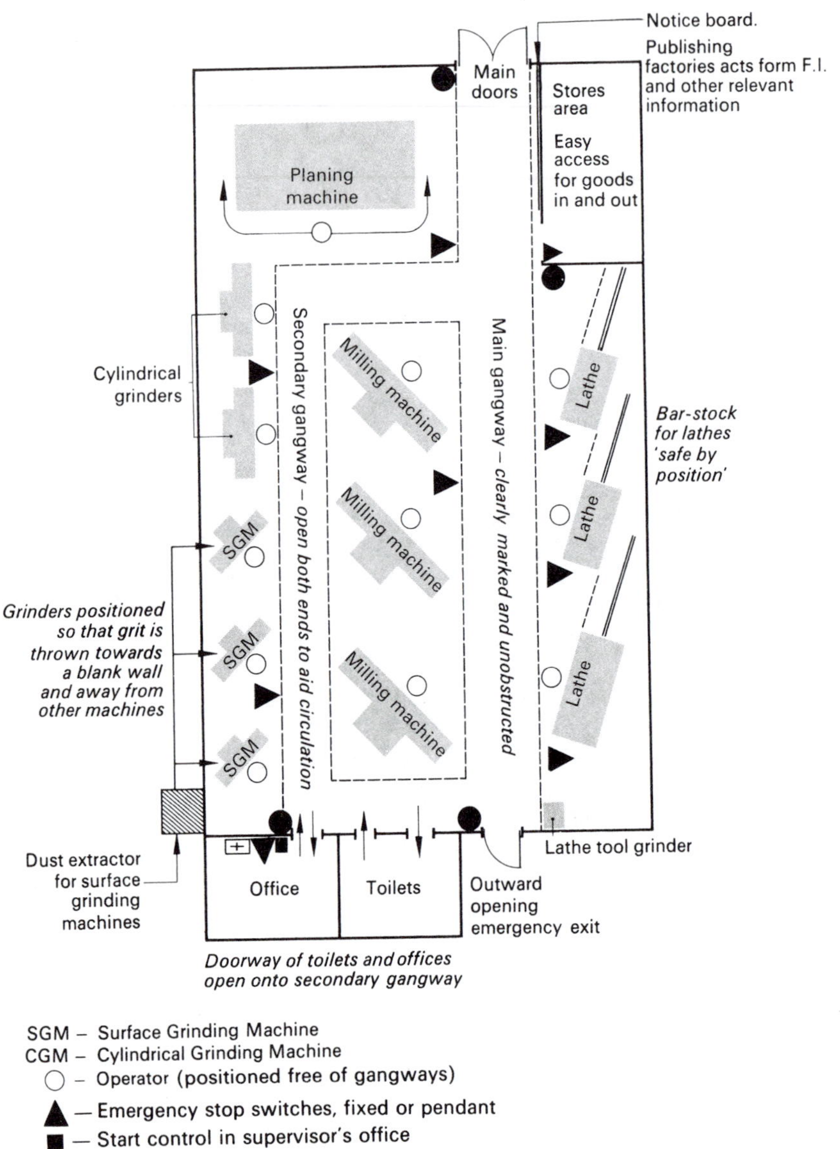

Fig. 6.19 'Process' or 'functional' machine shop layout

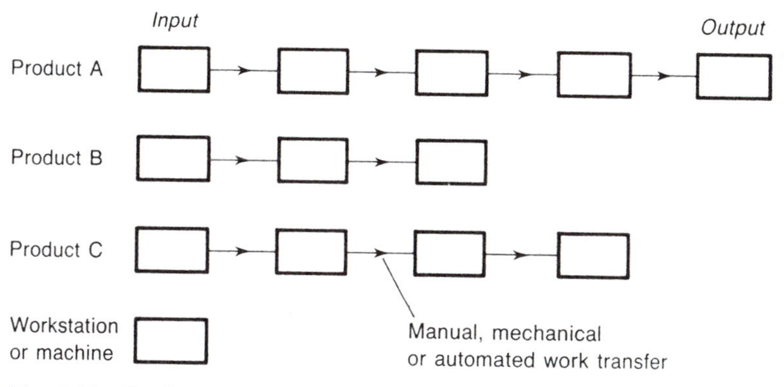

Fig. 6.20 'Product' machine shop layout

suggests that the advantages in the product layout are only obtainable when product volumes are very high, in order to justify the cost of dedicated production equipment. It is perfectly true that the production volume of any one product in a typical batch production factory would not justify forming a product layout just for that one product. However, it is invariably found that within the product range of a batch production factory there are several different 'families' of parts which share similar production routings. If these part families can be identified, then the machines can be rearranged into groups, each one of these groups being capable of making every member of a different product family. These types of group-product layout typically use conventional equipment and manual work transfer to achieve the benefits of the product layout. This approach, developed in the 1960s and 1970s, was referred to as *group technology*.

The group technology approach is increasingly referred to as *cellular manufacture*. A typical low-technology cell is formed by rearranging conventional equipment from a traditional layout. Frequently a 'U'-shaped layout is adopted to ease visibility and aid communication. Cell personnel will often be cross-trained to operate a range of cell processes and strong feelings of pride of ownership and of identification with the cell are encouraged. Usually, not all the products made in a cell need every process every time. Therefore costing systems which encourage high levels of machine utilisation are inappropriate. Performance measures need to be targeted more on responding flexibly to customer demands. The management of flexible manufacturing (group technology) cells is discussed further in Chapter 11.

6.15 Cellular concepts in FMS

High-technology cells are the building blocks of FMS. A typical cell of this type is shown in Fig. 6.21. Cells such as this, using advanced

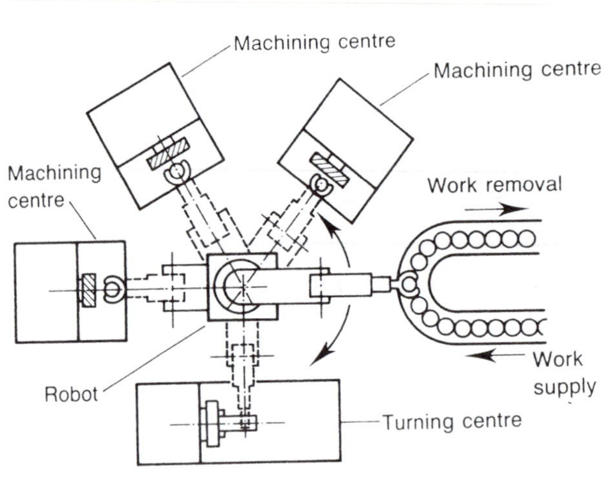

Fig. 6.21 Typical FMS cell

technology, are based upon the same general principles as the conventional technology cells described in section 6.14. Both low- and high-technology cells are flexible up to a point in that they can cope with the limited variety within a product family. Conventional technology cells cope with this variety by making use of the inherent flexibility of human operators and by having the full range of machine tools available even though some may be under-utilised by conventional efficiency standards. In contrast, advanced technology cells cope with the variety within a product family by making use of flexible, reprogrammable devices such as CNC machines and robots, and having a range of monitoring devices and systems to detect problems. Such cells can function without human aid and even under 'lights out' conditions.

6.16 Features of FMS

Flexible manufacturing systems consist of a single cell, or a series of linked cells, and have the following characteristics:

- They have the ability to process all members of a family of parts automatically and in random order under the control of a supervisory computer.
- They consist of a series of CNC machine tools with a materials-handling system made up of conveyors, pallets, shuttles, automatic guided vehicles (AGVs) and robots.
- They can store and download CNC and robot programs from master computers to individual machines and robots (DNC).
- They can control tooling from the CNC part program with machines needing large tool magazines to accommodate the range of tooling needed to machine all the members of the parts family.

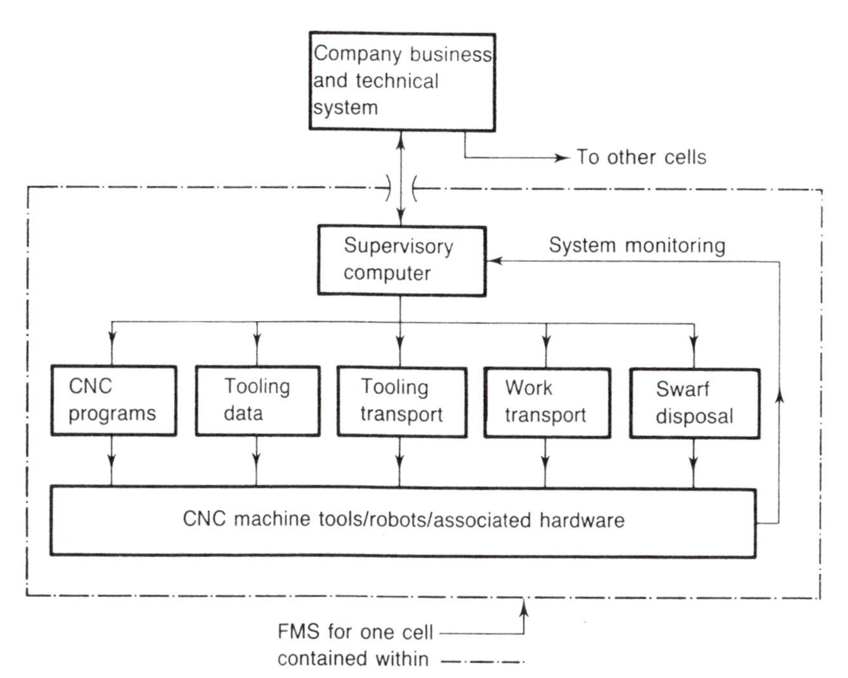

Fig. 6.22 Flexible manufacturing system

6.17 Control of FMS and its economic justification

Figure 6.22 shows the main features of a flexible manufacturing system and the lines of communication between the various computers. The supervisory computer analyses the management and technical data being downloaded to it from the company business and technical system. It then re-routes the data to the various machines and material-handling devices. These then use their own computers to interpret the messages being received from the supervisory computer and manufacture the components called for. The output from the cell and the whole system within the cell is monitored by the supervisory computer which also alerts the production management to any faults it cannot correct automatically.

The economic justification of full-scale FMS has always been difficult due to the enormous initial investment required to purchase, install and commission all the linked elements needed. There are relatively few clear-cut examples of flexible manufacturing systems that have met their technical specification and proved to be an undoubted financial success.

The most successful applications appear to be where the variety of work to be coped with is minimised. In this guise a flexible manufacturing system might be more accurately described as a slightly flexible transfer line.

On a smaller scale, there are many examples of successful small manufacturing cells consisting of a single process such as machining, presswork, welding, plastic moulding or die-casting combined with a robot for loading and unloading the machines.

What is evident is that cellular concepts lie at the heart of strategies to achieve internationally-competitive manufacturing and that successful cellular applications range from conventional technology cells using human operators and manual work transfer, to fully-automated high technology cells using state of the art CNC machine tools and robots.

6.18 Industrial robots

Industrial robots may be defined as computer-controlled, re-programmable mechanical manipulators with several degrees of freedom, capable of being programmed to carry out a variety of industrial operations. In order that a robot may be able to reach for, move, and position a workpiece or tool, it requires an arm, a wrist subassembly and a 'hand' or end effector. The sphere of influence of a robot depends upon the volume (envelope) into which the robot can deliver its wrist subassembly and end effector. A variety of geometric configurations have been developed and the most widely used will now be appraised.

Cartesian coordinate robots

Cartesian coordinate robots have *three orthogonal linear sliding axes*, as shown in Fig. 6.23(*a*). The manipulator hardware and control systems are the same as CNC machine tools. Therefore the arm positional resolution, accuracy and repeatability will be the same as for a CNC machine tool. An important feature of a cartesian robot lies in its spatial resolution which is equal and constant in all the axes of motion and throughout the work volume. This is not the case for the other configurations.

Cylindrical coordinate robots

Cylindrical coordinate robots have *two orthogonal linear sliding axes and one rotary axis*, as shown in Fig. 6.23(*b*). The horizontal arm telescopes in and out and moves vertically up and down the column which, in turn, rotates on its base. The working volume is, therefore, cylindrical. The resolution of a cylindrical robot is not constant and depends upon the radius of the wrist from the rotational axis of the column. With a standard resolution digital rotary encoder and an arm length of 1 m, the resolution of the wrist assembly will be of the order of 3 mm which is poor compared with a cartesian robot where the resolution is constant at about 0.01 mm. Cylindrical geometry robots offer (in theory) the advantage of higher linear velocity at the wrist end of the arm, as a result of having a rotary axis. In practice, this is limited by the moment of inertia of the arm and the work load. In fact, it is quite difficult to obtain good dynamic performance from rotary base robots. The moment of inertia reflected at the base drive depends not only upon the mass of the

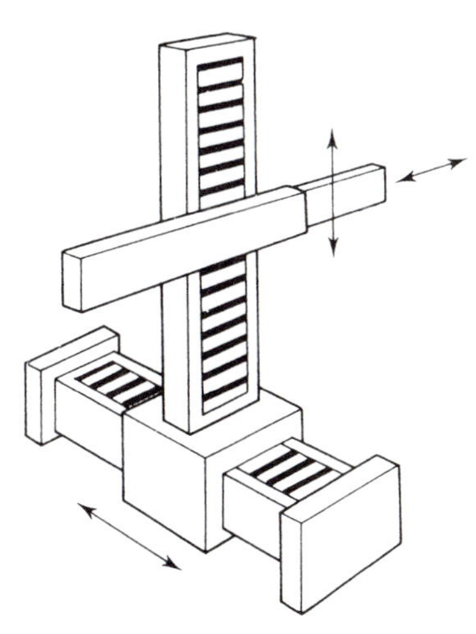

(a) Cartesian coordinate robot

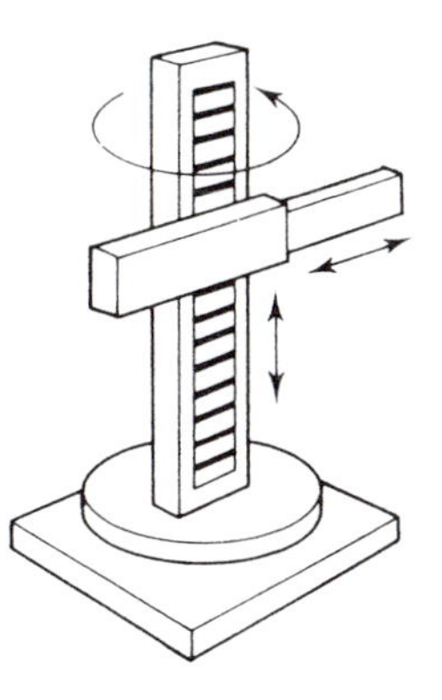

(b) Cylindrical coordinate robot

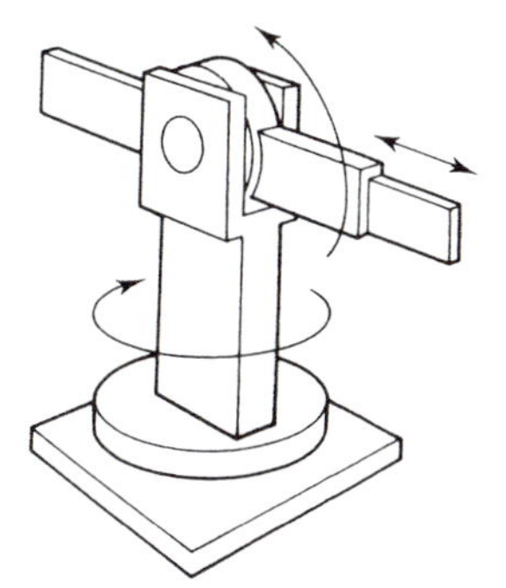

(c) Spherical (polar) coordinate robot

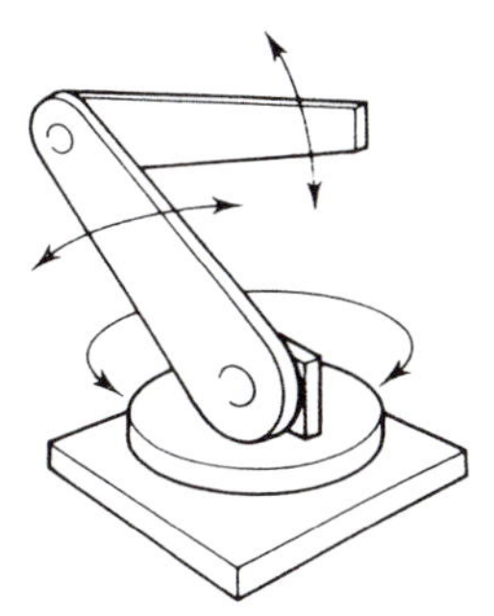

(d) Revolute (angular) coordinate robot

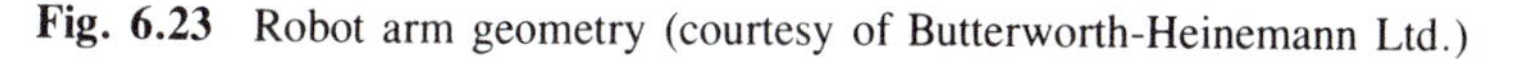

Fig. 6.23 Robot arm geometry (courtesy of Butterworth-Heinemann Ltd.)

arm and the work load but also its distance from the pivot point. This is one of the main drawbacks of robots using revolute joints.

Spherical (polar) coordinate robots

Spherical coordinate robots have *one orthogonal linear sliding axis and two rotary axes*, as shown in Fig. 6.23(*c*). The arm can move in and out and it can be tilted up and down by a horizontal pivot. The whole assembly can pivot about a vertical axis since it is mounted on a rotary base. The magnitude of the rotaional movement is measured by encoders built into the pivots. The working envelope is a spherical shell. Again, the resolution is limited by the vertical axis encoder and the length of the

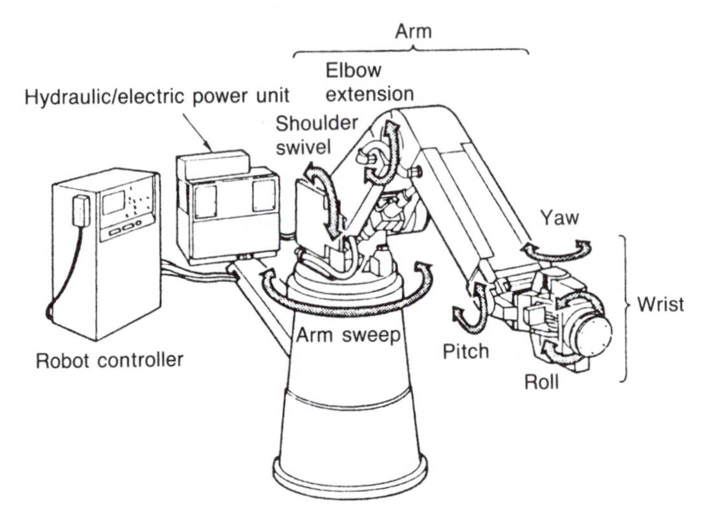

Fig. 6.24 Revolute robot, complete installation (courtesy of Butterworth-Heinemann Ltd.)

robot arm, the resolution becoming poorer as the distance of the wrists from the vertical axis increases. On the other hand, the movements of a spherical coordinate robot tend to be quicker and more flexible compared with a cartesian robot.

Revolute (angular) coordinate robots

Revolute coordinate robots have *three rotary axes and no linear axes*, as shown in Fig. 6.23(*d*). The revolute, angular or articulated robot (different names for the same thing) has two rotary joints and a rotary base. Its range of movements closely resembles those of the human arm and it is fast and extremely flexible. Unfortunately, having three revolute movements, its resolution is very poor compared with a cartesian coordinate robot. However, this configuration is the most popular with small and medium sized robots.

Figure 6.24 shows a complete revolute-type robot installation and indicates the additional movements of the wrist subassembly. Various types of end effectors can be mounted on the wrist to hold a variety of workpieces and tools. Remember that, the greater the number of joints and movements, the lower the overall resolution will be. Further, as wear occurs in service, the greater the fall off in accuracy will be.

6.19 Robot end effectors

As has previously been stated, robots can be programmed for a wide variety of tasks and are available in a variety of sizes, shapes and physical capabilities which are reflected in a correspondingly wide variety of end effectors. However, end effectors can be categorised into *grippers* and *special tools*. Examples of these are:

- Grippers:
  mechanical clamping
  magnetic or electro-magnetic
  suction
- Special tools:
  screw-drivers
  nut runners
  assembly tools
  welding tools
  inspection probes
  various types of sensor
  pouring ladles
  portable power tools
  spray-painting guns

6.20 Robot programming methods

As has already been stated robots share a number of hardware and software similarities with computer-controlled machine tools. They use the same linear and rotary encoders, the same stepper and servo drives, they can have open- or closed-loop control systems, and they are controlled by a dedicated computer which can, in turn, be linked with other computer-controlled devices to build up an automated manufacturing cell. However, the method of programming can be substantially different to that used with CNC machine tools.

'Lead-through' programming

'Lead-through' programming is done by a skilled operator who leads the robot end effector through the required pattern manually. For example, he/she may hold the spray gun on the end of the robot arm and guide it through the sequence of movements necessary to paint a car body panel. The robot arm joint movements needed to complete this operation are automatically recorded in the computer memory of the robot and can be repeated when required.

'Drive-through' programming

The robot is programmed by driving it through the required sequence of movements, under power, with the operator controlling speed and direction, etc., by means of a *teaching pendant* which is a small hand-held key pad connected to the robot controller by a trailing lead. The motion pattern is recorded in the computer memory and can be repeated when required.

'Off-line' programming

'Off-line' programming is rather like CNC manual programming but uses a specialist programming language. The following example based on a Unimation Robot uses a language called VAL. The robot is required to pick up a component from the unload chute of a machine tool and place that component in a box. The sequence required is:

(1) move to a position above the part in the chute;
(2) move close to the part, with the gripper jaws open;
(3) close the gripper to hold the part;
(4) lift the part from the chute;
(5) move to position the part above the box;
(6) move to position the part within the box;
(7) open the gripper to release the part;
(8) withdraw the gripper from the box.

The equivalent VAL program is as follows (note use of comment lines to explain the program — identified by REM):

```
(1) APPRO PART, 75      REM move to 75 mm above part
(2) MOVES PART          REM move along a straight line to part
(3) CLOSE I             REM close gripper (immediate — not in
                        following time)
(4) DEPARTS, 250        REM withdraw 250 mm in a straight line
(5) APPRO BOX, 250      REM move to 250 mm above box
(6) MOVE BOX            REM move into box in a straight line
(7) OPEN I              REM open gripper — immediate
(8) DEPARTS, 250        REM withdraw 250 mm above box
```

Location of 'PART' and 'BOX' are position variables defined by driving the robot to the required positions and using the monitor — immediate — command 'HERE PART' to define the position variable 'PART' as the current robot position and 'HERE BOX' to define the other required positions. 'BOX' and 'PART' can also be defined directly using X, Y and Z coordinates from the robot zero datum. However, in practice, the difficulties of measuring the actual positions of the component on the chute and also the box location are considerable. Therefore it is easier, and more often preferable, to use the drive-through technique to define position variables.

6.21 Robot simulation

The ability to program a robot 'off-line' and to carry out a graphical simulation to check the interaction of the robot with the work place is a useful feature which can be carried out at the work cell design stage.

GRASP is an example of this type of software package. It was developed by the Department of Production Engineering and Production Management at the University of Nottingham. The program has a 3D solid modeller which allows the user to develop models of robots from a standard library of geometrical shapes. The user can program the robot movements using the system and then play back the sequence. In this way the best layout for the robot application can be developed, the best robot choice made, and checks can be carried out to detect possible collisions. Finally, the GRASP robot program can be post-processed to obtain a control program for the particular robot to be used. Some examples of robot simulation are shown in Fig. 6.25.

6.22 Typical robot applications

- *Load and unload operations* — particularly in hostile environments, e.g. diecasting machines, plastic moulding machines, and furnaces.
- *Flame cutting* — the robot can be programmed to cut the required profile.
- *Fettling* — the robot can be programmed to manipulate a grinding head to remove runners and risers from castings and remove any unsightly excrescences.
- *Pouring* — the robot can be programmed to pour molten metal and other dangerous substances to avoid health hazards to human operators.
- *Palletising* — loading components to a programmed pattern onto pallets or into boxes.
- *Assembly* — the robot can be programmed to assemble components together, and to fit and tighten screwed and other fasteners. Parts often have to be redesigned to suit robotic assembly to avoid them tangling or seizing.
- *Manipulation* — the manipulation of X-ray and radiation sources which would be hazardous to human operators at close quarters.

6.23 Reasons for using robots

- To release human operators from boring, repetitive work.
- To achieve unmanned (or lightly manned) operations during nightshifts and at weekends.
- To replace human operators in dangerous workplace environments.
- Where strength and/or reach requirements are beyond the capacity of human operators.
- Where errors due to tiredness and lack of concentration of human operators cannot be tolerated.
- Where flexibility — the ability to be reprogrammed to perform a different task — is important; a key advantage compared with 'dedicated' automation.

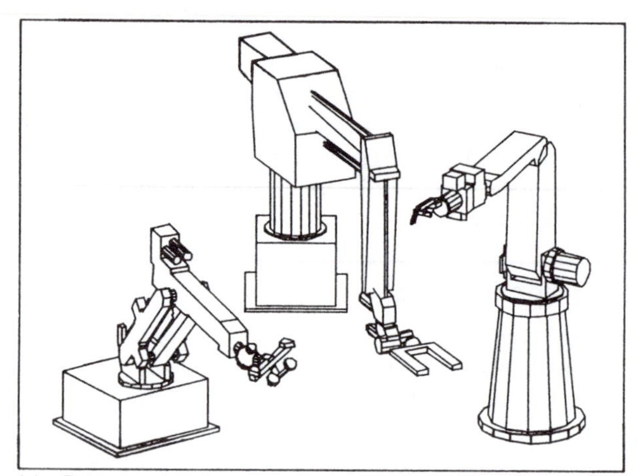

(a) Robot models from the GRASP library

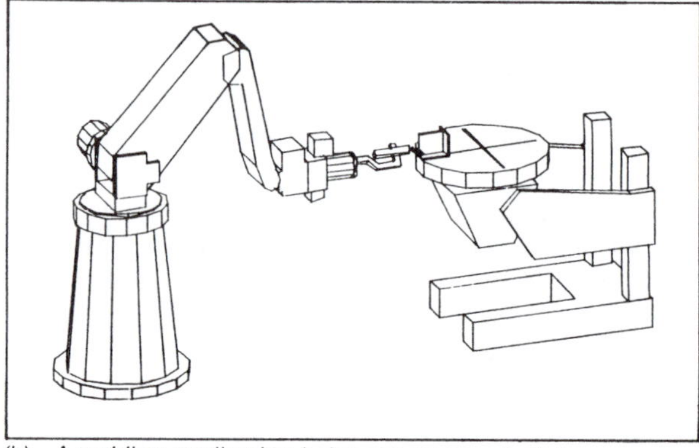

(b) A welding application being programmed off-line

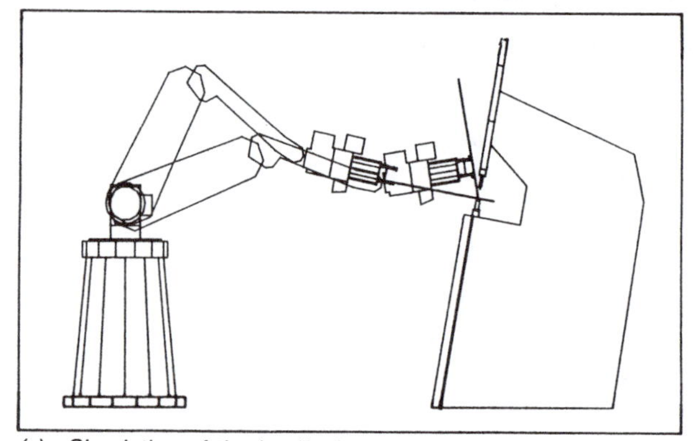

(c) Simulation of the load/unload configuration for assessing the position of the robot in relation to the press brake

Figure 6.25 Robot simulation using GRASP from Chapter 20 *Handbook of Industrial Robotics*

Assignments

1. (a) Describe the uses and benefits of sub-routines in CNC part programming.
   (b) Differentiate between macros and sub-routines in CNC programming.
2. Describe a typical machining situation where a macro-programming facility would be useful and explain in general terms how the macro would work.
3. Compare the advantages and limitations of computer aided CNC part programming with manual part programming.
4. List the main stages in the preparation of a CNC part program using the PEPS computer aided part programming system and briefly explain what happens at each stage.
5. With the aid of simple sketches, explain the following CNC part programming features:
   (a) zero shift;
   (b) scaling;
   (c) rotation;
   (d) mirror imaging.
6. Describe the different methods of ensuring that the replacement of a worn or broken tool can be accomplished without loss of machining accuracy.
7. Outline the advantages to a CNC machine tool user company of adopting a standard tooling system.
8. Briefly describe the different methods of industrial robot programming.
9. (a) Compare and contrast the advantages and limitations of industrial robots with human operatives in a manufacturing situation.
   (b) Describe three workplace situations where a robot might be used in preference to a human operator.
10. With the aid of a sketch, describe the basic layout of a flexible manufacturing 'cell' and discusss the criteria for justifying the high investment involved in setting up such a 'cell'.

7 Assembly processes

7.1 Introduction

The principles of assembly were introduced in *Manufacturing Technology: volume 1*, including various methods of joining components and subassemblies together. Also discussed was the need for tolerancing component dimensions so that the components are interchangeable, and the advantages to be found in using standard components.

It is important that assembly and dismantling is considered carefully at the design stage. Some of the questions that must be asked are:

- Is the assembly process to be manual or automated?
- Can the components and subassemblies be placed in position easily?
- Do the components need support to keep them in position while being fastened together?
- If they do need support, what work-holding devices need to be designed and manufactured and how will these affect tooling costs and lead time?
- Do the work-holding devices need to be indexable and, if so, do they need to be power operated and do they need to be linked with the assembly robot control system?
- Does the assembly operator have to make decisions concerning the positioning of components? For example, the electronic solid state device shown in Fig. 7.1 can be placed in one of two positions because of the symmetry of its legs. The designer of the device has

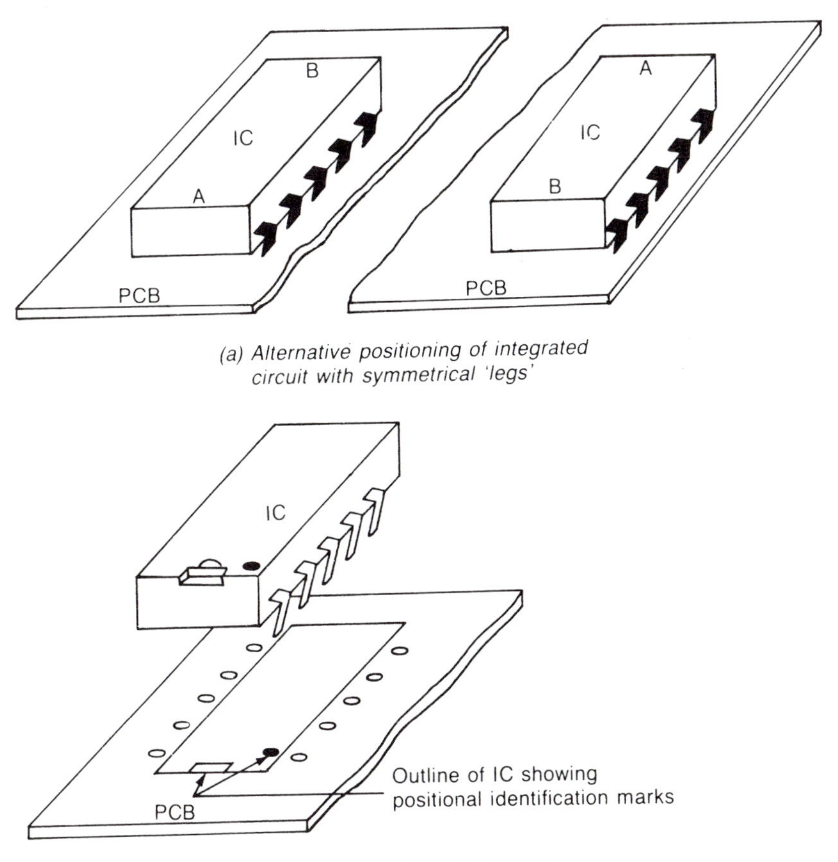

(a) Alternative positioning of integrated circuit with symmetrical 'legs'

(b) Positioning an IC using identification marks

Fig. 7.1 Designing to avoid assembly errors

included a notch and spot on the case to help the assembly operative decide which way round it goes. The designer of the printed circuit board completes the process by showing the outline of the device together with its notch and spot. By matching the device to the outline, decision making and potential error on the part of the assembly operative is reduced.

- Can the fastenings be readily positioned and are they accessible to standard spanners, nut runners, screwdrivers, etc., and is there room to turn these devices? It must be remembered that special tools available in the factory may not be available to the service engineer in the field.
- Can key components, which have to be changed during routine servicing, be removed readily without having to strip off other components? It should not be necessary to dismantle a vehicle engine

to reach a sparking plug, or to remove the engine to change an oil filter.

At one time, ease of assembly and access for repairs and routine servicing was given scanty consideration, but with ever-increasing labour costs and the increasing use of automated assembly, it is now a major design consideration.

To attempt to build an automated assembly cell to reproduce manual assembly processes is asking for trouble. No automated system can provide the dexterity, sensibility and thought processes of a human being. The cost and complexity of trying to use state-of-the-art technology to approach the skills of human assembly operatives greatly outweighs the results. Therefore, if it is intended to use automated assembly, the product must be designed to suit automated assembly and not the other way round. The aim should be to keep the assembly cell as simple and as reliable as possible and capable of being set up and maintained at a cost that can be justified.

The limitations of robots in terms of speed and accuracy are discussed in Chapter 6, and these limitations must be kept in mind. There are no 'brownie points' to be won for using a three-axis robot if a simpler, cheaper and more rigid two-axis machine will do the job. Further, it should be remembered that a simple transfer mechanism will operate more quickly and reliably than a robot in many cases. The robot comes into its own mainly when flexiblity is required.

7.2 Planning assembly

Modern manufacturing systems thinking suggests that production rates should be matched as closely as possible to customer usage rates (see section 11.27), so that both the supplier and the customer are relieved of holding stock ahead of need. This means that each product variant should be assembled in levelled quantities over the shortest possible time scale. For example, Table 7.1(*a*) shows an old style, large-scale, monthly production schedule, while Table 7.1(*b*) shows a typical weekly small-batch levelled schedule.

The main barriers to the achievement of this type of levelled scheduling are lack of cellular factory organisation and long set-up times whenever a new variant is to be made. Cellular organisation is needed because levelled scheduling is only feasible within the small-scale environment of a cell where product variants are few. At the overall factory level, the number of product variants means that the level of complexity is too great for levelled scheduling to be achieved.

Levelled scheduling entails frequent resetting as machines are changed from job to job. Where long set-up times are involved, levelled scheduling can result in too much capacity being lost from production to changeover. Thus long set-up times and levelled scheduling are largely incompatible. However, changeover analysis can invariably be applied to

Table 7.1 Assembly scheduling

Product type	Monthly requirement	Weekly production quantity			
		Week 1	Week 2	Week 3	Week 4
A	1100	500	500	100	—
B	300	—	—	300	—
C	200	—	—	—	200
D	100	—	—	—	100

(a) Large-batch monthly production schedule

Product type	Monthly requirement	Weekly production quantity			
		Week 1	Week 2	Week 3	Week 4
A	1100	275	275	275	275
B	300	75	75	75	75
C	200	50	50	50	50
D	100	25	25	25	25

(b) Small-batch weekly levelled schedule

reduce changeover times to insignificant levels. In the longer term, companies aspiring to world-class manufacturing standards seek to move to daily or even hourly levelled schedules where mixed-mode production is the norm.

Required rate of production

As stated in the previous section, the customer usage rate determines the required rate of production and hence the desired 'cycle time'. For example, a company supplies car rear-axle assemblies to a car-manufacturing company who sell 15 000 cars per month. Assuming that there are 20 working days in a month and an 8-h working day, then:

Number of rear axle assemblies required per day
$= 15\,000/20 = 750$ per day.

Therefore the cycle time $= (8 \times 60\ \text{min})/750$ assemblies
$= 0.64$ min/assembly.

Assembly methods

The main factors which determine the choice of assembly method are:

- the cycle time required;
- the anticipated product life cycle (and, by extension, the total number of products to be made);
- the design of the product, i.e. whether it is simple or complex and its suitability for assembly by automated methods.

For example, consider a high-volume product such as the rear-axle assembly mentioned previously. This type of product would clearly be more likely to justify investment in dedicated and automated assembly equipment than a low-volume long-cycle time product. This latter type of product would probably be most economically assembled using general purpose equipment and manual techniques.

Estimation of assembly times

At the planning stage, the assembly system is designed to match the required cycle time as closely as possible. Alternative assembly system designs will be considered at this time and reliable estimating techniques are needed to predict the cycle times that would be achieved.

For manual assembly systems, the established work study techniques of predetermined motion time systems, analytical estimating, synthetic time data, and data from existing operations may all have a useful part to play. In addition, desk-top manual simulation and computer simulation packages may help to explore the potential interaction and queuing problems between assembly operations.

Automated assembly operations can be estimated on the basis of fixed, machine-paced, transfer and operating cycles. Prototype assemblies, using mock-ups, may also be valuable to confirm assembly cycle times.

7.3 Types of assembly systems

A convenient way to classify assembly systems is as follows:

- manual assembly,
- dedicated automated assembly,
- programmable assembly (flexible automation).

Manual assembly

Manual assembly encompasses a wide variety of operations. At its simplest, operators may transport and position the parts and then perform the assembly using hand or power tools. On a more sophisticated level, a basic unit is delivered in a part carrier to the operator on a conveyor or other transfer device and the parts to be added are delivered by automatic feeders or placed in conveniently-positioned containers within easy reach of the operator. The final positioning and fixing of at least some of the parts is carried out by the operator, as shown in Fig. 7.2.

Dedicated assembly

A dedicated assembly system is dedicated to the assembly of a single product. The assembly system is built around either a linear or a rotary transfer machine. This transfer machine delivers the base units mounted on work carriers to the assembly station. The parts are delivered, positioned on the base unit and fixed in place by automated assembly devices. Figure 7.3 shows, diagrammatically, a rotary assembly system.

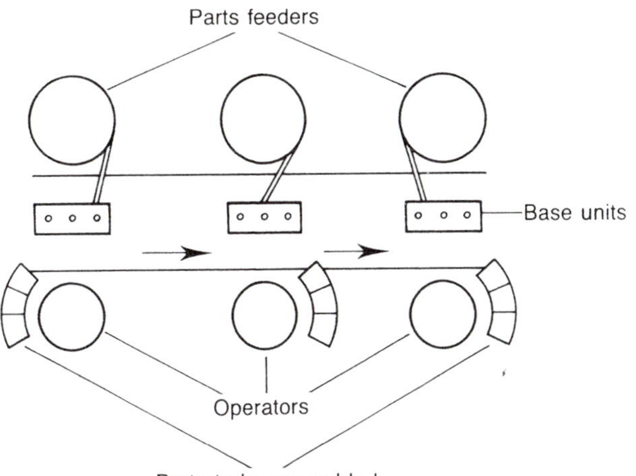

Fig. 7.2 Manual assembly layout

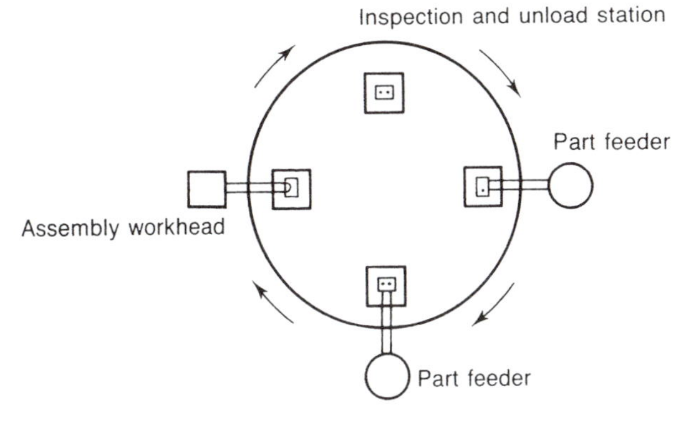

Fig. 7.3 Rotary automatic assembly layout

Programmable assembly

Programmable assembly uses a programmable robot to transfer and position the parts onto the base unit. This gives a degree of flexibility since the robot can be reprogrammed to cope with different parts, locations and assembly sequences. Figure 7.4 shows, diagrammatically, a typical programmable assembly cell.

7.4 Advantages of automated assembly

The actions involved in assembly — including visual selection and sorting of parts, orientation and positioning, compensating for awkward access and less than perfect fit — are all relatively easy for a human operator to

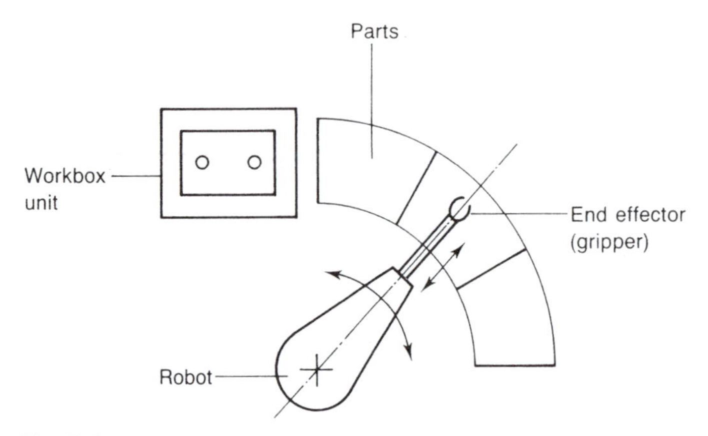

Fig. 7.4 Programmable assembly cell

accomplish. To achieve the same flexibility in an automated system requires a great deal of attention both to the design of the product for ease of assembly and to the design of the system to compensate for the absence of human intelligence, problem-solving ability and versatility. However, when used with well-designed products, in volumes which justify the investment needs, the following advantages can be claimed for automated assembly:

- lower overall cost,
- increased productivity,
- improved quality arising from less process variation, and 'in-process' automated inspection,
- easier integration with other automated equipment,
- removal of operators from hazardous environments.

7.5 The elements of automated assembly equipment

The elements of automated assembly equipment include:

- part feeding devices;
- transfer and indexing devices to present the work carrier to the assembly station;
- part positioning devices (including robots);
- workhead mechanism to tighten, rivet, peen, apply adhesive, weld, snap in, or otherwise fix the part to the base unit.

Part feeding devices

A typical part feeder uses vibratory action to transfer components from a mass storage hopper onto a feed track which delivers the parts to the workhead with the correct orientation. Figure 7.5 shows a length of feed track that has been fed from a vibratory hopper. It can be seen that

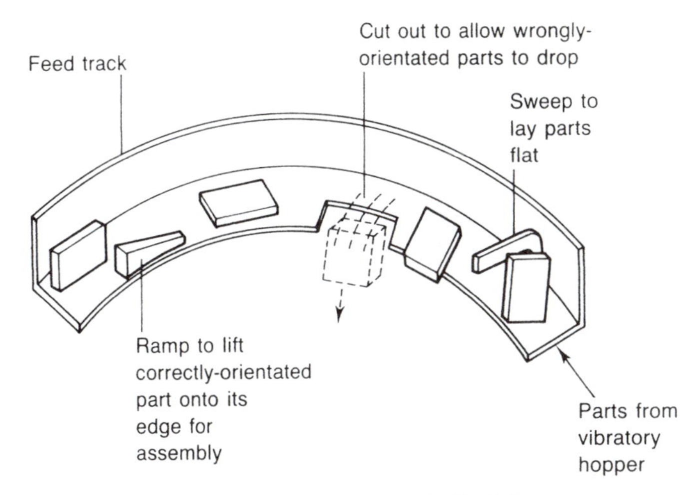

Fig. 7.5 Sorting and orientating automatically fed parts

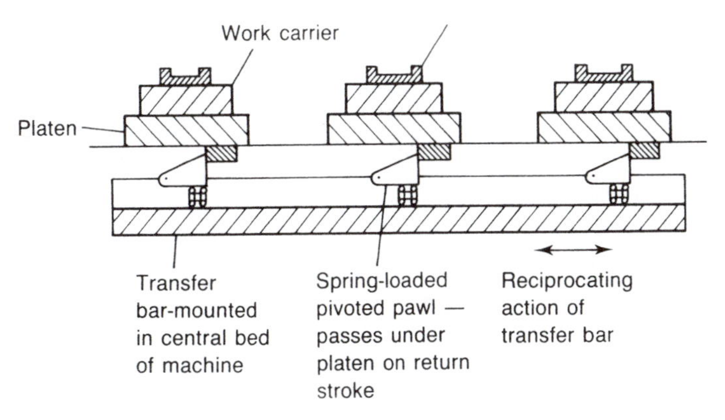

Fig. 7.6 Linear transfer device

simple sorting and orientating devices on the track can ensure that only parts with correct orientation reach the workhead. Wrongly-orientated parts are either corrected or fall off back into the hopper to be recirculated.

Transfer and indexing devices

Most automated assembly is done with the major components of the assembly (called the base unit) being delivered to the assembly station mounted on a work carrier. The other parts are then added to the base unit in sequence and secured. Therefore, the purpose of the transfer or indexing device is to deliver the work carrier to the workhead. A transfer device is shown in Fig. 7.6. On each forward stroke of the feed bar, all

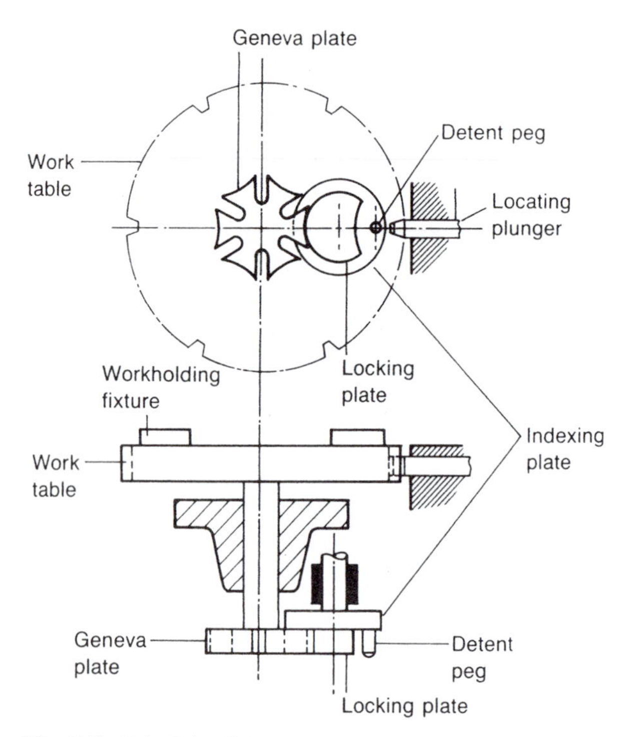

Fig. 7.7 Principle of Geneva indexing mechanism

the work carriers are moved forward by one increment. On the return stroke the spring-loaded pawls are depressed and slide under the work carriers which remain stationary. The cycle is repeated for each forward and return stroke of the feed bar.

Rotary indexing devices are basically rotary tables which transport the work carriers to the workheads. A typical intermittent drive is the Geneva mechanism shown in Fig. 7.7. This indexing plate rotates with constant velocity and the Geneva plate indexes through a prescribed angle intermittently. Having six slots, the Geneva plate in Fig. 7.7 would index through increments of 60° for each revolution of the indexing plate. The locking plate prevents movement of the Geneva plate when it is not being indexed.

Part-positioning devices

Part-positioning devices may be of the escapement type, fed by a feed track, or of the robotic pick-and-place type. There are many types of escapement devices successfully in use. In general, they need to be individually designed to suit the component to be fed. An example is shown in Fig. 7.8(*a*), while Fig. 7.8(*b*) shows a typical pick-and-place arrangement. The pick-and-place arrangement uses a programmable robot

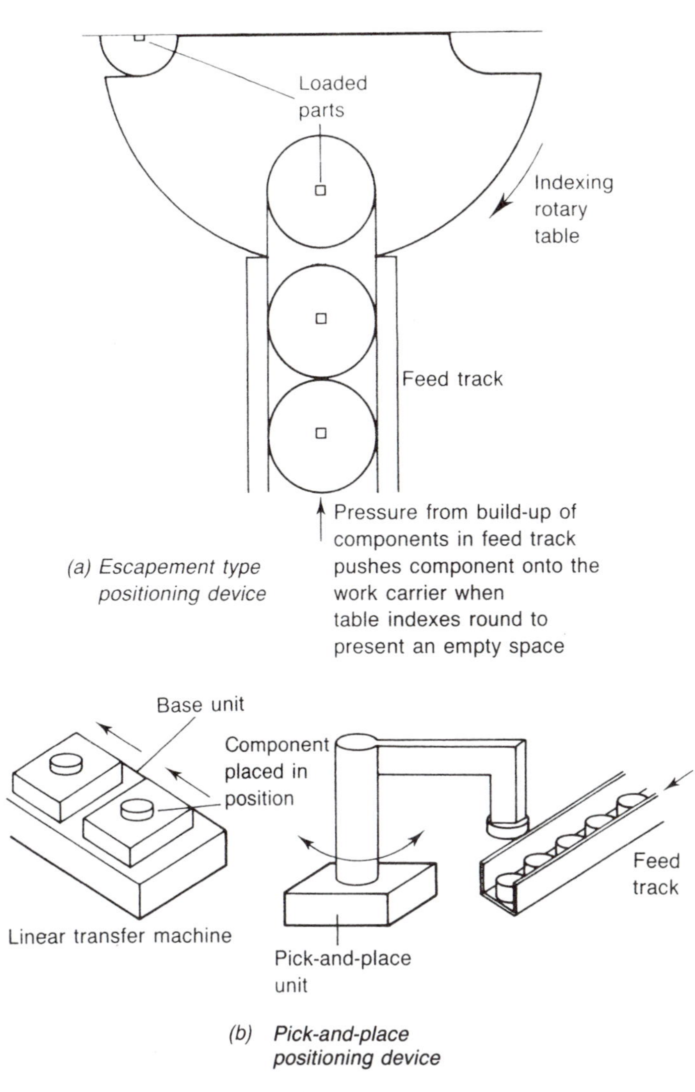

Fig. 7.8 Part-positioning devices

to transfer the components from the feed track and place them on the work carrier or the base unit.

Workhead mechanisms

Workhead mechanisms vary enormously, depending upon the application and the fixing action required. They may be operated mechanically, pneumatically or hydraulically. Figure 7.9 shows a typical workhead being used to secure components together by peening over the stem of the

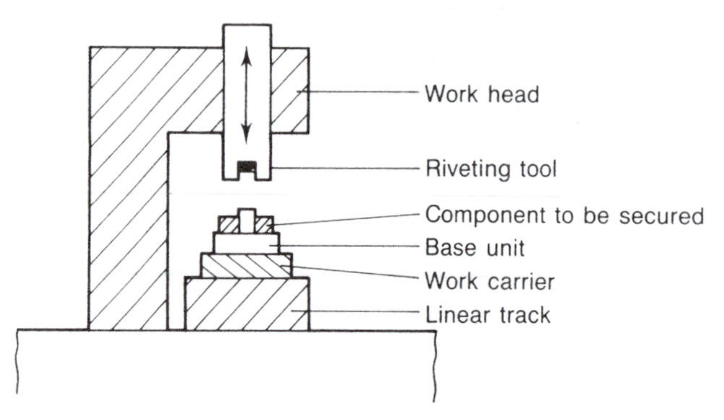

Fig. 7.9 Assembly workhead

component after the previous assembly stations have positioned the components on the base units.

Assignments

1. Discuss the criteria influencing assembly and dismantling which must be considered at the design stage in order to achieve efficient manual assembly and routine maintenance.
2. Discuss the citeria which must be considered at the design stage when automated assembly is to be used.
3. Discuss the advantages of levelled scheduling when planning assembly.
4. Briefly compare:
   (a) manual assembly;
   (b) dedicated automated assembly;
   (c) programmable assembly.
5. Compare the advantages and limitations of automated assembly with manual assembly.
6. (a) With the aid of sketches describe the elements of automated assembly equipment.
   (b) For a simple assembly of your choice, outline an automated assembly 'cell'.

8 Measurement and inspection

8.1 Kinematics of measuring equipment

The measurement and inspection of engineering components was introduced in *Manufacturing Technology: volume 1*. In this chapter linear and angular measurement will be developed further, together with the inspection of screw threads and gear teeth, the manufacture of which is discussed in Chapter 5. However, it is first necessary to consider the kinematics of measuring instruments themselves.

The word 'kinematics' is derived from the Greek work 'kinema' meaning movement. Hence the word kinematics, when applied to mechanisms, relates to movement and position without reference to mass or force. Machine tools and measuring instruments are dependent upon kinematic principles for the location and controlled movement of their various components and assemblies. Thus the aim of kinematic design is to allow a component to move in a specified manner with the utmost freedom, while restraining it in the remaining degrees of freedom with the utmost rigidity (stiffness).

As a reminder, Fig. 8.1 shows the six degrees of freedom of a body in space. The body is free to slide along or to rotate about any of the three axes. Figure 8.2 shows Kelvin's coupling which is the classic solution to prevent movement in all six degrees of freedom without duplicated or redundant restraints. It can be seen that the top plate is supported on a tripod of balls and is dependent only upon gravity to keep it in place.

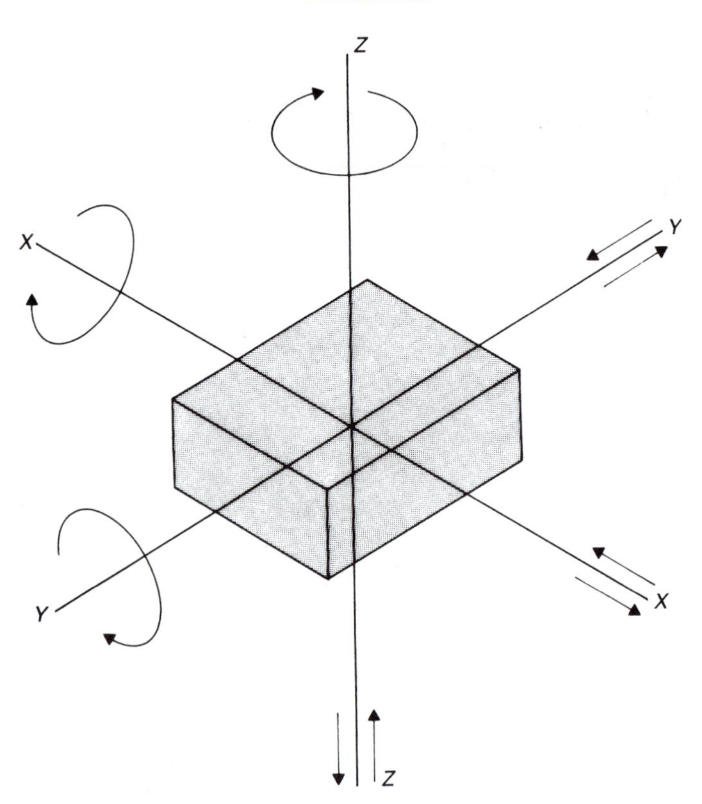

Fig. 8.1 Six degrees of freedom

Providing that the gravitational force, and any extraneous forces, act within a triangle joining the centres of the balls, the plate will remain firmly in position. The balls rest in a trihedral hole, a vee-groove, and a plain recess. The coupling satisfies the following conditions:

- The plates are mutually restrained without the application of external forces other than the force of gravity.
- The plates are free to expand or contract without distortion and without affecting the integrity of the coupling.
- The top plate is supported without distortion.

For most instrument applications only five degrees of freedom need constraint, while total freedom of movement is required in a single specified direction. An adaptation of Kelvin's coupling to satisfy these criteria is shown in Fig. 8.3. No redundant or duplicated constraints are present and the minimum number of constraints have been applied. This type of slideway arrangement is used in the floating carriage diameter measuring machine (see section 8.8).

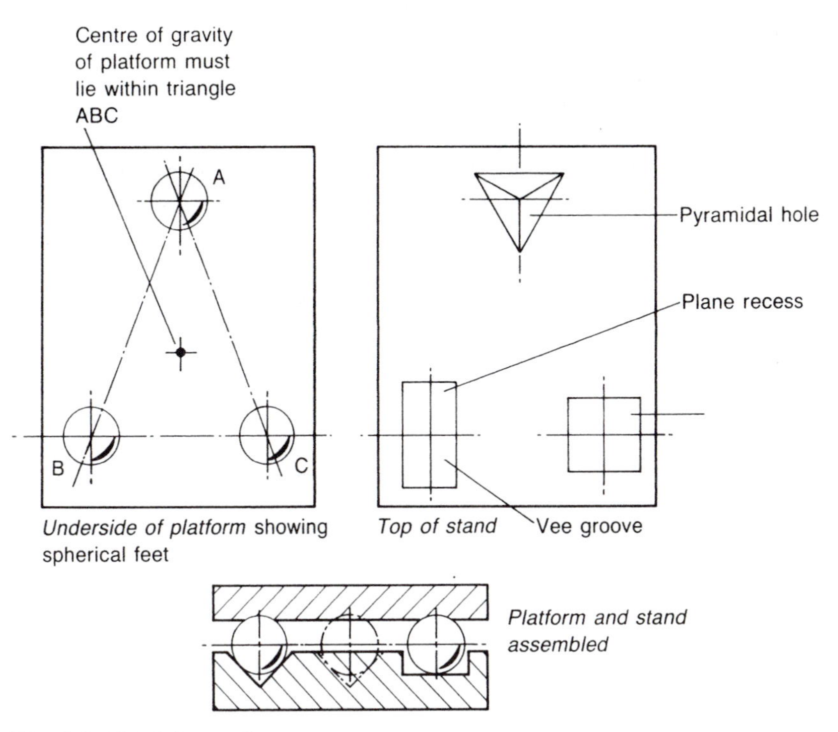

Fig. 8.2 Kelvin's coupling

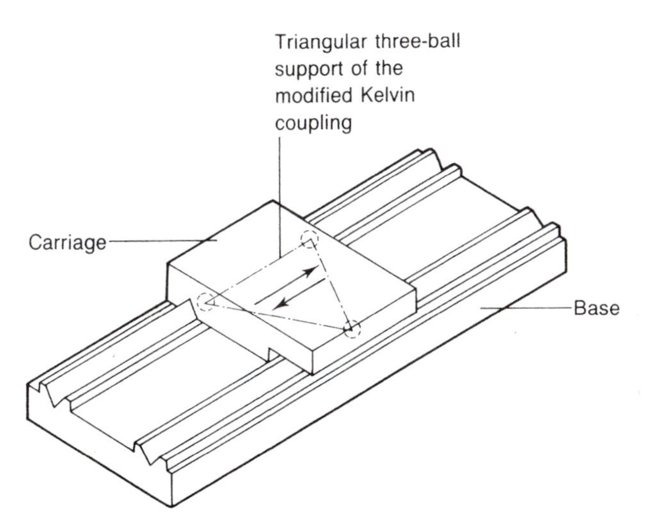

Fig. 8.3 Application of Kelvin's coupling

8.2 Instrument mechanisms

Where relatively small movements and deflections are required in instrument mechanisms, conventional pivots, hinges, and slides can be dispensed with and replaced by simple, robust mechanisms which are free from wear and backlash. For example, Fig. 8.4(*a*) shows how two reinforced spring steel laminae can provide parallel movement free from

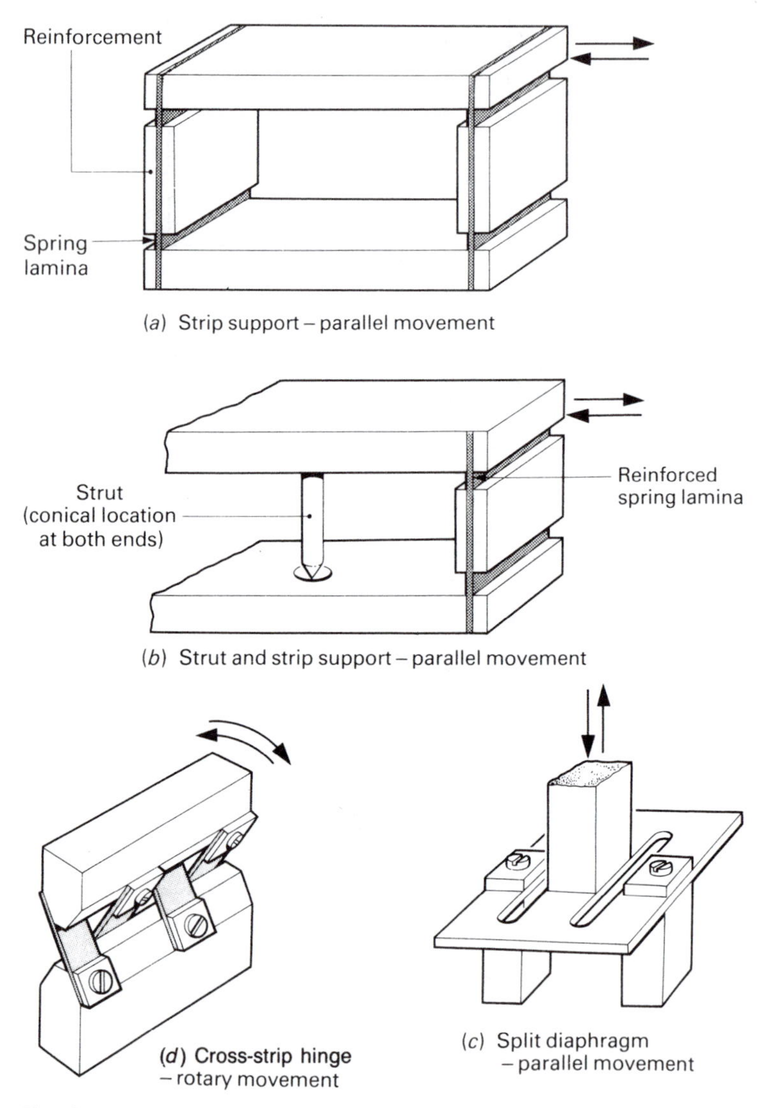

Fig. 8.4 Instrument mechanisms (principles)

sliding and wear. Although the distance between the members varies, this is of little practical importance over the small deflections encountered. Figure 8.4(*b*) is a variation on this mechanism and is to be found in the screw-pitch measuring machine described later in this chapter. Figure 8.4(*c*) shows how a split diaphragm can be used to provide parallel motion. Finally, Fig. 8.4(*d*) shows a cross-strip hinge to provide angular or rotary motion. Since there is no pivot pin, there is no wear in this type of hinge. Since all the above devices use spring steel strips and laminae, they are self-restoring when used in measuring instruments and can provide the necessary measuring pressure to keep the stylus in contact with the workpiece.

8.3 Magnification

In the dial test indicators discussed in *Manufacturing Technology: volume 1*, magnification was provided by gear trains or lever and scroll mechanisms. Such mechanisms are subject to wear and develop inaccuracies. Where greater magnifications are required alternatives must be sought. Many ingenious mechanisms have been developed and these can be categorised as: mechanical, optical, pneumatic and electrical/electronic. Some of these will now be considered.

Mechanical

Figure 8.5 shows a simple, but effective, mechanism using a cross-strip hinge and slit diaphragms to provide high magnification, high accuracy

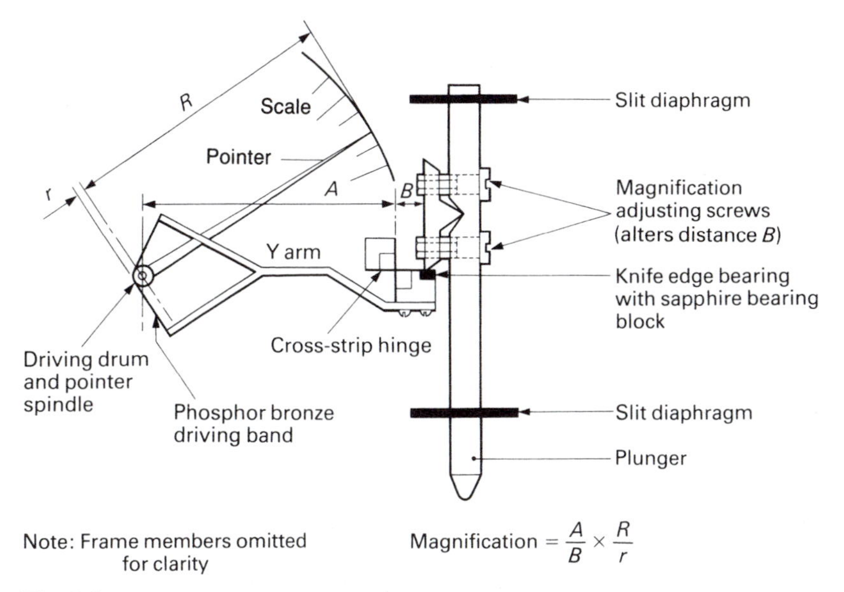

Fig. 8.5 Mechanism of the Sigma comparator

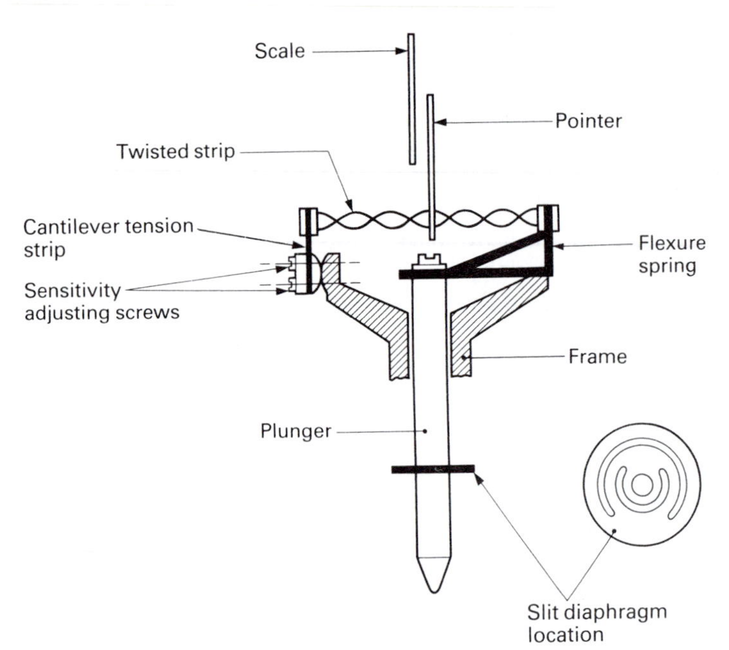

Fig. 8.6 Mechanism of the Johannson Mikrokator

and repeatability, and negligible wear and wear-related errors. Figure 8.6 shows how tension on a twisted strip can be used to provide magnification. The pull of the flexure spring on the twisted strip tends to unwind the strip and this, in turn, moves the pointer across the scale. A slit diaphragm is used to locate the plunger and provide the measuring pressure. Again, there are no pivots or gears which may wear and introduce errors.

Optical

Figure 8.7 shows one method of using optical systems for magnification. The object is to minimise the use of mechanical moving parts and replace them with inertia-less light beams. Unfortunately such instruments have to be viewed in subdued light and they also tend to be bulky because of the optical systems. Note how the reinforced spring diaphragm mechanism is used to provide parallel movement between members A and B. Movement of A relative to B (which is fixed) causes movement of the graticule.

Pneumatic

Pneumatic comparators can be divided into back-pressure types and flow types. The back-pressure types can be further subdivided into high-pressure and low-pressure types. Like electrical comparators, their measuring heads can be remote from the display. The main advantage of

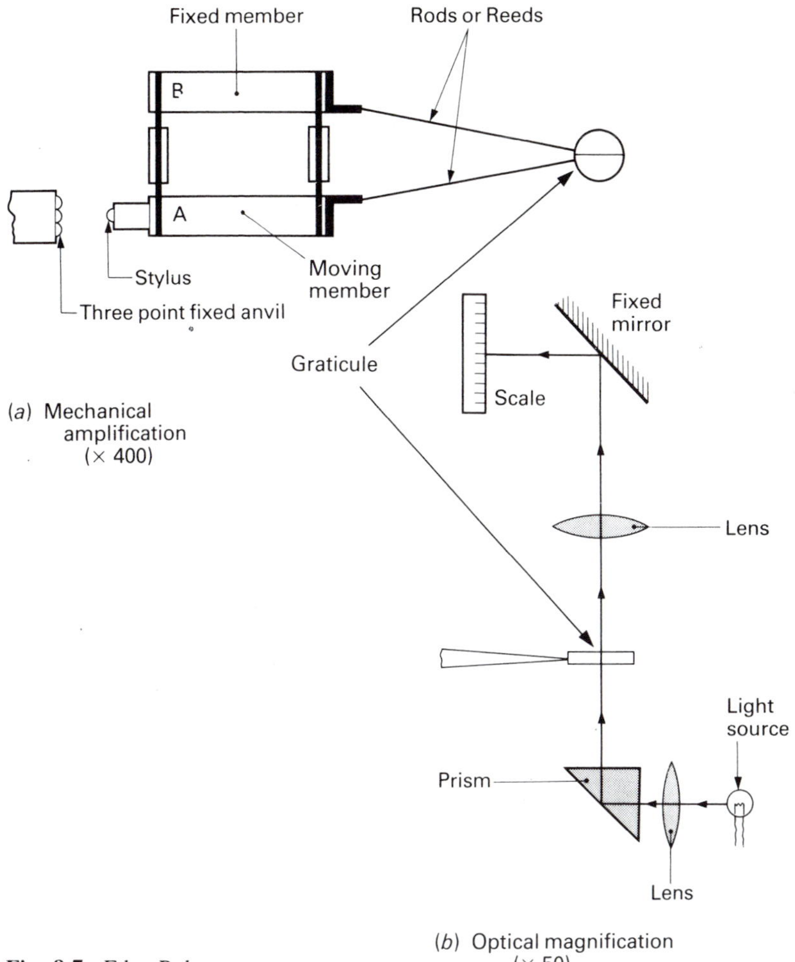

Fig. 8.7 Eden-Rolt comparator

this type of device is the lack of mechanical contact with the surface being measured (proximity gauging).

The principle of operation of a back-pressure instrument is shown in Fig. 8.8(*a*). Air is passed into the measuring head at a controlled pressure P_c. It then passes through the control orifice O_1, into the intermediate chamber. While the control orifice has a constant size, the effective size of the measuring orifice O_2 will vary according to its proximity to the work surface (distance D). Variation in the distance D produces corresponding variations in the back pressure P_b. An indicating device such as mamometer (low pressure) or a bourdon-tube pressure gauge (high pressure) is used to measure the back pressure and is calibrated to read directly the distance D in units of linear measurement.

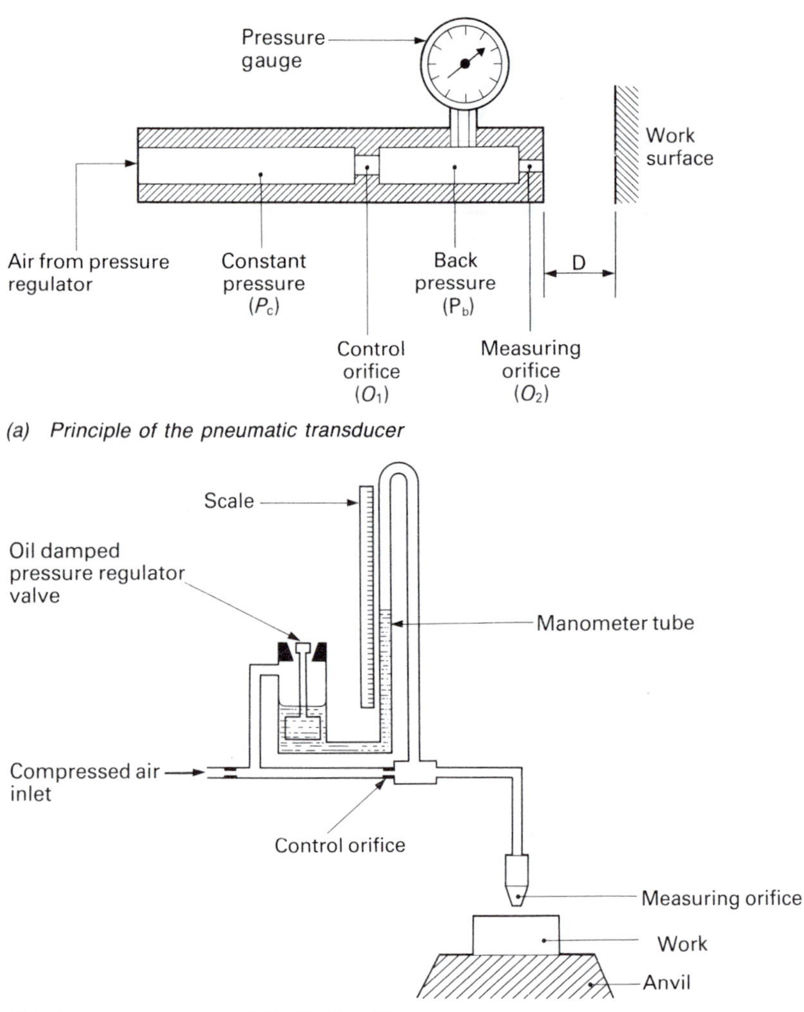

Fig. 8.8 Solex-type low-pressure pneumatic comparator with oil dashpot pressure regulator

Figure 8.8(*b*) shows a typical low-pressure pneumatic comparator with an oil dashpot pressure regulator. Figure 8.9 shows some typical measuring heads.

Electrical/electronic

Electrical and electronic measuring devices have largely displaced the devices previously described for many applications. The use of digital electronics and integrated circuits allows instruments to be built relatively

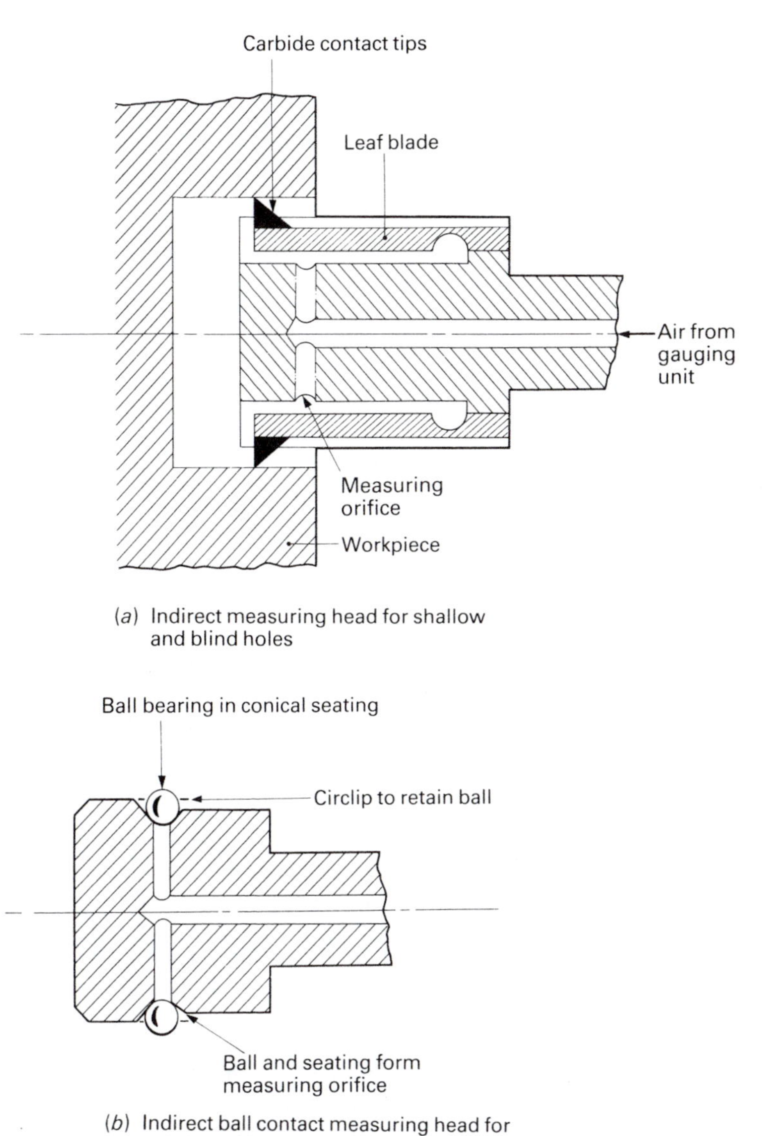

(*a*) Indirect measuring head for shallow and blind holes

(*b*) Indirect ball contact measuring head for rough surfaces

Fig. 8.9 Pneumatic comparator measuring heads

cheaply, yet be more accurate, reliable and compact than their predecessors.

One of the most simple approaches for electrical comparators is based upon some form of *bridge* circuit. Figure 8.10 shows the circuit for a simple bridge using resistors for the ratio arms. When the bridge is in a

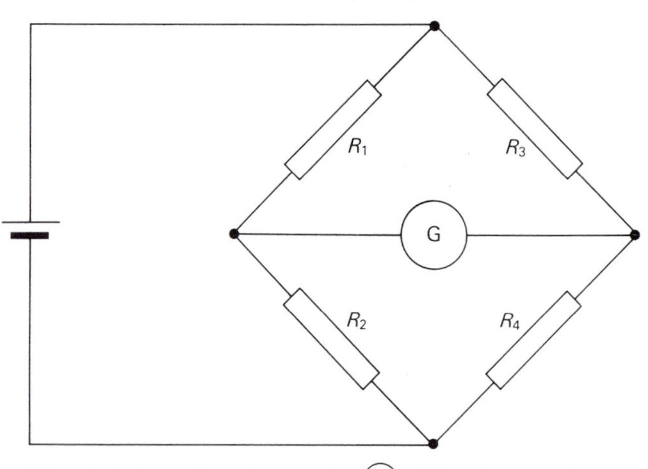

When in balance the galvanometer (G) will read zero, and:

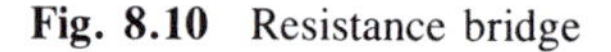

$$\frac{R_1}{R_2} = \frac{R_3}{R_4}$$

Fig. 8.10 Resistance bridge

state of balance, no current will flow through the galvanometer and it will have zero deflection. The relationship between the resistors in the ratio arms is $R_1/R_2=R_3/R_4$. The galvanometer will show a reading either side of zero should the value of any one of the resistors be changed so as to unbalance the current flow in the circuit. It would appear that the insertion of a measuring head containing a variable resistor into one of the ratio arms would produce a practical comparator. Unfortunately this is not the case because of the difficulty of producing reliable variable resistors with adequate sensitivity. Most practical electrical comparators use an alternating current bridge in which the ratio arms are inductors (coils of wire which impede the flow of an alternating current). The inductance (ability to impede alternating current flow) of the inductors can be increased by winding the coil of wire around a soft iron (ferrite) core.

Figure 8.11(*a*) shows the circuit of a practical comparator based upon an inductance bridge, and a suitable measuring head (transducer) is shown in Fig. 8.11(*b*). When the plunger and core assembly rises, more of the core is inserted into L_3 and its inductance increases. Correspondingly, less of the core is inserted into L_4 and its inductance decreases. This unbalances the bridge and a reading is indicated by the galvanometer, G. Note that since an alternating current is being used, a rectifier must be inserted into the galvanometer circuit and the instrument calibrated accordingly to compensate for the effects of rectification. Unlike a variable resistor, there are no sliding contacts in a variable inductor and this is where the a.c. bridge gains its reliability and sensitivity.

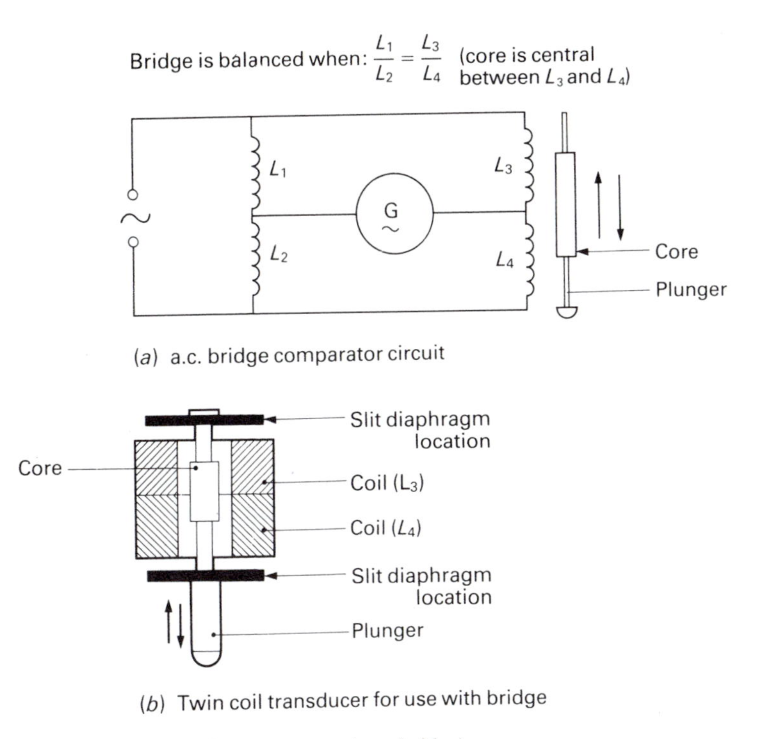

Fig. 8.11 Electronic comparator (a.c. bridge)

The differential inductance transducer shown in Fig. 8.11(*b*) contains only two external coils to form the ratio arms of the bridge. With the introduction of reliable high gain amplifiers in the form of integrated solid state circuits, it has become a practical proposition to use the arrangement shown in Fig. 8.12(*a*). The measuring head is shown in Fig. 8.12(*b*), and it can be seen that it has three coils and is called a linear variable differential transformer (LVDT). This type of transducer does not require a bridge circuit. In operation, a steady high-frequency alternating current from the frequency-stabilised oscillator is applied to the centre (primary) winding of the LVDT and induces equal and opposite alternating currents in the two secondary windings. As the core moves from the point of equilibrium, it changes the magnetic flux pattern. This results in the induced current in one secondary winding being increased at the expense of the current in the other winding which is decreased. This provides a signal for the amplifier which can vary in both magnitude and direction when compared with the steady state signal from the oscillator. By using a high-frequency alternating current the sensitivity of the instrument is increased while, at the same time, the size of the transducer is reduced by the use of smaller coils and a core of lower mass.

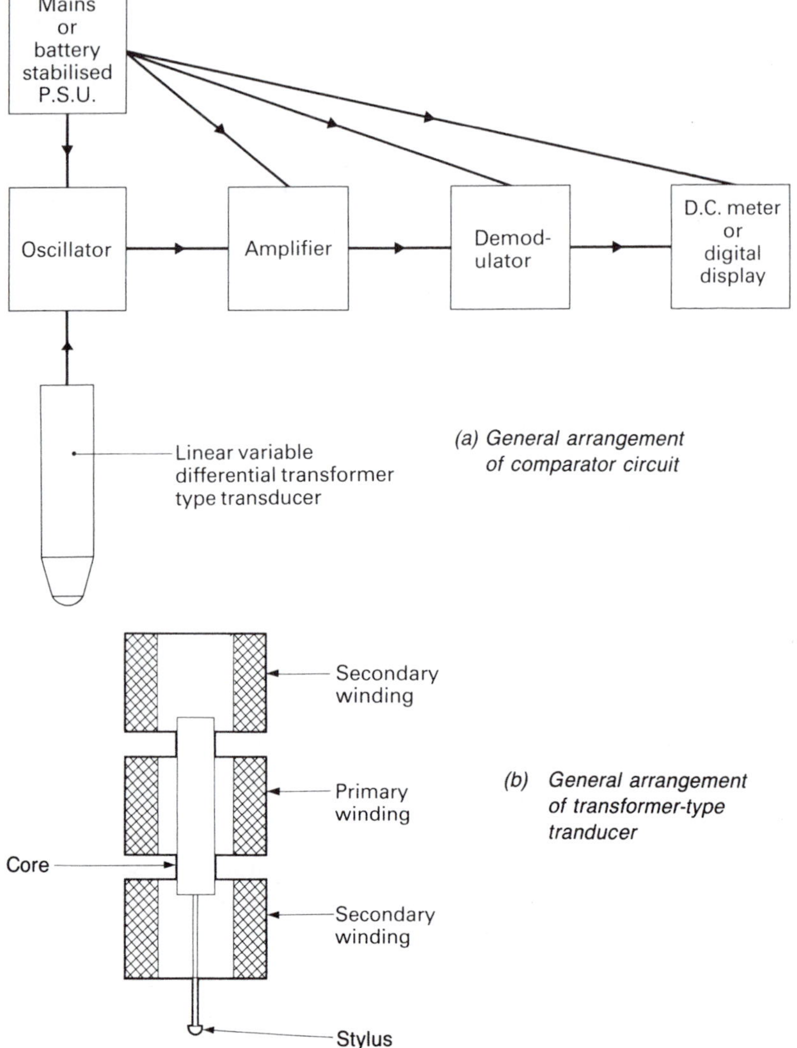

Fig. 8.12 Electronic comparator (LVDT)

Although more complex than mechanical and pneumatic instruments, electrical and electronic measuring devices are becoming ever more widely used as they become lower in cost, increasingly reliable, more and more compact, and as portable as mechanical instruments now that they can be battery powered. Some of the advantages of electronic comparators are:

- they do not contain mechanisms that are liable to introduce friction, backlash, and inertia effects;

- the magnification can be changed by simply switching the electrical circuit;
- zero adjustment is also affected electrically, thus avoiding the need for fine mechanical adjustments to the position of the measuring head;
- the system is energised externally and the movement of the plunger/stylus does not have to drive the magnification mechanism. This allows extremely low measuring pressures to be used, resulting in negligible component distortion, which is important when measuring soft and thin-walled components;
- the output of the device can be used in conjunction with analogue or digital displays. It can also be used to operate machines, recording instruments, sorting gates, light signals, etc.;
- remote readings with the measuring head up to 30 m from the rest of the equipment are possible. This is particularly true with electronic instruments using LVDT measuring heads as they are less susceptible to the capacitance effects of long interconnecting leads.

8.4 Angular measurement

Angular measurement by such devices as plain and vernier protractors and, where greater precision is required, a sine bar were discussed in *Manufacturing Technology: volume 1*. Angles can also be measured by comparison. Just as comparative measurements of linear dimensions can be made using slip gauges, so can comparative measurements of angles be made using *combination angle gauges*. These are made from hardened, stabilised, and lapped blocks of alloy steel in the same manner as slip gauges. The blocks are made to precise angles and the surface finish of the measuring faces enables them to be wrung together. The sizes of a typical set of blocks and the manner in which they may be combined so that their angles may be added or subtracted is shown in Fig. 8.13. A set of 20 blocks of the sizes shown enables any angle to be built up in steps of 3 seconds of arc. An additional 9° block extends the range to 90°, and the use of the precision square block enables the complete range of 360° to be obtained. Alternatively, the combination angle gauges are also available in decimal notation.

These gauge blocks are normally used in conjunction with optical measuring devices such as the Angle Dekkor. Alternatively, they may be used in the workshop for direct measurement and setting in a similar manner to the sine bar.

8.5 Collimation

Figure 8.14(*a*) shows that if a point source of light is placed at the principal focus of a collimating lens system, the rays of light will be projected as a parallel beam. If these rays strike a reflecting surface perpendicular to the optical axis, they will be reflected back along the same path and refocused at the point of origin.

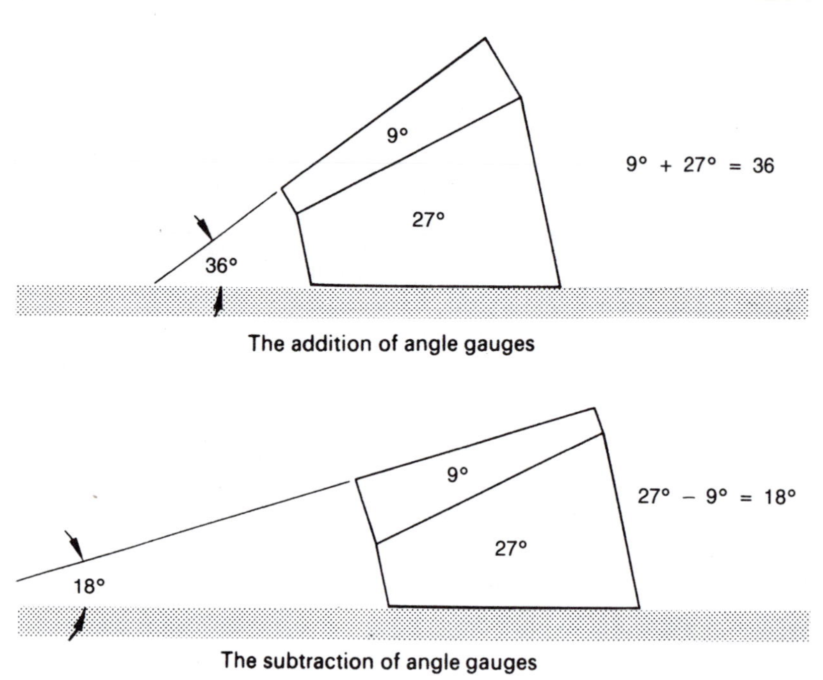

(a) Use of combination angle gauges

Degrees	Minutes	Seconds
1	1	3
3	3	9
9	9	27
27	27	
41		
A square block is also provided		

Degrees	Minutes	Decimal minutes
1	1	0·05
3	3	0·1
9	9	0·3
27	27	0·5
41		
A square block is also provided		

(b) Range of sizes

Fig. 8.13 Combination angle gauges

Figure 8.14(*b*) shows that if the reflective surface is tilted through some small angle $\theta°$, the reflected parallel rays will be inclined at $2\theta°$ to the optical axis and will be brought to a focus at a point in the focal plane of their origin, but displaced by a small distance x. The following characteristics of such an arrangement should be noted since they are of supreme importance in the optical devices to be described in the next two sections.

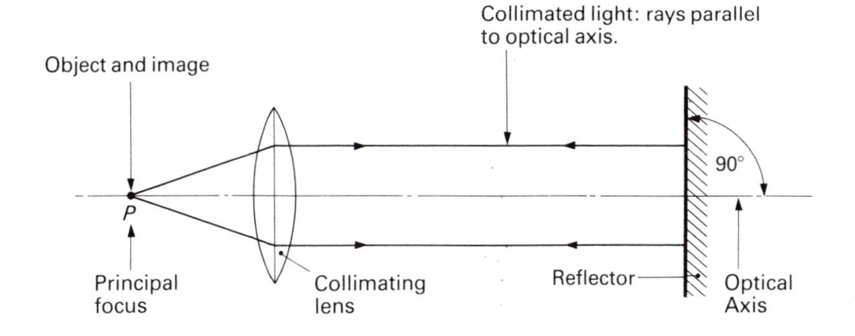

(*a*) Collimated light reflected from perpendicular surface

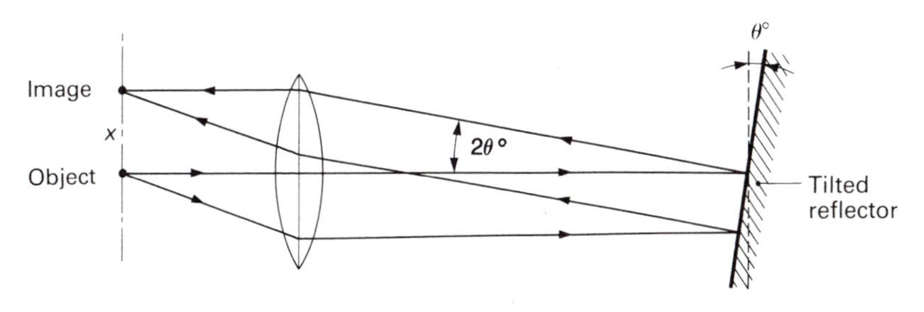

(*b*) Collimated light reflected from tilted surface

Fig. 8.14 Reflection of collimated light

- The distance between the reflector and the lens has no effect on the separation of the source and the image (distance x). However, too great a distance will result in the rays missing the collimating lens altogether and no image will be formed.
- For high sensitivity a long focal length collimating lens is required to give a large value of x for a small value of $\theta°$.

8.6 The Angle Dekkor

The Angle Dekkor does not measure angles directly, but by comparison with a known standard such as a combination angle gauge. Thus it is a comparator with a scale of limited range, capable only of giving a difference reading.

The optical system and the scale as seen in the eyepiece is shown in Fig. 8.15. The magnifying eyepiece views both the fixed datum scale and the reflected image of the illuminated scale at right angles to each other, as shown. The image does not fall across a simple datum line, but across a similar fixed scale at a right angle to the image. Thus the reading on the reflected scale measures angular deviations in one plane at 90° to the

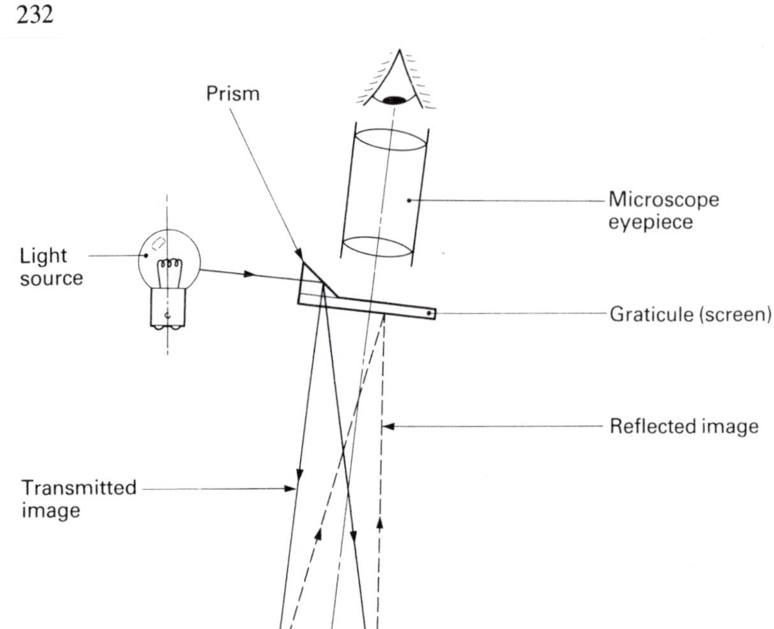

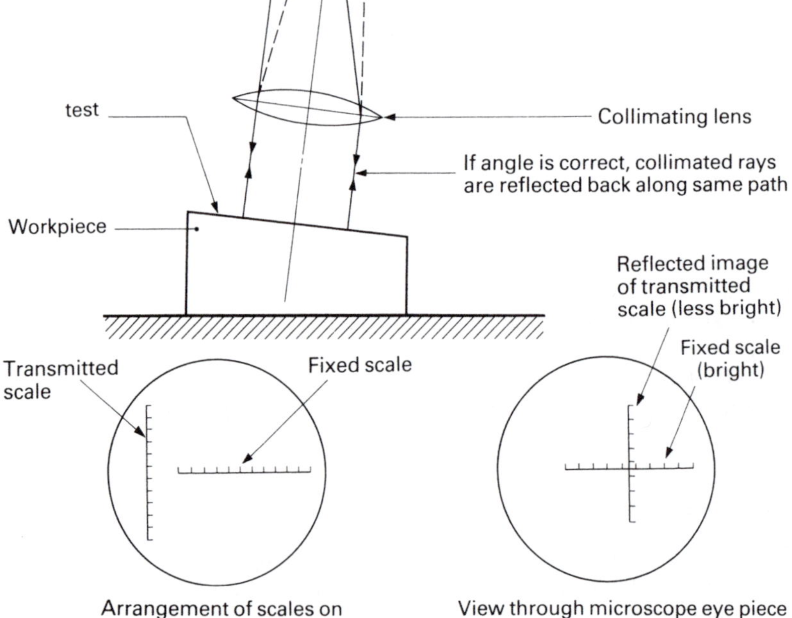

Fig. 8.15 Angle Dekkor optical system

optical axis, and the reading on the fixed scale gives the deviation in a plane that is perpendicular to the former plane. This feature enables angular errors in two planes to be dealt with at the same time and, more importantly, to ensure that the reading on the setting gauge and on the workpiece is the same in one plane, while the error is read in the other plane. If there is no error the two scales should cross at the same reading as would have been shown when the instrument was set up against the setting gauge. Normally it would be set with the scales appearing to cross in the middle of the eyepiece. Figure 8.16 shows the displacement of the scales for various errors in the reflected surface of the workpiece. Although the Angle Dekkor requires considerable skill to set it and read it accurately, it is a very useful instrument for a wide range of angular measurements at short distances. Direct readings can be obtained down to 1 min of arc over a range of 50 min and, by estimations, readings down to 0.2 min of arc are possible with practice. Figure 8.17 shows an Angle Dekkor and its stand.

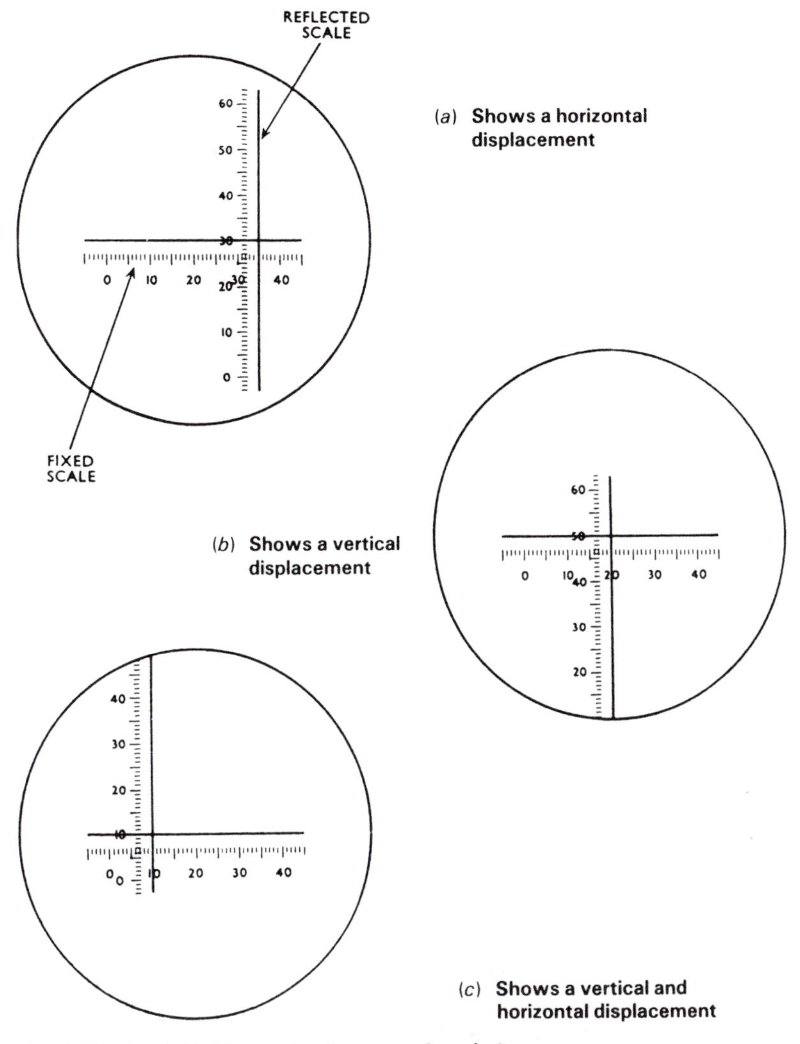

Fig. 8.16 Angle Dekkor scales (compound angles)

Fig. 8.17 Angle Dekkor and stand

8.7 The Autocollimator

Unlike the Angle Dekkor, the Autocollimator is a direct-reading angular measuring device. It does not require to be used in conjunction with a setting master; it is very much more sensitive, and it can be used over very much longer distances than the Angle Dekkor. Figure 8.18(*a*) shows a typical Autocollimator, Figure 8.18(*b*) shows a reflector suitable for checking machine tool alignments, and Fig. 8.18(*c*) shows the principles of its optical system. It can be seen from Fig. 8.18(*c*) that the image of the target wires are projected onto the reflective surface of the workpiece through a collimating lens system and returned, after reflection, back through the same lens system in a similar manner to the Angle Dekkor. The target wires and their reflected images are viewed simultaneously through a magnifying eyepiece, as shown in Fig. 8.19 after passing through the semi-reflector.

The eyepiece contains an adjustable graticule engraved with setting lines and a scale divided into minutes and half-minutes of arc. The setting lines are adjusted by the micrometer control until they straddle the reflected image of the target wires. The scale is then read to the nearest $\frac{1}{2}$ minute of arc plus the micrometer reading. The divisions of the micrometer scale each represent a $\frac{1}{2}$ second of arc. Instruments are available with the scales in decimal notation. The Autocollimator has two sets of scales and two micrometer adjustments perpendicular to each other. This is so that compound angles may be measured or that setting errors may be reduced. Figure 8.20 shows two uses of the Autocollimator for testing machine tool alignments.

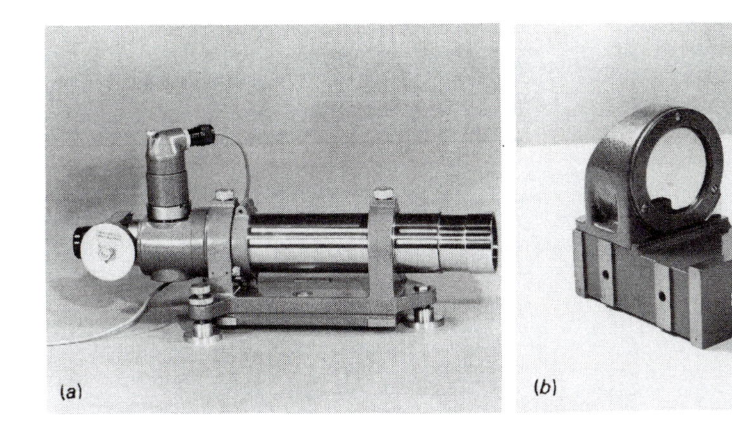

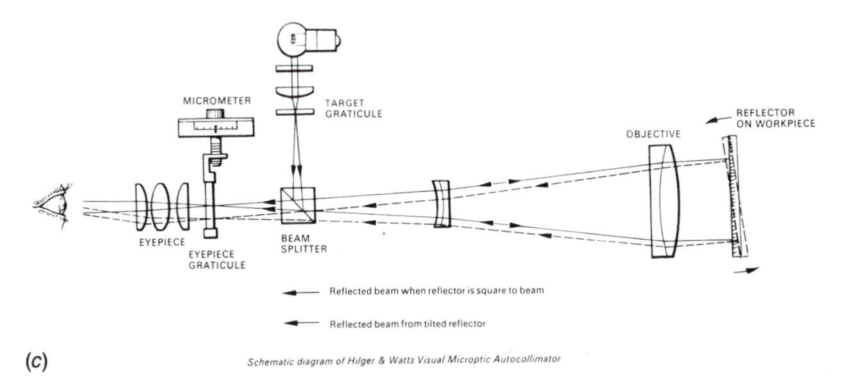

Fig. 8.18 Microptic Autocollimator

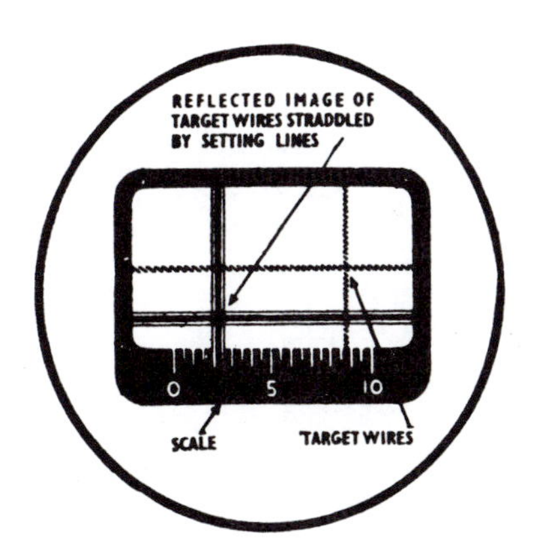

Fig. 8.19 Autocollimator field of view

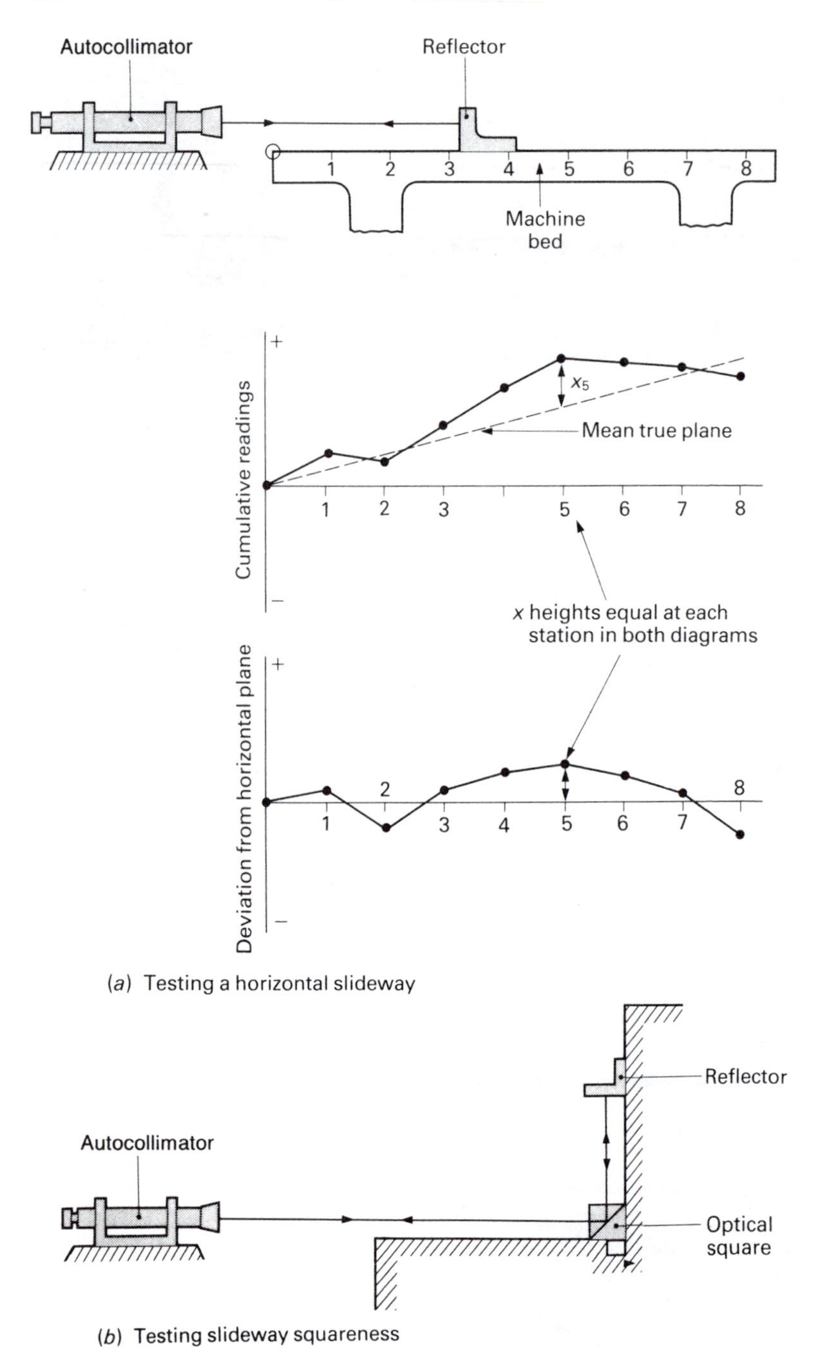

Fig. 8.20 Testing machine alignments with the autocollimator

8.8 Screw-thread diameter measurement

Figure 8.21 shows the elements of a screw thread. It can be seen that in order to completely inspect a screw thread it is necessary to measure the major, minor, and effective (pitch) diameters. It is also necessary to measure the pitch and check the thread profile — particularly the flank angle.

Figure 8.22 shows a floating-carriage diameter measuring machine. The workpiece to be measured is supported between centres, and the micrometer and fiducial indicator are mounted on a carriage. This carriage maintains the measuring axis perpendicular to the workpiece axis and ensures that the measurement is across the diameter of the workpiece. The carriage is free to move parallel to the axis of the workpiece, and the top slide, which carries the micrometer and fiducial indicator, is free to move perpendicularly to the axis of the workpiece. The slides follow the principles described in section 8.1 to give maximum freedom (minimum friction) in the required direction and maximum stiffness in all other directions. The micrometer has a large diameter barrel and thimble and has a reading accuracy of 0.001 mm over a range of 25 mm. Its spindle does not rotate and this reduces the wear on the contact face. The fiducial indicator ensures a constant measuring pressure each and every time the instrument is used. Adjustable arms are provided for suspending prisms and cylinders, and the importance of these will be described later.

8.9 Measurement of the major diameter

Measurements taken with a diameter-measuring machine are comparative. The initial reading is taken over a known standard as near to the major diameter as possible, as shown in Fig. 8.23(*a*). The thread to be measured is then substituted for the setting standard and a reading is taken over the major diameter of the workpiece as shown in Fig. 8.23(*b*). The major diameter is then calculated as follows:

$$M = D_s \pm (\text{difference between } R_1 \text{ and } R_2)$$

where: M = major diameter of screwthread;
D_s = actual diameter of setting standard;
R_1 = reading over setting standard;
R_2 = reading over screw thread.

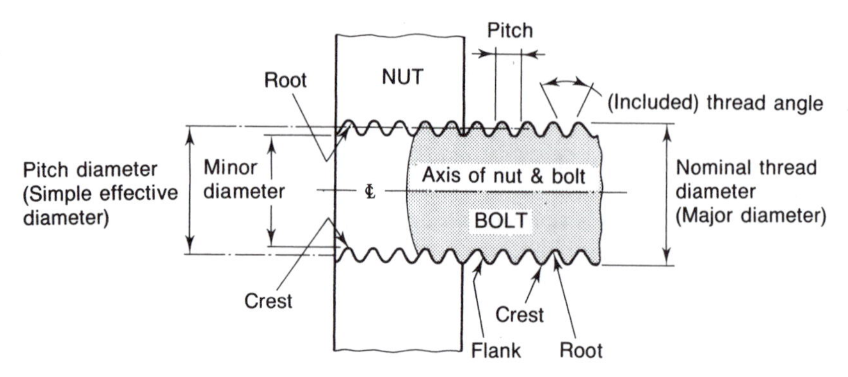

Fig. 8.21 ISO screw-thread form

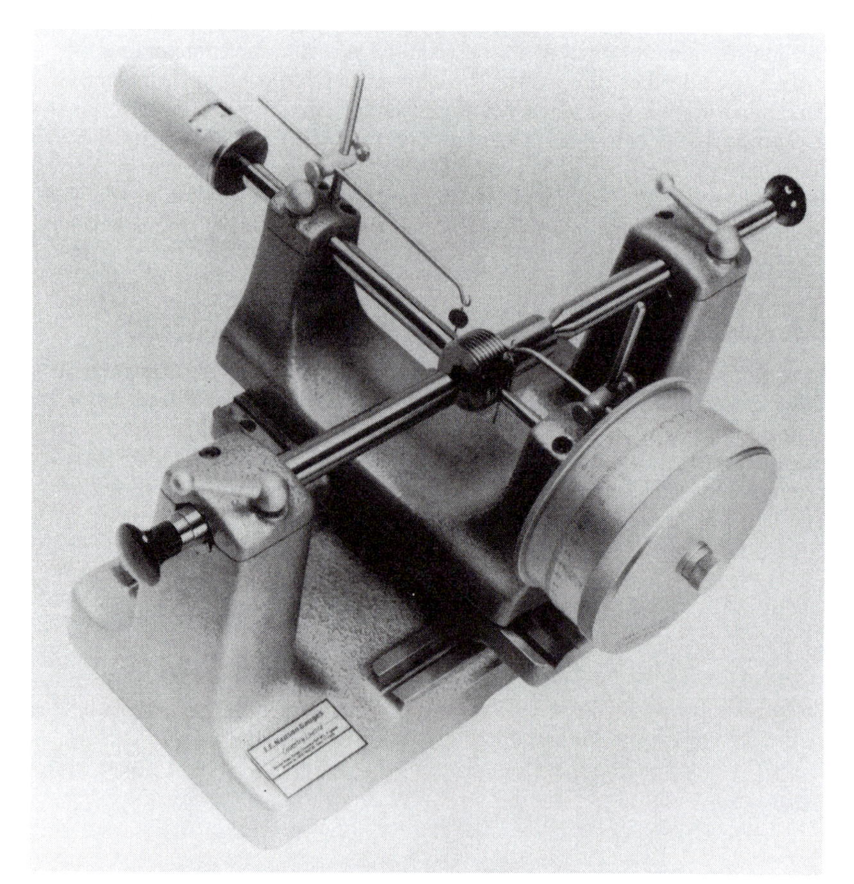

Fig. 8.22 Floating-carriage diameter measuring machine (courtesy of J.E. Nanson Gauges, Coventry, Ltd.)

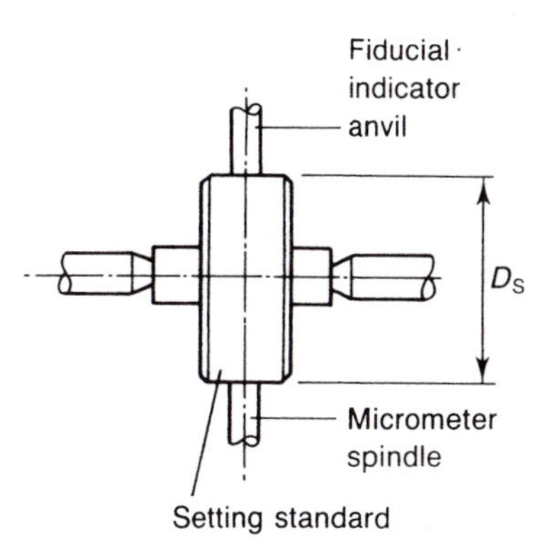

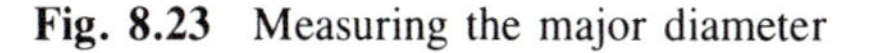
(a) *Calibrating the diameter measuring machine*

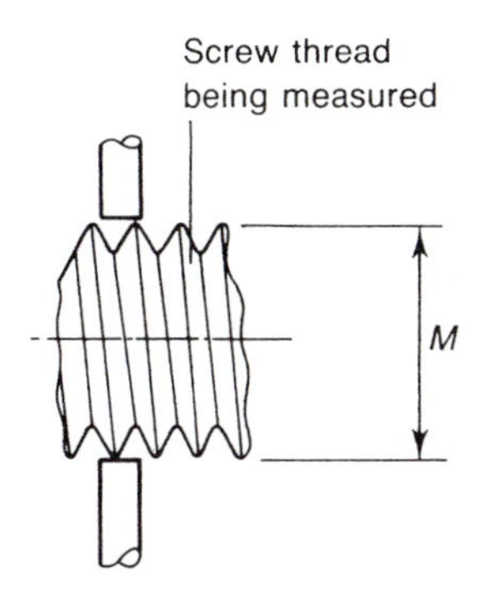

(b) *Measuring the major diameter*

Fig. 8.23 Measuring the major diameter

Example 8.1 Calculate the major diameter of a screw thread with a nominal diameter of 28 mm, given that:

- calibrated diameter over setting standard = 25.001 mm
- reading over setting standard = 17.908 mm
- reading over screw thread = 20.898 mm

Since the major diameter is larger than the setting standard, use:

$M = D_s + (R_2 - R_1)$ where: $D_s = 25.001$

$M = 25.001 + (20.898 - 17.908)$ $R_1 = 17.908$

$\underline{M = 27.991 \text{ mm}}$

8.10 Measurement of effective diameter

To measure the effective or pitch diameter of a screw thread, contact must be made with the straight flanks of the thread. To do this, precision cylinders or 'wires' are introduced between the micrometer anvils and the screw thread, as shown in Fig. 8.24(*a*). If a floating-carriage measuring machine is not available, then three wires have to be used (as shown in Fig. 8.24(*b*)) to align the micrometer and prevent it from twisting. Figure 8.24(*c*) shows the geometry of a cylinder in contact with a thread. It can be seen that:

$E = T + 2x$ where: E = effective diameter;

T = the measured size over the cylinders minus twice the cylinder diameter;

$2x$ = a constant (P) whose value depends upon the cylinder diameter, the pitch of the thread and the semi-angle of the thread form.

For all practical purposes when measuring ISO threads where the included flank angle of the thread is 60°, the following expression may be used to calculate P:

$P = 0.86603p - d$ where: p = pitch
d = cylinder diameter

Dimensions are in inches when unified threads are being measured, and in millimetres when metric threads are being measured.

Again, the measurements taken are comparative and the procedure is as follows. An initial reading is taken over the setting standard with the cylinders or 'wires' in place, as shown in Fig. 8.25(*a*). A second reading is then taken over the screw thread with the wires in place as shown in Fig. 8.25(*b*), and dimension T can be calculated as follows:

$$T = D_s \pm (\text{difference between } R_3 \text{ and } R_4)$$

where: D_s = actual diameter of setting standard;
R_3 = measurement over setting standard with cylinders in place;
R_4 = measurement over the screw thread with cylinders in place.

If the effective diameter is *larger* than the diameter of the setting standard use D_s + ...

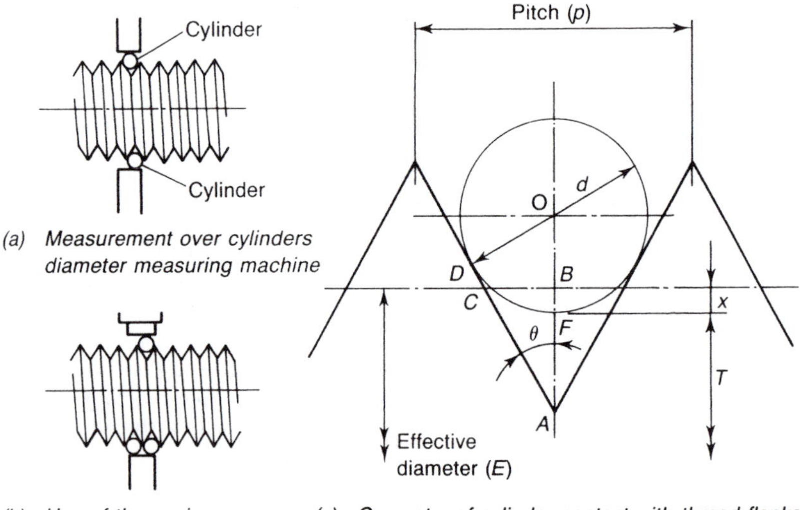

(*a*) *Measurement over cylinders diameter measuring machine*

(*b*) *Use of three wires micrometer caliper*

(*c*) *Geometry of cylinder contact with thread flanks*

Fig. 8.24 Use of measuring wires

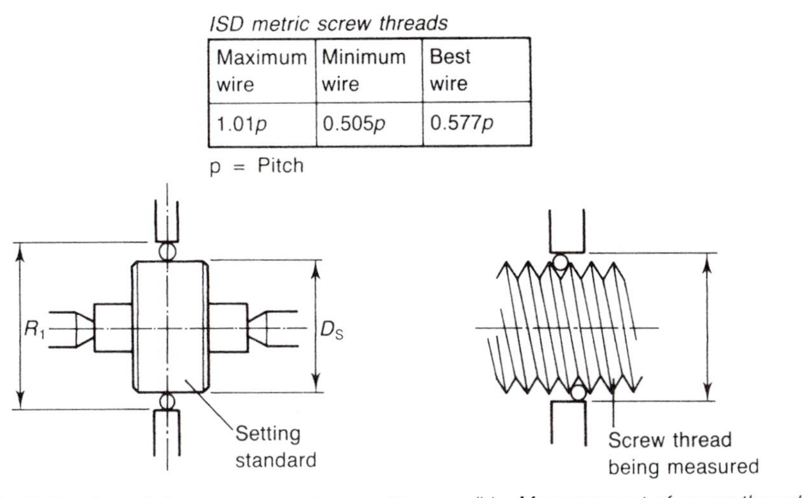

ISD metric screw threads

Maximum wire	Minimum wire	Best wire
1.01*p*	0.505*p*	0.577*p*

p = Pitch

(a) Calibration of diameter measuring machine with cylinders in place

(b) Measurement of screw thread

Fig. 8.25 Measuring the effective diameter

If the effective diameter is *smaller* than the diameter of the setting standard use D_s − . . . . .

Example 8.2 Calculate the effective diameter for an external screw thread given the following data:

Pitch = 3.500 mm
Diameter of calibrated standard = 30.000 mm
Measurement over standard = 15.377 mm
Measurement over screw thread = 14.343 mm
Cylinder diameter = 2.000 mm

Since the effective diameter is smaller than the setting standard, use:

$T = D_s - (R_3 - R_4)$ where: $D_s = 30.000$ mm
$T = 30 - (15.377 - 14.343)$ $R_3 = 15.377$ mm
$\underline{T = 28.966 \text{ mm diameter}}$ $R_4 = 14.343$ mm

$P = 0.86603p - d$ where: $p = 3.5$ mm
$P = 0.86603 \times 3.5 - 2.0$ $d = 2.0$ mm
$\underline{P = 1.0311 \text{ mm}}$

Since:

The effective diameter $E = T + P$
$E = 28.966 + 1.0311$
$\underline{E = 29.971 \text{ mm diameter}}$

The above calculations hold good providing the cylinders touch the threads somewhere on the straight flanks and proving that the *flank angle is correct*. The maximum and minimum wire sizes for a given thread are shown in Fig. 8.25 for metric and unified threads. The *best cylinder size* is also shown. This is used if there is a possibility of the flank angle being incorrect.

8.11 Measurement of the minor diameter

The minor or root diameter is measured in a similar way to the effective diameter, except that *prisms* are used instead of cylinders. These are inserted between the micrometer anvils and the screw thread so that contact can be made with the minor diameter of the thread. Again, the measurement is comparative and the procedure is as follows:

- A reading R_5 is taken over the setting standard with the prisms in place as shown in Fig. 8.26(*a*).
- A reading R_6 is taken over the screw thread with the prisms in place as shown in Fig. 8.26(*b*).
- The minor diameter is then calculated from the expression:

$$m = D_s \pm (\text{difference between } R_5 \text{ and } R_6)$$

  If the minor diameter is *larger* than the diameter of the setting standard use D_s + ...

  If the minor diameter is *smaller* than the diameter of the setting standard use D_s − ....

Note that the size of the prisms does not affect the calculation, the only proviso being that the prisms will fit between the minor diameter of the thread and the micrometer anvils without touching any other part of the thread.

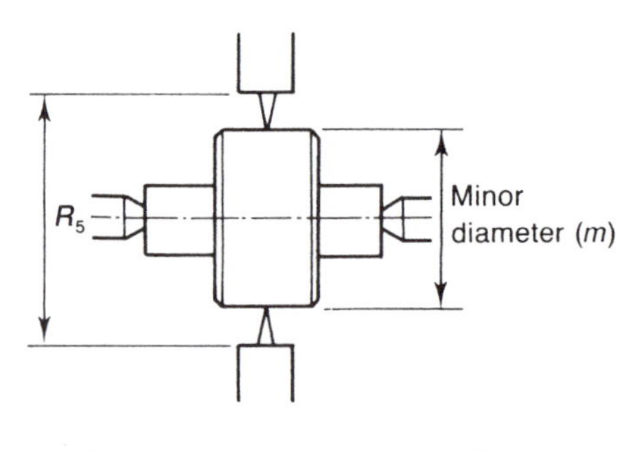

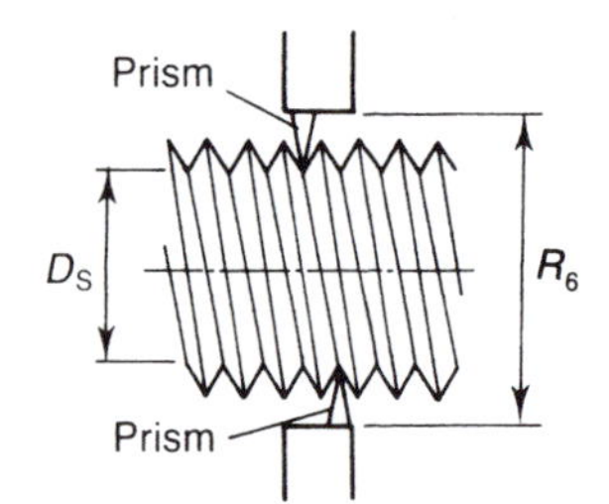

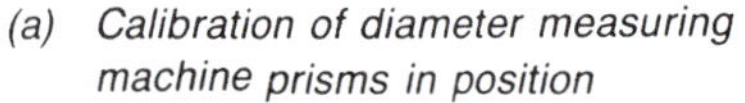
(*a*) *Calibration of diameter measuring machine prisms in position*

(*b*) *Measurement of screw thread*

Fig. 8.26 Measuring the minor diameter

Example 8.3 Calculate the minor diameter of a screw thread given the following data:

D_s = 30.005 mm diameter; R_6 = 12.377 mm diameter; R_6 = 14.513 mm diameter.

Since the minor diameter is larger than the setting standard diameter use:

$m = D_s + (R_5 - R_5)$
$m = 30.005 + (14.513 - 12.377)$
$m = 32.141$ mm diameter

8.12 Pitch measurement

Figure 8.27 shows a screw-thread pitch measuring machine. The carriage is moved parallel to the axis of the thread being measured by a lead screw and micrometer. A fiducial indicator mounted on the carriage indicates when corresponding points have been reached on adjacent threads. The difference between the readings of the micrometer dial indicates the pitch of the screw being measured.

Figure 8.28(a) shows the principle of the mechanism of a fiducial indicator suitable for use on a screw-thread pitch measuring machine. A round-nosed stylus engages the thread approximately on the pitch line. As the carriage is moved, the stylus rides up the thread and causes the reinforced diaphragm spring and strut to deflect in a radial direction. This does not register on the pointer. However, should the side forces on the stylus (F_1 and F_2) be unequal, the diaphragm spring will twist and the pointer will be deflected.

To measure the pitch of the thread, the stylus is engaged in the thread groove with sufficient force for the pointer (2) in Fig. 8.28(*b*) to lie

Fig. 8.27 Pitch measuring machine (courtesy of J.E. Nanson Gauges, Coventry, Ltd.)

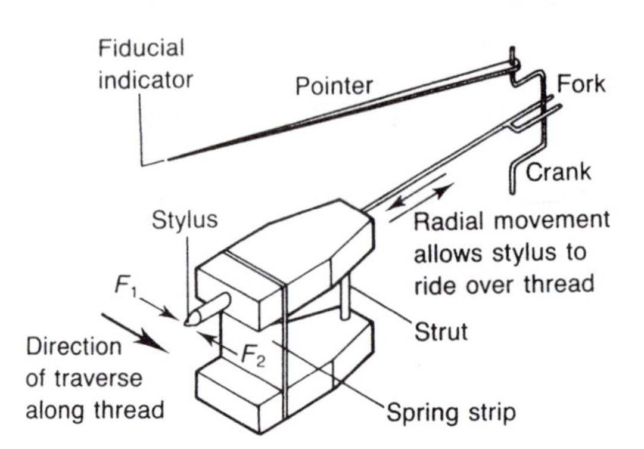

(a) Mechanism (principle)

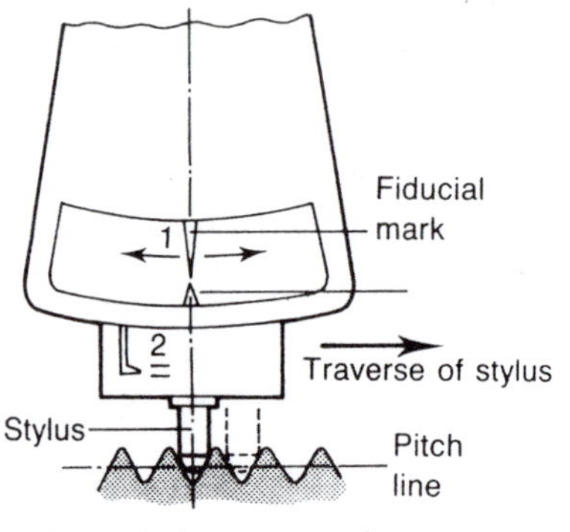

(b) Indicator scales

Fig. 8.28 Use of fiducial indicator; (a) Mechanism (principle); (b) Indicator scales

between its datum lines. The micrometer is adjusted until the pointer (1) in the same figure lies opposite its datum mark and the initial reading of the micrometer is noted. Continuing to rotate the micrometer in the same direction, so that no backlash is introduced, the carriage is moved along by one thread until the pointer (1) is again opposite its datum mark and the micrometer is read for the second time. The difference between the readings is the measured pitch of the thread.

8.13 Thread form

Finally it is necessary to check the flank angles, the root and crest radii, and the profile in general. This is usually done by means of an optical projector. The projection of screw threads presents a special problem due to the helix of the thread which interferes with the passage of the light rays, as shown in Fig. 8.29(*a*). There are two ways of overcoming this problem.

- One method, as shown in Fig. 8.29(*b*), is to incline the axis of the thread so that the collimated light beam grazes the thread without appreciable interference. Since the thread is inclined to the focal

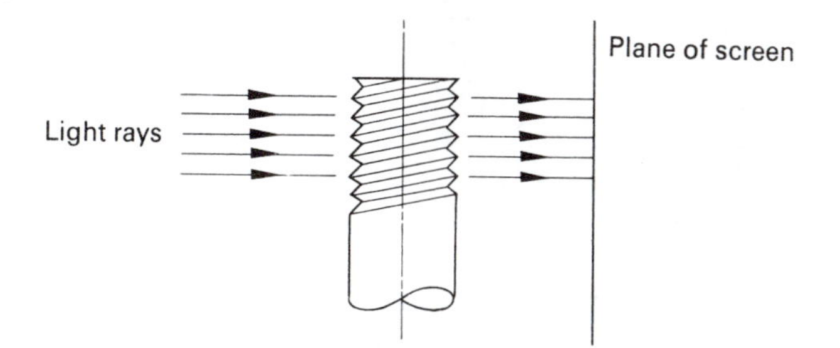

(*a*) **Light rays perpendicular to thread axis**

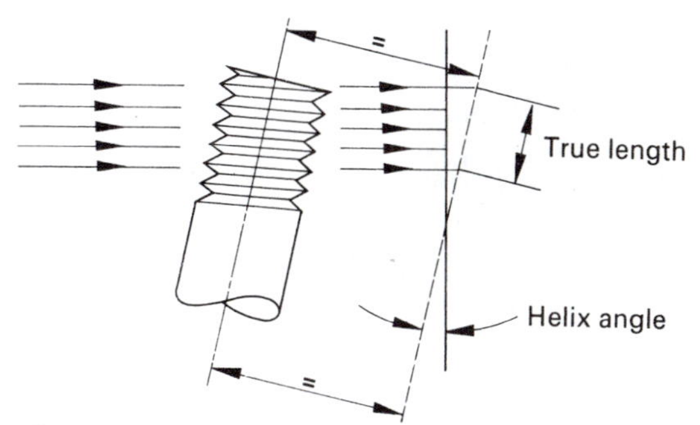

(*b*) **Thread inclined at helix angle**

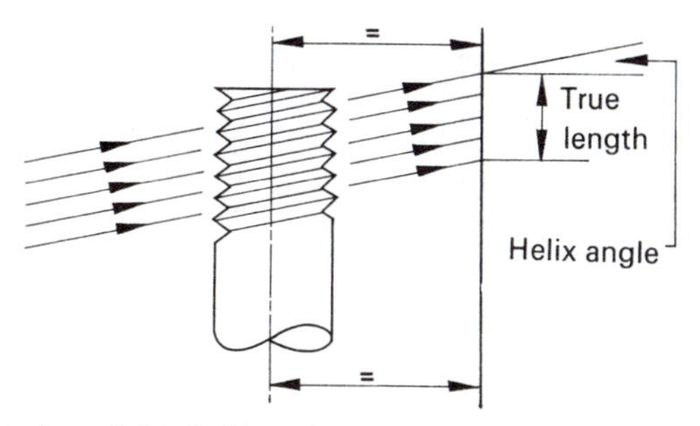

(*c*) **Light rays inclined parallel to helix angle**

Fig. 8.29 Projecting a screw thread

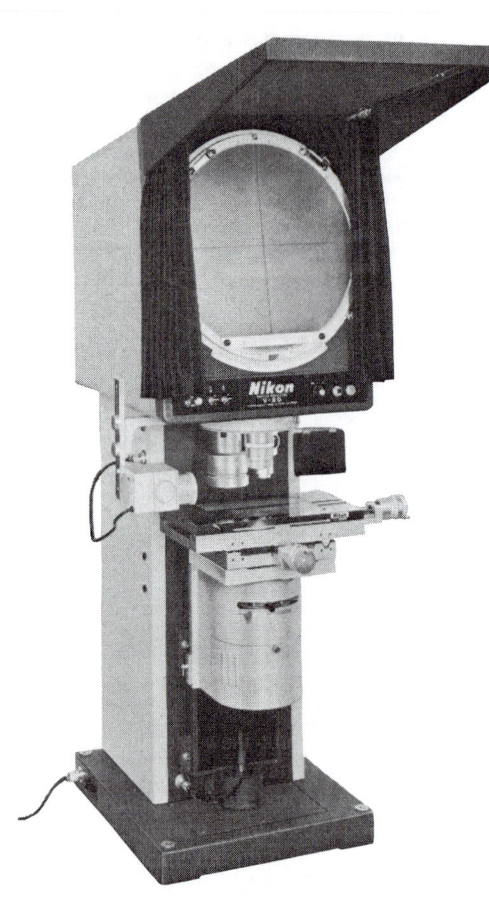

(a)

Fig. 8.30 Optical projector (a) Profile projector; (b) Light path of profile projector

plane of the projection lens, the image formed at the screen is that of a plane cutting the thread normal to the helix and, therefore, lying at an angle to the axis. Although a sharply focused image will result, this image will be foreshortened as shown.

- Alternatively, the thread axis can be left parallel to the screen and the collimated beam of light can be inclined to the helix angle, as shown in Fig. 8.29(*c*). Although the collimated beam of light is inclined, the axis of the screw thread lies in the focal plane of the projection lens and, therefore, the projected image will be that of a section plane passing through the axis of the screw thread. This results in a sharp image with a minimum deviation from the true form. The projection lens has to be specially designed for this

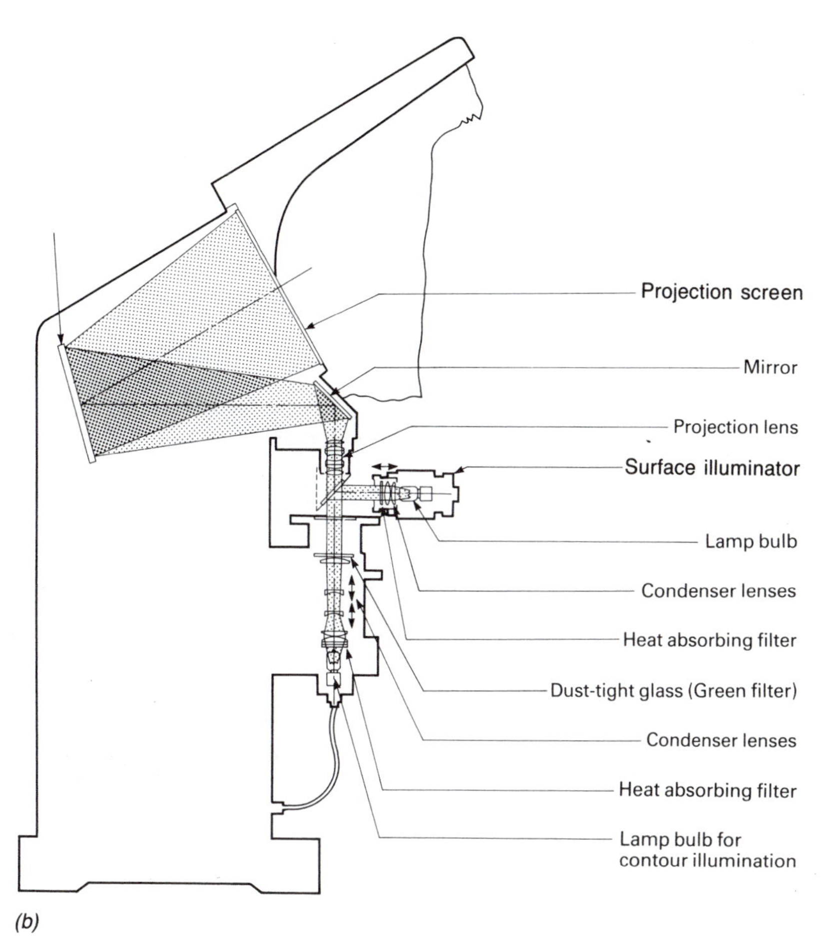

(b)

application, otherwise only very small helix angles could be projected without considerable loss of field and definition. This second method is the one normally used in practice. Figure 8.30(*a*) shows an optical projector suitable for checking screw threads, and Fig. 8.30(*b*) shows its optical path.

When checking screw-thread forms by projection, two elements are usually measured simultaneously; one is the thread profile and the other is the flank angle. The angle is, of course, part of the profile and is treated as such for production screw threads. However, for gauge measurement it is usual to measure the thread angle directly by a protractor fitted to the rear projection screen of the projector. A protractor for use with a projector is shown in Fig. 8.31(*a*). The scale is calibrated in intervals of 0.5° and the micrometer scales on the adjusting

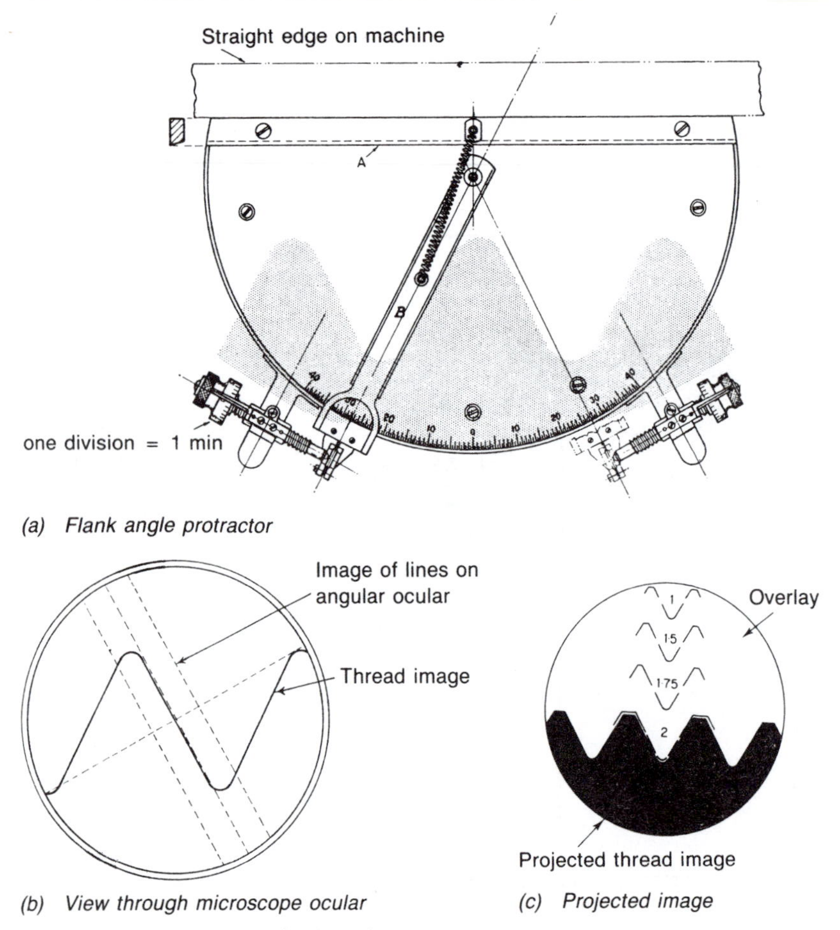

(a) Flank angle protractor

(b) View through microscope ocular

(c) Projected image

Fig. 8.31 Measuring the flank angle

screws are calibrated in intervals of 1 minute of arc. Protractors with decimal notation are also available.

Root and crest radii together with pitch errors can be checked against templates or overlays, as can the flank angle of production screw threads. The templates or overlays are made to a precise magnification and are used as shown in Fig. 8.31(*b*). Pitch errors can be measured by means of the micrometer slide on which the screw thread is mounted.

8.14 Screw-thread gauging

It has been shown in the preceding sections that the measurement of screw threads is a complex process involving skilled personnel and costly precision instruments. Except for gauges and special high-precision screwed components made on a 'one-off' basis, it would be too costly to use measurement techniques for checking production screw threads. For screwed components produced on a batch or continuous production basis, thread gauging is used.

As for any other component, or features of components, screw threads cannot be made to an exact size. BS 3643 provides a system of tolerancing for male and female screw threads in a similar manner to the system for plain shafts and holes. In the ISO system, the fundamental deviation is designated by capital letters for nuts and lower-case letters for bolts. A number is used to designate the tolerance grade. Thus the *tolerance class* is a combination of fundamental deviation letter and the tolerance grade number, as shown in Fig. 8.32(*a*). It can be seen that just as a hole basis system is the most practical for plain shafts and holes, a nut basis system is the most practical for screwed components. This is because the internal thread is invariably cut with a fixed-size tap, while the external thread can be varied in size by adjusting the thread-cutting and thread-forming tools. Figure 8.32(*b*) shows how a metric screw thread should be specified.

Providing the designer has applied correct limits to the dimensions in accordance with BS 3643, the screw will function satisfactorily if its manufactured size lies within those limits. Therefore, when checking a screw thread, it is not necessary to measure its size exactly but only to determine if its elements lie within the prescribed limits. Provided that the number of threaded components being manufactured warrants the outlay, the most economical method of checking screw threads is to use limit gauges. Taylor's Principle of Gauging was discussed in *Manufacturing Technology: volume 1* and screw-thread gauges follow these principles closely since they check geometrical form as well as dimensional accuracy.

External threads

Because of the complexity of a screw-thread form, more than one gauge is required to check it thoroughly. In the case of an external thread, current practice favours the use of the following gauges:

- *A screw-thread caliper gauge* has full-form GO anvils and form-relieved NOT GO anvils in accordance with Taylor's Principle of Gauging. This enables the gauge to check the effective diameter and, over the length of the anvils, the pitch of the thread.
- *A plain caliper gauge* is used to check the major diameter over the crests of the threads. Only the NOT GO element is required in this instance. It is not usual to check the minor diameter separately, as this feature is checked for interference by the screw-thread caliper gauge at the same time that it checks the effective diameter.

Internal threads

As for the external screw threads discussed above, internal screw threads also require more than one type of gauge to be used for their inspection

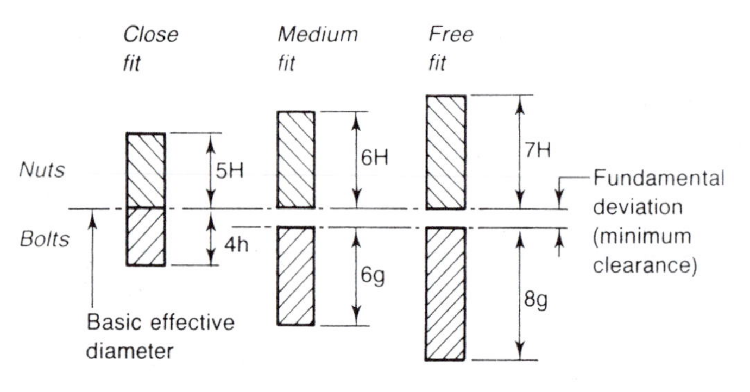

(a) *Screw-thread tolerances*

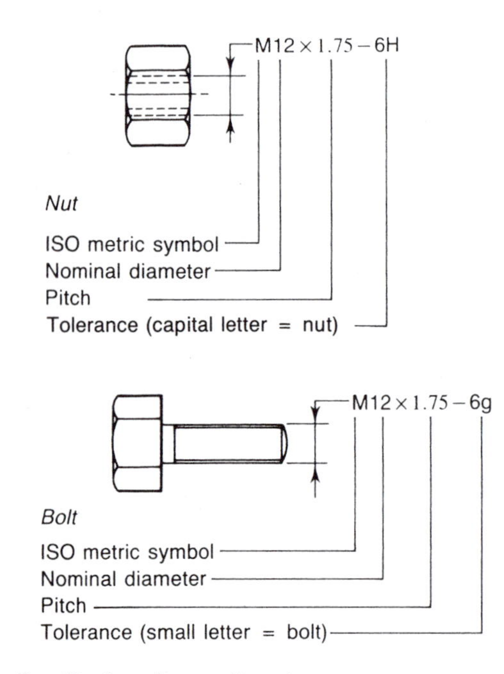

(b) *Specification of screw threads*

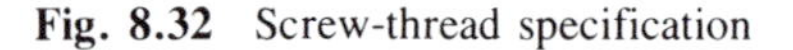

Fig. 8.32 Screw-thread specification

and to check the thread form thoroughly. Current practice favours the use of the following gauges:

A screw-thread plug gauge, as shown in Fig. 8.33(*a*). In compliance with Taylor's Principles, it has a full-form, full-length GO element and a shortened, form-relieved, NOT GO element. The NOT GO element has a truncated thread with the crest removed, and a heavily-relieved minor

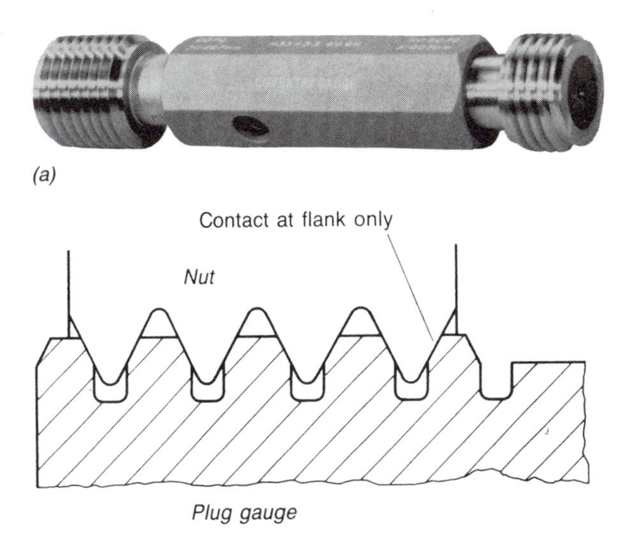

Fig. 8.33 Plug gauge for internal thread

diameter as shown in Fig. 8.33(*b*). This permits flank contact only so that it can check the effective diameter.

A plain single-ended plug gauge. This is used to check the minor diameter. Only the NOT GO element is required to check that the core diameter is not *over-size*, thus reducing the strength of the threads. If the core diameter was *under-size*, then the screw-thread gauge would not enter, hence there is no need for a GO element on the plain plug gauge. It is not usual to check the major diameter of an internal thread, since the screw-thread plug gauge checks this for interference while checking the effective diameter.

8.15 Screw-thread gauge applications

The quality and type of screw-thread gauge used depends upon the class of work being checked. Inspection can vary from 'just trying a nut on the screw' to complete gauging and even individual measurement for critical components.

For most general-purpose production purposes it is not necessary to use a full set of gauges as set out in section 8.14. Thread-cutting equipment is manufactured to high standards of accuracy and, normally, only the effective diameter has to be strictly controlled. Hence the inspector usually relies solely upon the use of a screw-thread caliper gauge for external threads or a screw-thread plug gauge for internal threads. Screw-thread caliper gauges are usually adjustable so that they can be set to a master gauge. This allows for adjustment when wear takes place.

For general production internal threads, once the tap being employed has been fully inspected, it is not necessary to use a double-ended screw-thread gauge. Only a gauge with a NOT GO element is required, and this speeds up the inspection process, with a corresponding lowering in costs.

8.16 Gear tooth inspection

The inspection of spur gears falls into two distinct areas of investigation.

- The *measurement* of individual gear-tooth elements to determine whether or not they have been manufactured to the specified design tolerances. The measurement of gears is a very complex subject, therefore, within the confines of this chapter, only the following basic measurements will be considered:
  - — tooth thickness at the pitch line;
  - — constant chord;
  - — pitch measurement;
  - — base tangent;
  - — involute form testing.
- The *dynamic testing* of gears to ensure that they are concentric and that they run quietly and smoothly. Such tests include the use of:
  - — rolling gear testing machines;
  - — noise and vibration analysing machines.

8.17 Measurement of tooth thickness at the pitch line

Figure 8.34 shows the elements of involute gear teeth and their relationships when meshing together. One way to measure the tooth thickness at the pitch line is to use a gear-tooth caliper, as shown in Fig. 8.35. It can be seen that this instrument has two sets of scales perpendicular to each other. This enables the tooth thickness to be measured at any given point from the root to the top of the tooth by adjusting the auxiliary slide controlling the height dimension (h). In practice, the auxiliary slide is set so that the main scale shows the thickness of the tooth (t_c) at the pitch circle. This is called the *chordal thickness* (t_c). The theoretical value of the chordal width and the setting height of the auxiliary slide can be calculated from the expressions:

$$t_c = m \times N \times \sin (90°/N)$$

and

$$h = m + \{m \times N/2\ [1 - \cos (90°/N)]\}$$

Where: t_c = chordal thickness
h = setting height
m = modular pitch
N = number of teeth

as shown in Fig. 8.36

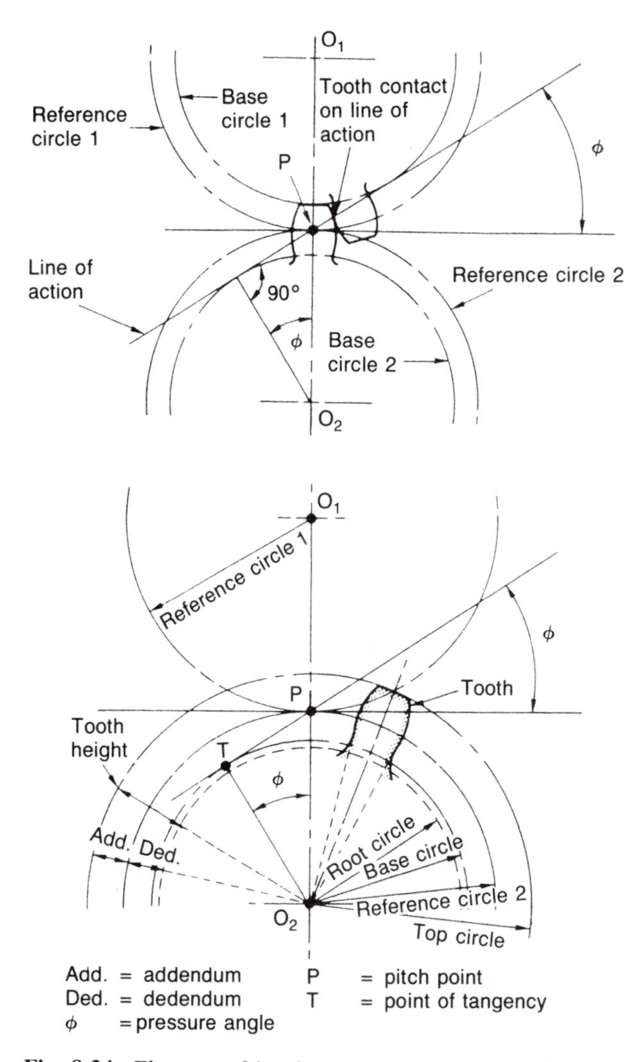

Fig. 8.34 Elements of involute gear teeth (courtesy of Butterworth-Heinemann Publishers Ltd.)

Thus the auxiliary slide of the gear-tooth vernier caliper is set to the calculated value of the setting height (h), and the actual reading of the chordal thickness (t_c) is read from the main scale of the caliper. This is compared with the calculated value of (t_c) and the difference is the error. In practice, allowance must be made for variations in tooth form (e.g. stub teeth), and also for designed clearance (backlash) which will reduce the chordal thickness (t_c).

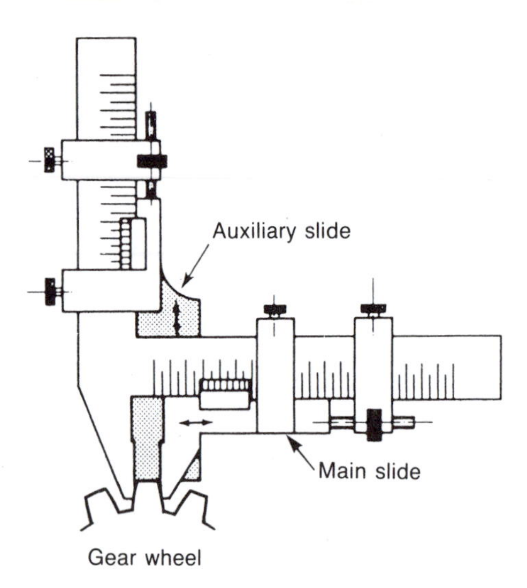

Fig. 8.35 Gear tooth vernier caliper

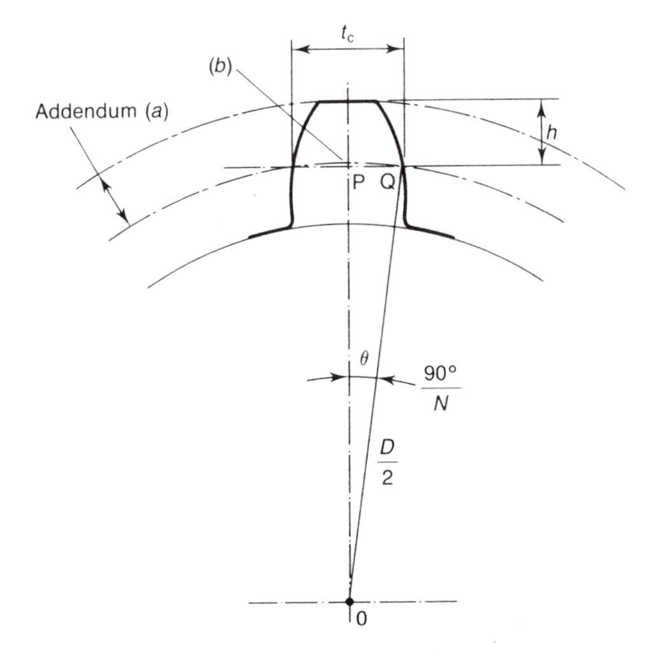

Fig. 8.36 Tooth thickness at the pitch line

Example 8.4 A spur gear has 25 teeth and a modular pitch of 3 mm. Calculate the setting height (h) and the chordal thickness (t_c) of an ideal involute form tooth.

$t_c = m \times N \times \sin(90°/N)$ where: $m = 3$ mm
$t_c = 3 \times 25 \times \sin(90°/25)$ $N = 25$ teeth
$t_c = 4.71$ mm

$h = m + \{m \times N/2\ [1 - \cos(90°/N)]\}$
$h = 3 + \{3 \times 25/2\ [1 - \cos(90°/25)]\}$
$h = 3.074$ mm

8.18 The constant chord

In the previous section the chordal thickness (t_c) and the setting height (h) were dependent upon the number of teeth on the gear as well as the modular pitch. The *constant chord* is independent of the number of teeth and this is a useful measure where a large number of gears have the same modular pitch and pressure angle. Figure 8.37 shows an involute gear tooth in mesh with an involute rack. The tooth thickness (t) is measured at the points of contact at a dimension (H) from the tip of the tooth, and the theoretical values can be calculated from the following expressions:

$$t = m\pi/2\ (\cos^2\phi)$$
$$H = m\ (1 - \pi/4\ \sin\phi\ \cos\phi)$$

where: m = modular pitch
ϕ = pressure angle

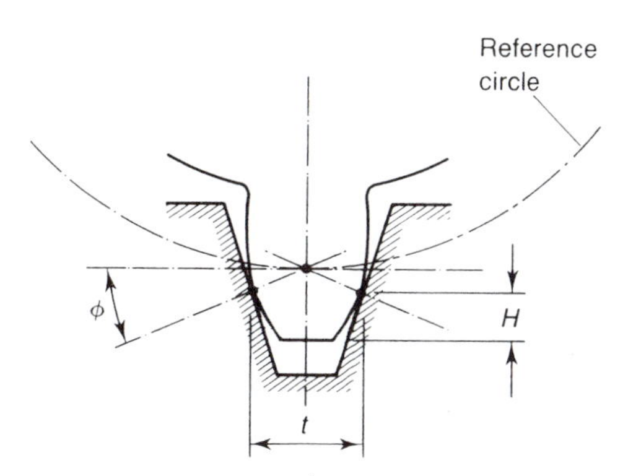

Fig. 8.37 The constant chord

Example 8.5 Calculate the constant chord tooth thickness (t) and its height (H) for a gear with a modular pitch of 3 mm and a pressure angle of 20°.

$t = m\pi/2\ (\cos^2\phi)$ where: $m = 3$ mm
$t = 3 \times \pi/2\ (\cos^2 20°)$ $\phi = 20°$
$\underline{t = 4.161 \text{ mm}}$
$H = m\ (1 - \pi/4 \sin\phi \cos\phi)$
$H = m\ (1 - \pi/4 \sin 20° \cos 20°)$
$\underline{H = 2.243 \text{ mm}}$

8.19 Pitch measurement

The circular pitch of the teeth of a spur gear when measured on the base circle is called the *base pitch* (P_B). The circular distance EF in Fig. 8.38(a) is the base pitch for the involute tooth form shown. The geometrical properties of an involute curve ensure that any tangents to the base circle drawn as shown in Fig. 8.38(a) gives the following relationship:

$$\text{base pitch } (P_B) = EF = CD \ldots\ldots \text{ etc.,}$$

also

$$\text{number of teeth } (N) \times P_B = \text{base circle circumference,}$$

thus

$$N \times P_B = \pi D_B$$

and

$$P_B = \pi D_B / N \qquad (1)$$

The following relationship can be deduced from Fig. 8.38(b):

$$(D_B/2) \div (D/2) = \cos\phi$$

thus

$$D_B = D \cos\phi \qquad (2)$$

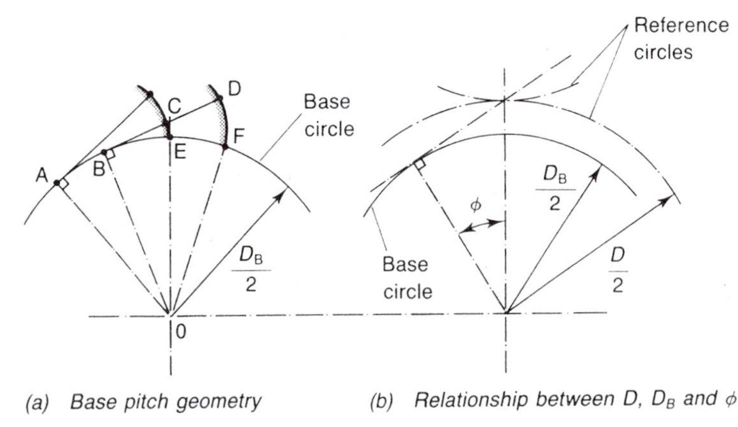

(a) Base pitch geometry *(b) Relationship between D, D_B and ϕ*

Fig. 8.38 Base pitch

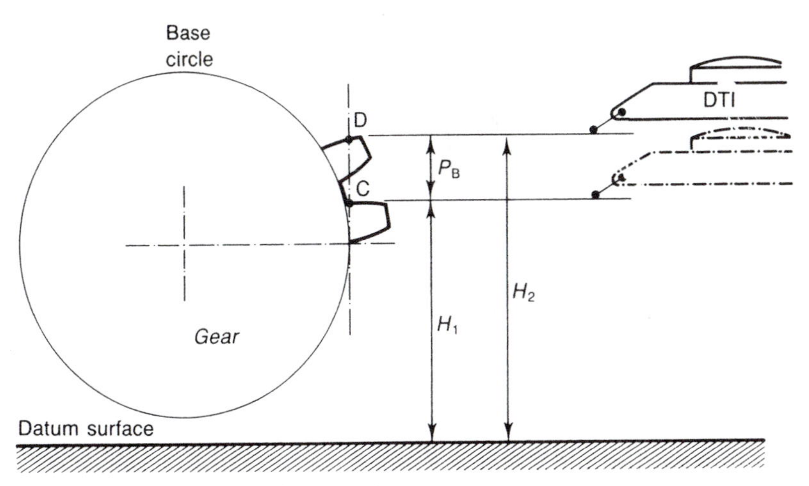

Fig. 8.39 Measurement of base pitch

Combining equations (1) and (2)

$$P_B = \pi(D/N)\cos\phi \qquad (3)$$

where: modular pitch (m) = D/N

thus

$$P_B = m\pi\cos\phi \qquad (4)$$

Thus the base pitch (P_B) can be calculated from the known data of modular pitch and the pressure angle (ϕ), since DCB is also tangential to the base circle. Although difficult to gauge, base pitch can easily be measured, as shown in Fig. 8.39. The DTI is used as a fiducial indicator and the vernier height gauge, on which it is mounted, is adjusted so that the DTI gives the same reading for each height setting. Thus:

$$P_B = H_3 - H_1$$

For greater accuracy, the DTI could be set over slip gauges.

Example 8.6 Calculate the base pitch for a spur gear having a modular pitch of 3.5 mm and a pressure angle of 20°.

$P_B = m\pi\cos\phi$ where: $m = 3.5$ mm
$P_B = 3.5 \times \pi \times \cos 20°$ $\phi = 20°$
$\underline{P_B = 10.332 \text{ mm}}$

8.20 Base tangent

The base tangent is a useful measure of gear-tooth spacing since it can be made with an ordinary vernier caliper as shown in Fig. 8.40(a). Since

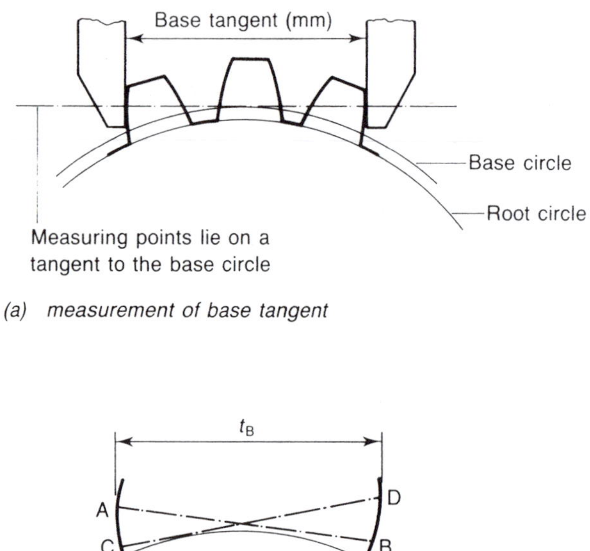

(a) measurement of base tangent

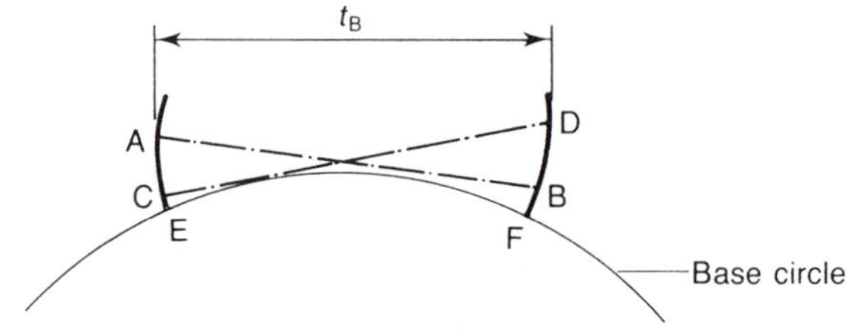

(b) Geometry of the base tangent

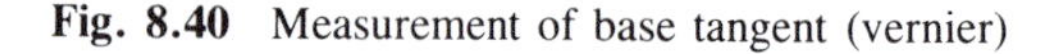

Fig. 8.40 Measurement of base tangent (vernier)

this measurement is taken over several teeth, it also indicates cyclical errors as the caliper is stepped around the gear.

The base tangent measurement can be taken at various positions AB, CD, etc., as shown in Fig. 8.40(*b*). It should be apparent, from previous discussion in this chapter, that any base tangents across opposed involute flanks are equal. Thus:

$$\text{base tangent } (t_B) = AB = CD = \text{arc } EF$$

where the base tangent is measured with a vernier caliper.

The base tangent can also be calculated using the following empirical expression, providing the standard pressure angle of 20° is used.

$$t_B = m \cos 20° \, [(N/\pi)\text{inv } 20° + k - 0.5)$$

where: inv 20° = involute function of 20°
k = number of teeth spanned by t_B
m = modular pitch

N	10–18	19–27	28–36	37–45	46–54	55–63	64–72	73–81	82–90
k	2	3	4	5	6	7	8	9	10

Example 8.7 Calculate the base tangent for a 50 tooth gear with a modular pitch of 3 mm. Pressure angle = 20°.

$t_B = m \cos 20° \, [(N/\pi)\text{inv } 20° + k - 0.5)]$ where: $m = 3$ mm
$t_B = 3 \times 0.9397 \times [(50/\pi) \times 0.014\,904 + 6 - 0.5]$ $k = 6$ (from the above table)
$\underline{t_B = 16.17 \text{ mm}}$

8.21 Measurement over rollers

Rollers can be placed between the jaws of a measuring instrument and a surface when it is necessary to obtain line contact only. For this reason, measurement over rollers can be employed to check gear teeth. The geometrical relationship between a roller and an involute rack tooth is shown in Fig. 8.41. When taking measurements over rollers, it is essential to choose a suitable roller diameter so that the roller contacts the flanks of the tooth at the constant chord position (t_c). The roller diameter (d) can then be calculated by use of the expression: $d = (\pi/2)\, m \cos\phi$.

To check a straight-tooth spur gear, two rollers of equal diameter (d) have to be used, and the distance across them (M_1) has to be calculated and measured. Any difference between the measured size and the calculated size represents a manufacturing error. For a small gear that can be spanned by a vernier caliper or micrometer caliper, the rollers are spaced so as to be as nearly diametrically opposed as possible. Figure 8.42(*a*) shows the position of the rollers for gears with an *even* number of teeth so that the gears are diametrically opposite. In this case $M_1 = D + d$ where D is the base-circle diameter. For gears with an *odd* number of teeth it is not possible to arrange the rollers diametrically opposite to each other and allowance has to be made for their angularity as shown in Fig. 8.42(*b*). The previous expression is now modified to $M_2 = D \cos 90/N + d$.

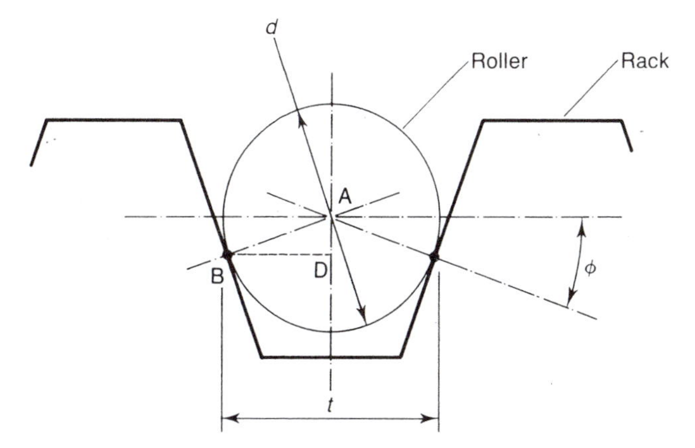

Fig. 8.41 Roller and rack tooth geometry

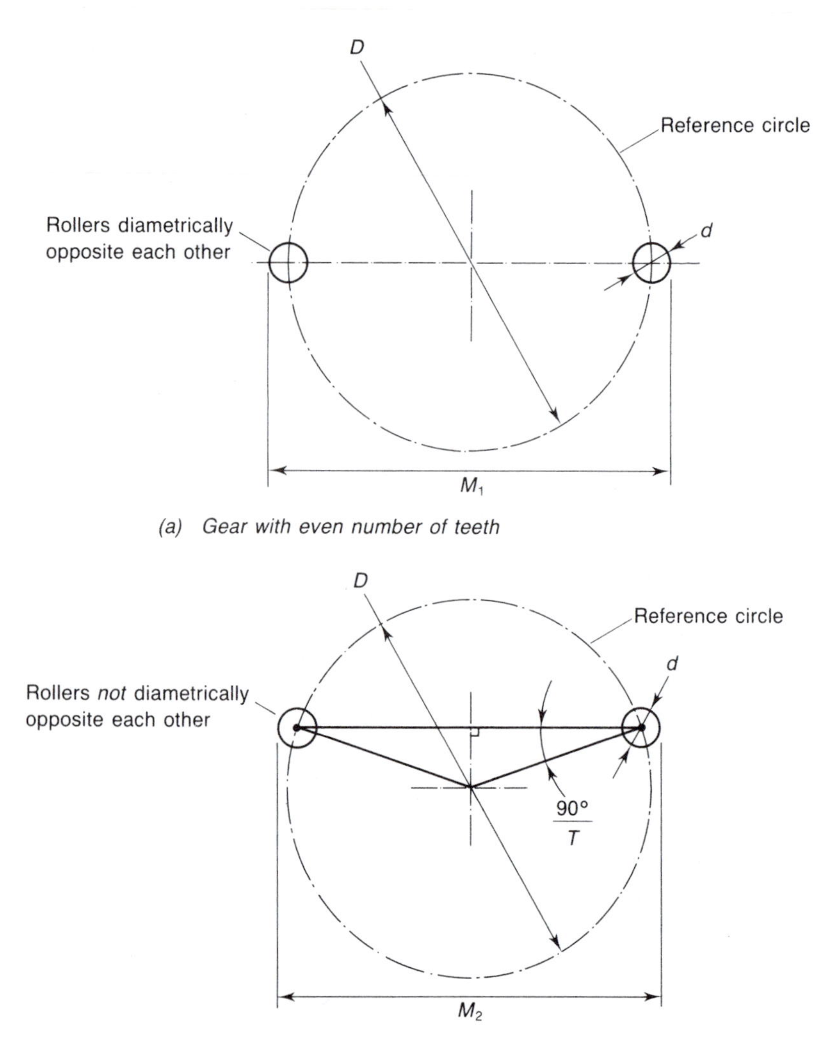

(a) Gear with even number of teeth

(b) Gear with odd number of teeth

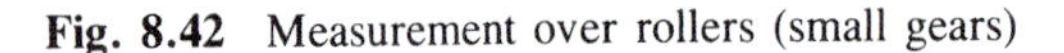

Fig. 8.42 Measurement over rollers (small gears)

When checking large gears it is not always possible or convenient to span the gear diametrically with the measuring instrument. In these instances, it is more usual to check the gear across a selected number of teeth, as shown in Fig. 8.43. The angle $\theta°$ subtended by a selected number of teeth (N_s) can be determined from the expression

$$\theta° = (N_s/N) \times 360°$$

thus:

$$M_3 = (D \sin \tfrac{1}{2}\theta°) + d$$

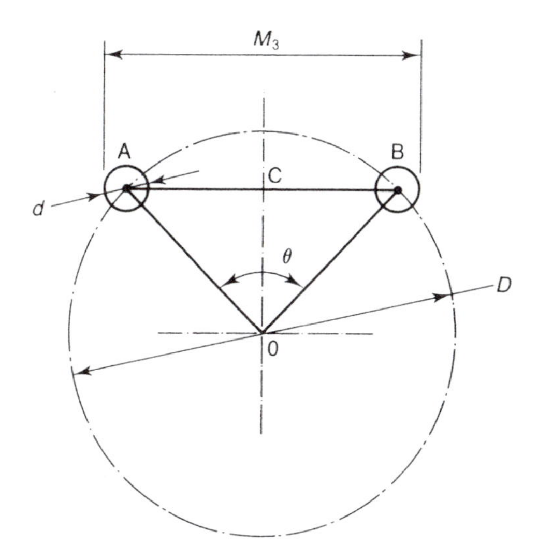

Fig. 8.43 Measurement over rollers (large gears)

Example 8.8 Calculate the diameter of the rollers and the dimension M_3 for a 180 tooth gear having a modular pitch of 8 mm and a pressure angle of 20°.

$d = (\pi/2)\ m \cos\phi$ where: $m = 8$ mm
$d = (\pi/2) \times 8 \times \cos 20°$ $\phi = 20°$
$\underline{d = 11.81 \text{ mm}}$

$M_3 = (D \times \sin\frac{1}{2}\theta°) + d$ where: $D = mN$
$M_3 = (8 \times 180 \times \sin 8°) + 11.81$ $\theta° = (N_s/N) \times 360° = 16°$
$\underline{M_3 = 212.22 \text{ mm}}$

8.22 Involute form testing

The form of involute gear teeth can be checked either by optical projection for comparison with a template or overlay as in screw-thread inspection or, more usually nowadays, by an electronic involute testing machine. Such machines use an electronic probe to trace the profile of each tooth and, after amplification, the signal is used to print out an enlarged profile.

8.23 Dynamic gear testing

The calculations and tests described previously in this chapter are intended to determine whether or not the gear is dimensionally correct. The dynamic tests are used to determine whether or not the gear will run smoothly and quietly.

Rolling test

The gear under test is mounted on a slide and meshed with a hardened and ground master gear. The two gears slowly rotate together and any errors due to eccentric running or irregular pitch will cause the gear under test and its slide to move to or from the master gear. This movement is measured and recorded. Spring tension keeps the gears in mesh.

Vibration and noise test

In a typical gear-tooth noise and vibration measuring machine, the finished gears are run in mesh at controlled speeds in both directions, with and without brake loads. This enables the character of the vibration and the volume and frequency of the sound to be ascertained and analysed under a variety of working conditions.

8.24 Geometrical tolerances (introduction)

Traditionally, inspection has been mainly concerned with the measurement and gauging of linear dimensions and angles. However, in the sections of this chapter dealing with screw threads and gear teeth the importance of the component profile has also been emphasised. The importance of correct geometry as well as correct dimensions is now well established.

Figure 8.44(*a*) shows a shaft whose dimensions have been given *limits of size*. The difference between the limits is called the *tolerance*. The reason for providing toleranced dimensions is that it is not possible to manufacture a workpiece exactly to size and, even if it were possible, there is no means of measuring such a workpiece exactly. For many applications, if the process used can achieve the dimensional tolerance, it will also produce a geometrically correct form. For example, the shaft shown in Fig. 8.44(*a*) would most likely be cylindrically ground to achieve the dimensional accuracy specified. Cylindrical grinding between centres should give a satisfactory degree of *roundness*. However, the shaft could also be mass produced by centreless grinding and some *out-of-roundness* may occur such as ovality and lobing. Figure 8.44(*b*) shows how the shaft can be out-of-round yet still be within its limits of size.

If cylindricity is important for the correct functioning of the component, then additional information must be added to the drawing to emphasise this point. This additional information is a *geometrical tolerance* which, in this case, would be added as shown in Fig. 8.45(*a*). The symbol in the box indicates cylindricity, and the number indicates the geometrical tolerance. Figure 8.45(*b*) shows the interpretation of this tolerance.

8.25 Geometrical tolerance (principles)

It has just been shown that in some circumstances dimensions and tolerance of size, however well applied, do not impose the necessary

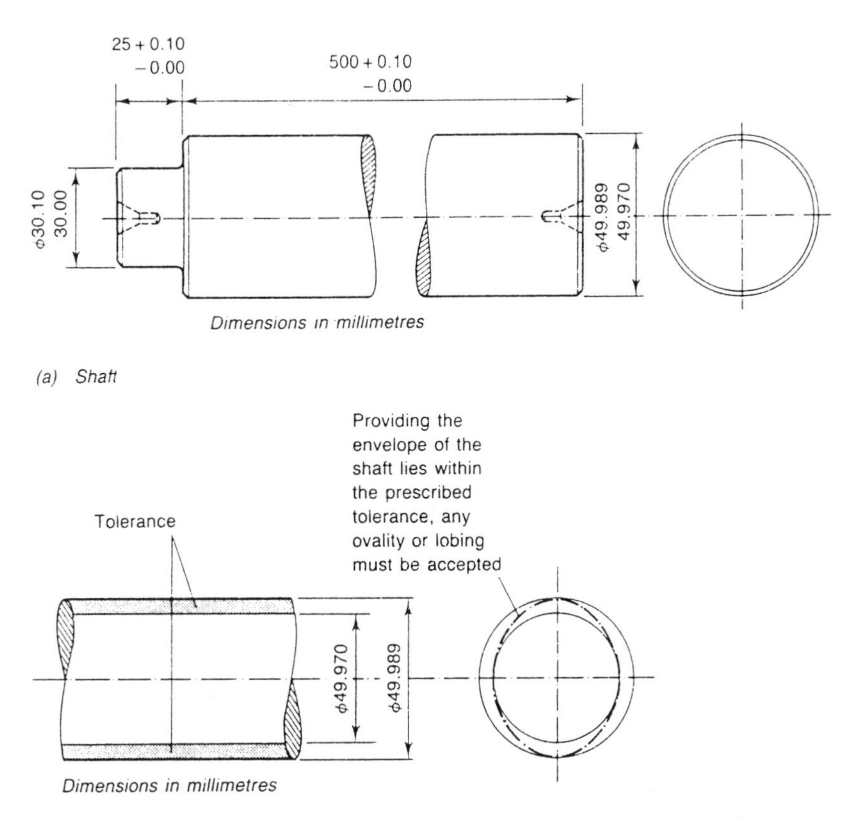

Fig. 8.44 Need for geometrical tolerancing

control of form. If control of form is necessary then geometrical tolerances must be added to the component drawing. Geometrical tolerances should be specified for all requirements critical to the functioning and interchangeability of components. An exception can be made when the machinery and techniques used can be relied upon to achieve the required standard of form. Geometrical tolerances may also need to be specified even when no special size tolerance is given. For example, the thickness of a surface plate is of little importance, but its accuracy of flatness is of fundamental importance.

Figure 8.46 shows the standard tolerance symbols as specified in BS 308: Part 3 and it can be seen that they are arranged into groups according to their function. In order to apply geometrical tolerances it is first necessary to consider the following definitions.

Geometrical tolerance

Geometrical tolerance is the maximum permissible overall variation of form or position of a feature. That is, it defines the size and shape of a

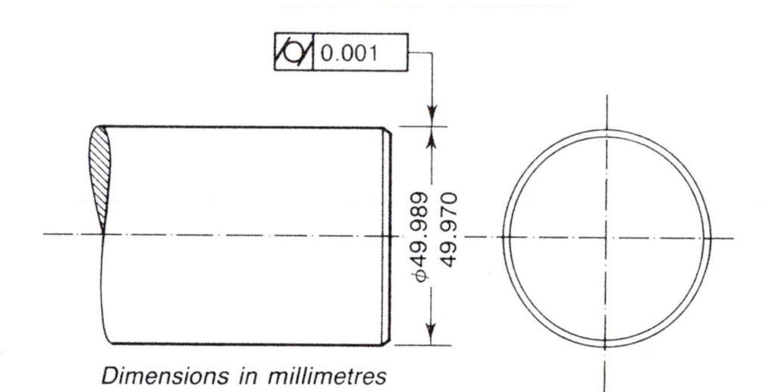

(a) Addition of geometric tolerance for cylindricity

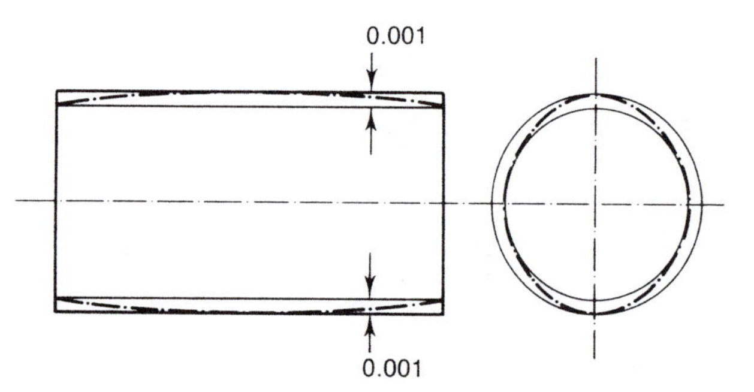

The curved surface of the shaft is required to lie within two cylinders whose surfaces are coaxial with each other, a RADIAL distance 0.001 mm apart.

The diameter of the outer coaxial cylinder lies between 49.989 and 49.970 mm as specified by the linear tolerance

(b) Interpretation of the geometric tolerance

Fig. 8.45 Geometrical tolerancing

tolerance zone within which the surface, median plane or axis of the feature is to lie. It represents the full indicator movement it causes where testing with a DTI is applicable, for example, the 'run-out' of a shaft rotated about its own axis.

Tolerance zone

The tolerance zone is the zone within which the feature is required to be contained. Thus, according to the characteristic which is to be toleranced and the manner in which it is to be dimensioned, the tolerance zone is one of the following:

- a circle or a cylinder;
- the area between two parallel lines or two parallel straight lines;
- the space between two parallel surfaces or two parallel planes;
- the space within a parallelepiped.

	Type of tolerance	Characteristic to be toleranced	Symbol	Clause ref.
For single features	Form	Straightness	⏤	3.1.1
		Flatness	⏥	3.1.2
		Roundness	○	3.1.3
		Cylindricity	⌭	3.1.4
		*Profile of a line	⌒	3.1.5
		*Profile of a surface	⌓	3.1.6
For related features	Attitude	Parallelism	//	3.2.1
		Squareness	⊥	3.2.2
		Angularity	∠	3.2.3
	Location	Position	⌖	3.3.1
		Concentricity	◎	3.3.2
		Symmetry	⌯	3.3.3
	Composite	Run-out	↗	4
Maximum material condition			Ⓜ	2.6
Boxed dimension (dimension which defines true position			□	2.7.1

*May be related to datum when it is necessary to control position in addition to form.

Fig. 8.46 Geometrical tolerance symbols

Note: The tolerance zone, once established, permits any feature to vary within that zone. If it is considered necessary to prohibit sudden changes in surface direction, or to control the rate of change of a surface within this zone, this should be additionally specified.

Geometrical reference frame

The geometrical reference frame is the diagram composed of the constructional dimensions which serve to establish the true geometrical relationships between positional features in any one group, as shown in Fig. 8.46. For example, the symbol ⌖ | $\phi 0.02$ indicates that the hole centre may lie anywhere within a circle of 0.02 mm diameter, which is itself centred upon the intersection of the frame. The dimension 40 indicates that the reference frame is a geometrically-perfect square of side length 40 mm.

8.26 Applications of geometrical tolerances

Figure 8.47 shows a plate in which four holes need to be drilled. For ease of assembly, the relationship between the hole centres is more

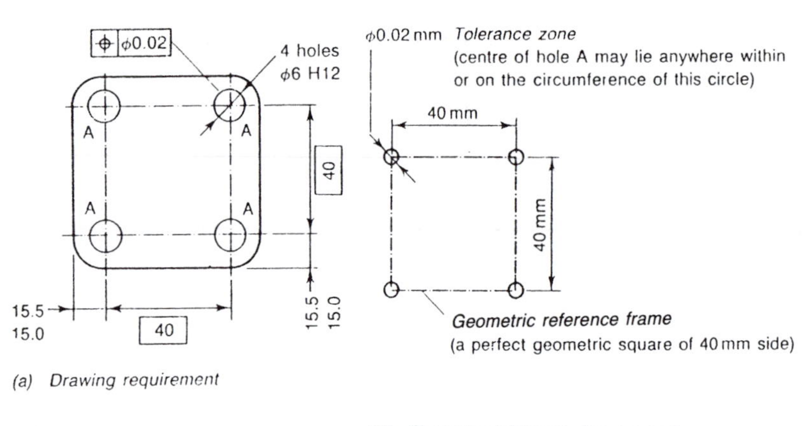

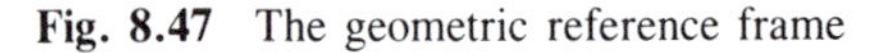

Fig. 8.47 The geometric reference frame

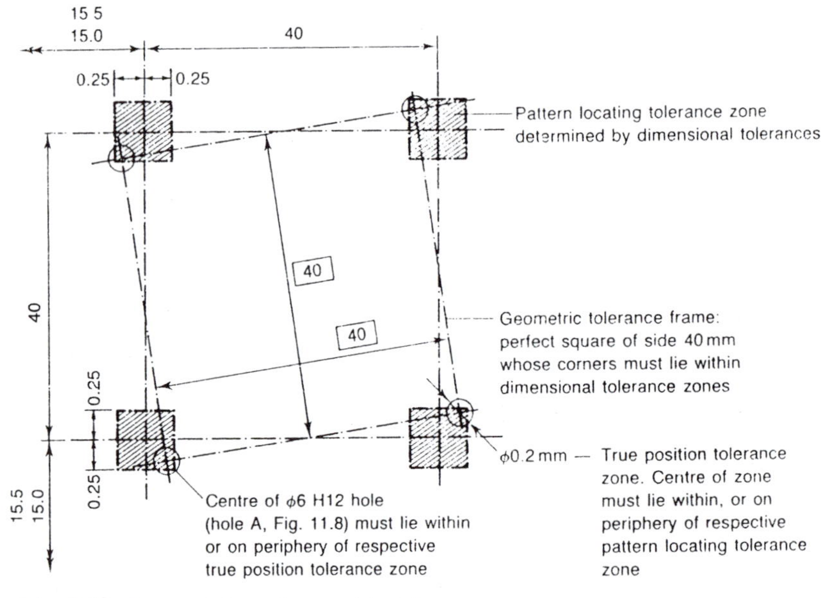

Fig. 8.48 Interpretation of geometric tolerances

important than the group position on the plate. Therefore a combination of dimensional and geometrical tolerances has to be used.

Using the dimensional tolerances and the boxed dimension that defines true position, it is possible to determine the pattern locating tolerance zone for each of the holes, as shown in Fig. 8.48. That is, the reference frame (which must always be a perfect square of 40 mm side in this

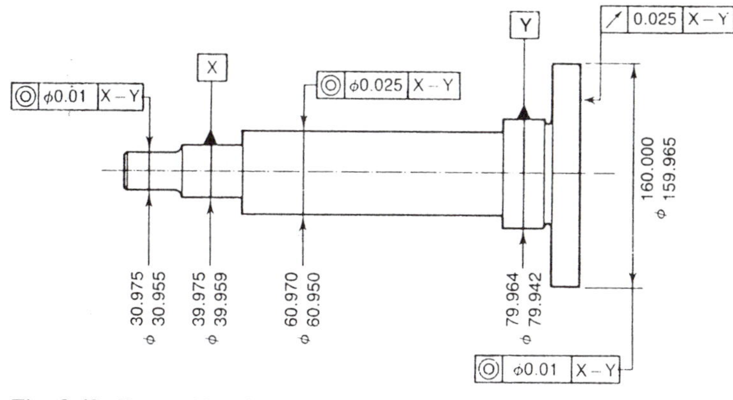

Fig. 8.49 Datum identification

example) may lie in any position provided that its corners are within the shaded boxes. The shaded boxes represent the dimensional tolerance zones. The hole centres may (in this example) lie anywhere within the 0.02 mm circles which are themselves centred upon the corners of the reference frame.

Figure 8.49 shows a further example. It is a stepped shaft with four concentric diameters and a flange which must run true with the datum axis. The shaft is to be located within journals at X and Y and these are identified as datums. Note the use of leader lines terminating in solid triangles resting upon the features to be used as datums. The common

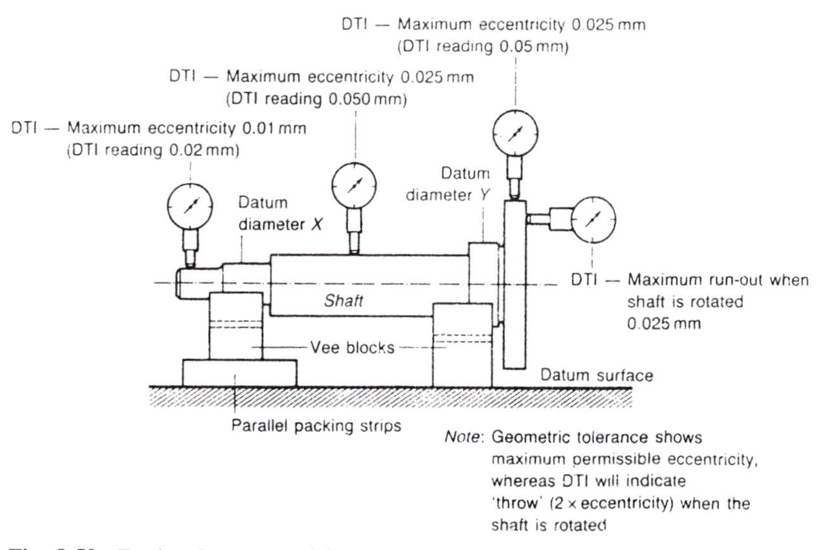

Fig. 8.50 Testing for concentricity

axis of these datum diameters is used as a datum axis for relating the tolerances controlling the concentricity of the remaining diameters and the true running of the flange face. Figure 8.49 shows that the dependent diameters and face carry geometric tolerances as well as dimensional tolerances. The tolerance frames carry the concentricity symbol, the concentricity tolerance and the datum identification letters showing that the concentricity is relative to the common axis of the datum diameters X and Y. The tolerance frame whose leader touches the flange face indicates that this face must run true, within the tolerance indicated when the shaft rotates in vee blocks supporting it at X and Y, as shown in Fig. 8.50. These two examples are in no way a comprehensive résumé of the application of geometric tolerancing but stand as an introduction to the subject and as an introduction to BS 308: Part 3. This British Standard details the full range of applications and their interpretation, and includes a number of fully-dimensioned examples.

8.27 Virtual size

To understand this section, it is necessary to introduce the terms *maximum metal condition* and *minimum metal condition*. The maximum

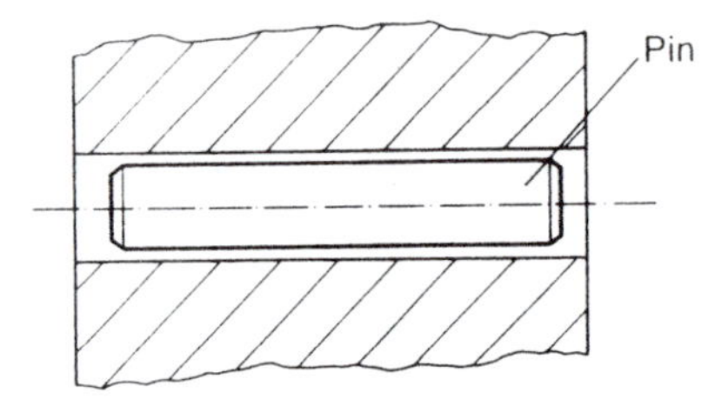

(a) *Pin with straight axis* — clearance fit in hole

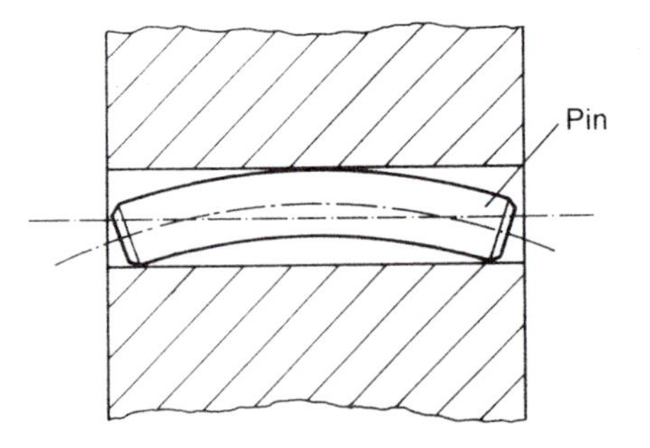

(b) *Pin with bowed axis* — tight fit in hole despite no change in size

Fig. 8.51 Effect of geometry on fit

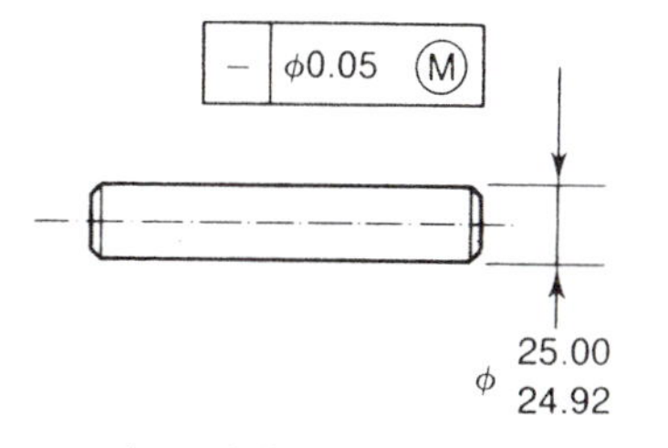

(a) *Pin with geometric and dimensional tolerances*

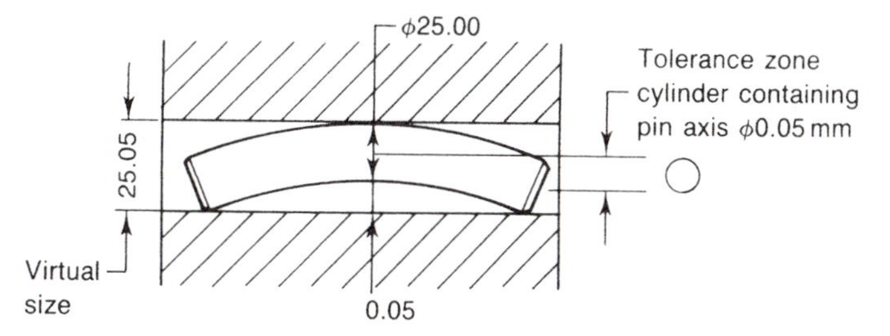

(b) *Virtual size under maximum metal conditions* — pin diameter = 25.00 mm

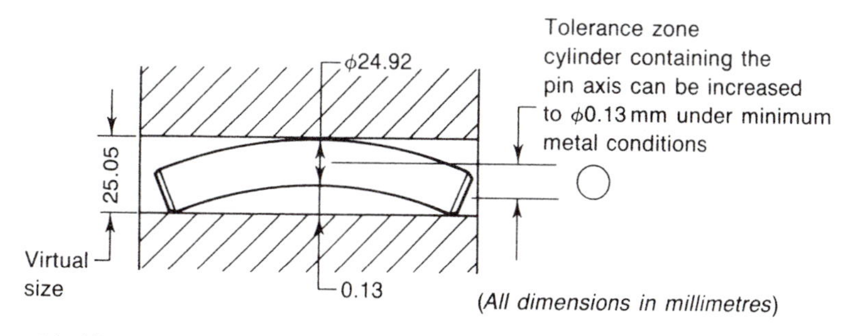

(c) *Virtual size under minimum metal conditions* — pin diameter = 24.92

Fig. 8.52 Virtual size

metal condition exists for both the hole and the shaft when the least amount of metal has been removed but the component is still within its specified limits of size, i.e. the largest shaft and smallest hole that will lie within the specified tolerance. The minimum metal condition exists when the greatest amount of metal has been removed but the component is still within its limits of size, i.e. the largest hole and smallest shaft.

Figure 8.51 shows the effect of a change in geometry on the fit of a pin in a hole. In Fig. 8.51(*a*), an ideal pin is used with a straight axis. It can be seen that, as drawn, there is clearance between the pin and the hole. However, in Fig. 8.51(*b*) the pin is distorted and now becomes a tight fit in the hole despite the fact that its dimesional tolerance has not changed.

Figure 8.52 shows the same pin and hole, only this time dimensional and geometrical tolerances have been added to the pin. For simplicity the hole is assumed to be geometrically correct. The tolerance frame shows a straightness tolerance of 0.05 mm. This means that the pin axis can bow within the confines of an imaginary cylinder 0.05 mm diameter. The dimensional limits indicate a tolerance of 0.08 mm. The worst conditions for assembly (tightest fit) occur when the pin and the hole are in their respective maximum metal conditions and, in addition, the maximum error permitted by the geometrical tolerancing is also present. Figure 8.52(*b*) shows the fit between the pin and the hole under these conditions.

Under such conditions the pin will just enter a truly straight and cylindrical hole equal to the pin diameter under maximum metal conditions (25.00 mm) plus the geometrical tolerance of 0.05 mm, i.e. a hole diameter of 25.05 mm. This theoretical hole diameter is referred to as the *virtual size for the pin.*

Under minimum metal conditions, the geometrical tolerance for the pin can be increased without altering the fit of the pin in the hole. As shown in Fig. 8.52(*c*), the geometric tolerance can now be increased to 0.13 mm without any increase in the virtual size or the change of fit.

8.28 The economics of geometrical tolerancing

Figure 8.53 shows a typical relationship between tolerance and manufacturing costs for any given component. The addition of geometrical tolerances on top of the dimensional tolerances aggravates the situation still further. Thus geometric tolerances are only applied if the function and interchangeability of the component is critical and absolutely dependent upon a tightly-controlled geometry. To keep costs down, it can often be assumed that the machine or process will provide adequate

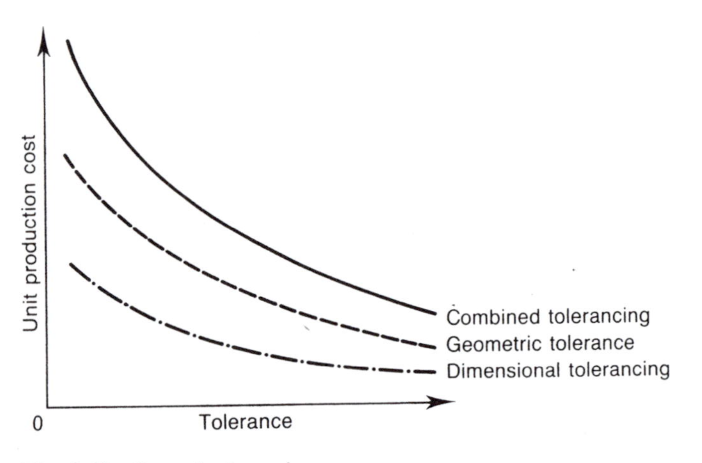

Fig. 8.53 Cost of tolerancing

geometric control and only dimensional tolerances are added to the component drawing. For example, it can generally be assumed that a component finished on a cylindrical grinding machine between centres will be straight and cylindrical to a high degree of geometrical accuracy. However, current trends in quality control and just-in-time (JIT) scheduling tend to increase the demand for closer dimensional and geometrical tolerancing to ensure no hold-ups occur during assembly and that quality specifications are met first time without the need for corrective action or rejection by the customer.

Assignments

1. Discuss the kinematics of measuring instruments and explain how backlash, and 'lost movement' can be avoided.
2. Explain how magnification is obtained in measuring instruments:
   (a) mechanically;
   (b) electrically;
   (c) optically;
   (d) pneumatically.
3. (a) Compare and contrast the advantages and limitations of the 'Angle Dekkor' and the 'Auto collimator' as devices for measuring angles.
   (b) Describe typical applications where these instruments would be used giving reason for your choice.
4. With the aid of a sketch, describe the measurement of the *major diameter* of a screw thread and calculate that diameter given:
   calibrated diameter over setting standard = 20.001 mm
   reading over setting diameter = 18.03 mm
   reading over screw thread = 17.509 mm
5. With the aid of a sketch, describe the measurement of the *effective diameter* of a screw thread and calculate that diameter given:
   pitch = 3.500 mm
   calibrated diameter over setting standard = 25.000 mm
   reading over setting standard = 28.003 mm
   reading over cylinders = 29.869 mm
6. With the aid of a sketch, describe the measurement of the *minor (root)* diameter of a screw thread and calculate that diameter given:
   D_s = 29.978 mm
   R_5 = 12.577 mm
   R_6 = 14.499 mm
7. With the aid of sketches explain how an optical profile projector can be used to measure thread pitch and flank angle.
8. (a) With the aid of sketches, explain how Taylor's Principles of Gauging are applied to screw thread gauges.
   (b) With the aid of sketches, explain how an external screw thread can be checked using gauges.

9. With the aid of sketches, and with typical calculations, explain how TWO of the following dimesions applicable to involute gears are calculated and checked:
   (a) tooth thickness at the pitch line;
   (b) constant chord;
   (c) pitch measurement;
   (d) base tangent.
10. Discuss the need for the geometrical tolerancing as well as the dimensional tolerancing of engineering components.
11. With the aid of sketches explain what is meant by the following terms as applied to geometrical tolerancing:
   (a) tolerance zone;
   (b) geometrical reference frame;
   (c) virtual size.
12. (a) Fully dimension a component of your choice, using dimensional and geometrical tolerancing
   (b) Discuss the economics of geometrical tolerancing.

Part B Management of manufacture

9 Health and safety

9.1 The Health and Safety at Work etc. Act

The Health and Safety at Work etc. Act was put on the Statute Book on 31 July 1974, and it was fully implemented by April 1975. It is an *enabling* Act, which means that it is a broad and general piece of legislation that does not go into specific detail in itself. However, it gives powers to the Secretary of State for Employment (acting through the Health and Safety Commission) to draw up detailed *regulations* and *codes of practice* without further reference back to Parliament.

The main purpose of the Act is to provide a comprehensive, integrated system of law dealing with the health, safety and welfare of work-people and the public as affected by work activities. The Act has six main provisions:

- to completely overhaul and modernise the existing law dealing with safety, health and welfare at work;
- to put *general duties* on employers ranging from providing and maintaining a safe place to work, to consulting on safety matters with their employees;
- to create a Health and Safety Commission (see section 9.2);
- to reorganise and unify the various inspectorates (see section 9.3);
- to provide new powers and penalties for the enforcement of safety laws (see section 9.4);

- to establish new methods of accident prevention, and new ways of operating future safety regulations.

The enabling provisions of the Act are superimposed on the existing legislation contained in the 31 relevant Acts and some 500 subsidiary regulations. It did not remove, cancel or affect existing legislation such as the Factories Acts, the Offices, Shops and Railway Premises Acts, and others. These and other existing pieces of legislation will continue side-by-side with the Health and Safety at Work Act until they are replaced by new legislation.

9.2 Health and Safety Commission

The Act provides for a full-time, independent chairman and between six and nine part-time commissioners. The commissioners are made up of three trade union members appointed by the TUC, three management members appointed by the CBI, two Local Authority members, and one independent member. The commission has taken over the responsibilities previously held by various government departments for the control of most occupational safety and health matters. The commission is also responsible for the Health and Safety Executive.

9.3 Health and Safety Executive

Since 1975, the unified inspectorate amalgamates the formerly independent government inspectorates such as the Factory Inspectorate, the Mines and Quarries Inspectorate, and similar bodies into one cohesive executive. This is called the Health and Safety Executive Inspectorate. The inspectors of the 'HSE' have wider powers than when working under previous legislation and their prime duty is to implement the policies of the Health and Safety Commission.

9.4 Enforcement

Should an inspector find a contravention of any of the provisions of previous Acts or Regulations still in force, or a contravention of any provision of the 1974 Act, the inspector has three possible lines of action available.

Improvement Notice

If there is a legal contravention of any of the relevant statutory provisions, the inspector can issue an *Improvement Notice*. This notice requires the fault to be remedied within a specified time. The notice can be served on any person deemed to be contravening the legal provision, or it can be served on any person on whom responsibilities are placed. This person can be employer, employee, or a supplier of equipment or materials.

Prohibition Notice

If there is a risk of serious personal injury, or hazard to health, the inspector can issue a *Prohibition Notice*. This immediately stops the activity giving rise to the risk until remedial action specified in the notice has been taken to the inspector's satisfaction. The Prohibition Notice can be served on the person undertaking the dangerous activity, or it can be served on the person in control of the activity at the time that the notice is served.

Prosecution

In addition to serving an Improvement Notice or a Prohibition Notice, the inspector can prosecute any person contravening a relevant statutory provision. Finally, the inspector can *seize, render harmless, or destroy* any substance or article that he considers to be the cause of imminent danger to health or serious personal injury.

9.5 Responsibility

The Act now places the responsibility for safe working equally upon the employer, the employee, and the manufacturers and suppliers of goods and equipment. Not only must an employer ensure the health, safety and welfare at work of all his or her employees; maintain the plant and equipment in a safe condition, and ensure that adequate instruction, training and supervision in safe working practices is provided, but it is an equal responsibility of every employee to cooperate in making proper and full use of the facilities provided. To ensure this cooperation, each company must have a health and safety committee representing both management and labour to examine working practices, actual and potential incidents, co-opt expert advisors when required, organise health and safety training, and enforce safe working practices. It must look at all aspects of the company's activities and ensure that it is a safe and healthy place in which to work.

Similarly the Act places duties on persons who design, import or supply articles for use at work. Any plant, machinery, equipment or appliance must be so designed and constructed as to be safe and without risk to health. Plant must be regularly tested and examined to ensure its continuing safety. It must also be erected or installed by competent persons to ensure its correct functioning.

Again, anyone who manufactures, imports or supplies goods and materials for use at work, has the responsibility for ensuring such goods and materials do not represent a hazard. The same applies to contractors hiring out plant and firms leasing plant and equipment. Such plant and equipment must be supplied in a safe condition and not represent a health hazard. The safety, health and welfare of visitors is also covered by the Act, as is the safety, health and welfare of persons living and working in the vicinity of any firm. This last point overlaps somewhat with environmental issues which are the subject of separate legislation and inspectorates.

9.6 European Machinery Safety Directive

The European Safety Directive became law in the UK on 31 December 1992. It requires that machinery manufactured and/or sold in the UK, including imported machinery, must satisfy wide-ranging health and safety requirements including those relating to noise levels. A new approach was used in preparing the Machinery Safety Directive by coupling it with European Standards as agreed by European Standards Committees CEN/CENELEC. The British Standards Institution participated in this activity, together with the professional engineering institutions.

Barriers to trade within European Community Countries have resulted from the difference in safety legislation that exists within the various member states. Therefore, in May 1985, European Community Ministers agreed on a new approach to harmonising the technical standards in order to overcome this problem. The Machinery Safety Directive (Directive 89/329/EEC) as passed by the European Community Ministers sets out essential safety requirements (ESRs) for machinery which must be met before machinery is placed on the market anywhere within the community. Thus compliance with the Directive will ensure the free movement of goods and services relating to machinery within the European Community. The Directive represents 'Best Practice Methods' and its implementation and enforcement is aimed at guaranteeing an adequate level of health and safety among those persons using machinery. Thus all manufacturers and importers of machinery and/or their agents will be required to certify that their products conform with the Directive. It must be remembered that, ultimately, it is the Courts of Law that will decide upon the meaning and interpretation of the Directive.

The ESRs are expressed in general terms and it is intended that *European Harmonised Standards* should fill in the detail so that designers, manufacturers and importers have clear guidance on how to conform with the Directive. To this end, the harmonised standards are those standards produced by CEN and CENELEC under a mandate from the Commission and which are subsequently approved by the 83.189 Directive Committee and published in the *Journal of European Communities*. The harmonised standards are, in themselves, not mandatory but are a way of showing conformity with the Directive. They form a 'benchmark' against which other ways of achieving conformity will be judged.

The Directive, which was first published in June 1989, is divided into three parts all of which have significant implications for persons and institutions responsible for drawing up standards.

- The first part sets out the overall objectives of the Directive and provides guidance on the interpretation of the subsequent Articles.

- The next part sets out the 14 Articles which lay down the scope, exclusions and duties contingent upon designers, manufacturers, importers and member states in implementing the Directive.
- The final part consists of a series of annexes detailing specific parts of the Directive and containing most of the technical details required for its implementation.

A detailed study of this comprehensive document is beyond the scope of this book. In fact, it would more than fill it. For more detailed information, readers are referred to the Directive itself and, for assistance in interpreting the Directive, to the proceedings published by the Institution of Mechanical Engineers, the Institution of Electrical Engineers and the Health and Safety Executive of the one-day symposium held in June 1991 which these three bodies co-sponsored.

9.7 Safe working practices

Safe working practices, safety clothing and equipment, and guarding were introduced, in some detail, in *Engineering Fundamentals* by Timings (Longman, 1988). The following is a summary of the main points to be observed, when engaged in the procsses listed, as a reminder that one can never become complacent where safety health and welfare is concerned and that the principles of good operating practice can never be reiterated too often.

Casting processes

As with all processes involving high temperatures, casting processes are potentially hazardous. This is particularly the case when large quantities of molten metal are involved. The structure and lining of all furnaces, crucibles and ladles should be regularly inspected to prevent failure and accidental spillage. It is essential to maintain the casting shop itself in good repair, since water dripping from a leaking roof or water pipes into molten metal can cause an erruption of violent and dangerous proportions. Further, the casting-shop floor should be so designed that any spillage should be contained locally and the flooring material should not break up and shatter if hot metal is spilt on it. (For this reason it must never be concrete.) Moulds are often placed on sand beds for pouring.

Casting-shop personnel must wear suitable protective clothing when preparing and pouring molten metal: a face visor, leather apron and gauntlet gloves, and leather spats and industrial footwear to protect the lower part of their legs and feet. Adequate ventilation is also required, not only to remove fumes associated with the fluxing and degassing of the molten metal and the burning out of the resin adhesive used in cores and shell-moulds, but also to provide a cool environment with an adequate supply of fresh air so that the personnel can remain alert and work comfortably.

Where automatic and semi-automatic processes, such as pressure die-casting, are used the machine should be fully guarded with the casting area totally enclosed during the injection cycle. This is necessary to contain any molten metal that could escape under very high pressure should the dies burst or should the machine fail to close the dies properly. Such guards should be interlocked with the machine so that it cannot function while the guard is open. This prevents the machine operating and trapping the operator while the casting is being removed from the dies.

Forging and extrusion processes

Usually forging presses and hammers are not guarded to the same extent as sheet-metal presses. This is because the operator has to load and unload the work zone with long-handled tongues since the workpiece is at or above red-heat during the process. However, the operator must be protected against accidental contact with the hot metal and protective clothing similar to that recommended for casting should be worn. In addition, ear protectors should be worn as the noise level in a forging shop is a potential hazard. As with all high-temperature processes adequate ventilation is essential to maintain a comfortable working environment and to remove any unpleasant or toxic fumes.

Impact extrusion has similar hazards to forging. However, because of the batch sizes involved, mechanical handling devices are frequently used to load and unload the presses and total guarding of the working zone is possible.

Plastic moulding

Most plastic-moulding operations are now heavily automated and this reduces operator risk. However, where manual loading and unloading of the press or the injection-moulding machine takes place suitable guards, interlocked with the machine operating mechanism, must be provided to prevent the operator becoming trapped and injured. Overalls are adequate protective clothing. Care has to be taken if moulding materials or synthetic adhesives have to be handled by the operator as this can cause skin sensitisation. Suitable barrier creams and gloves should be used. Ventilation of the working area is essential, particularly if solvents are being used as these can be both toxic and highly flammable.

Presswork (sheet metal)

Sheet-metal pressing operations are among the most hazardous processes used because the operator's hands have to enter the working zone of the tools to load and unload the machine. Thus presses must be fitted with comprehensive guarding which must be interlocked with the operating mechanism. This prevents the tools closing and trapping the operator while work is being loaded or unloaded. An example of the guarding of a small press is shown in Fig. 9.1. The guards and the presses to which they are fitted must be regularly inspected by an accredited inspector and

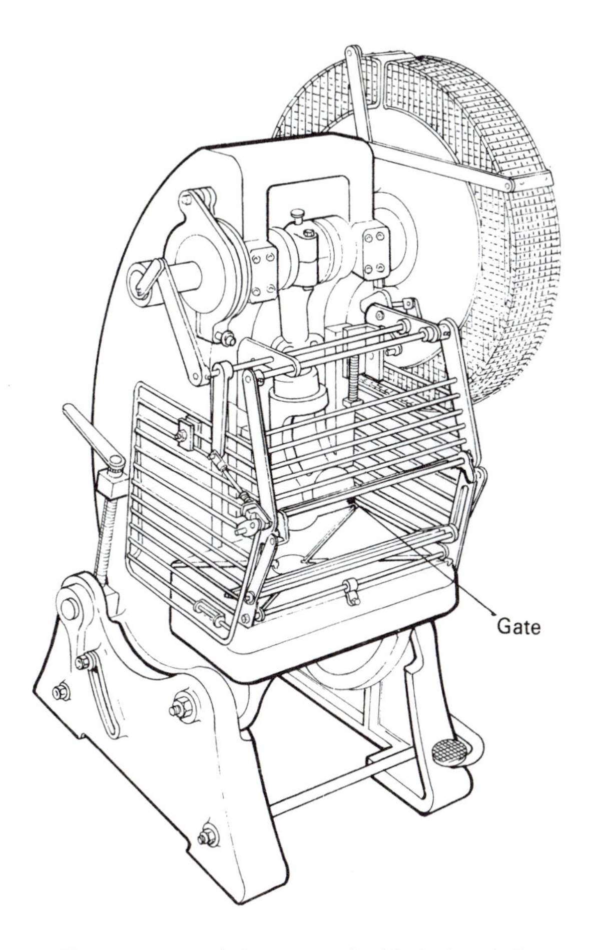

The press can only be operated with the 'gate' shut

Fig. 9.1 Interlocked guard for a power press

the guards can only be adjusted and set by a certificated fitter. The use of mechanical handling devices to load and unload a press greatly reduces the hazardous nature of the operation. No special protective clothing is required by the operator other than normal overalls and reinforced gloves as the raw edges of thin sheet metal can be very sharp.

Machining

It is the duty of machine manufacturers to ensure that their products have adequate transmission guards securely fencing the shafts, gears, belts and chains, etc. It is the duty of the employer and the employees to ensure that these guards are kept in place and in good condition.

In addition, it is the duty of the employer to ensure that such machines are provided with proper cutter, chuck and splash guards and that these guards are fitted and used correctly. It is the duty of the employee to make proper use of the guards provided at all times and not to tamper with them or remove them in order to speed up the process and earn increased bonus payments. The type of guard will depend upon the machine and the process. Milling machines are particularly hazardous and a selection of cutter guards for jobbing and small batch work are shown in Fig. 9.2. Automatic and CNC machine tools frequently have the work zone totally enclosed while cutting is taking place. This not only protects the operator but it prevents the coolant being splashed out of the machine. A considerable volume of coolant is usually flooded over the cutting tools and the workpiece in such machines throughout the cutting cycle. Loading and unloading can be by industrial robots but, where manual loading and unloading is resorted to, a separate work station is frequently provided remote from the cutting zone.

Grinding

The guards on grinding machines have two purposes:

- to prevent the operator coming into contact with the rapidly revolving abrasive wheel;
- to provide *burst containment*, i.e. to contain the wheel should it shatter while it is being used. For this reason, abrasive wheel guards should be made from steel and should be thicker than is necessary merely to keep the operator out of contact with the wheel. The thickness is specified in the regulations.

Under the grinding wheel regulations of 1970, only certificated persons may change and mount grinding wheels. Because of its relatively flimsy construction and high rotational speeds, a grinding wheel is potentially dangerous. A large wheel, rotating rapidly, stores a lot of kinetic energy which can be released with disastrous results if the wheel bursts. The usual causes of failure are misuse, incorrect mounting or overspeeding. The spindle speed of the machine must be prominently displayed. This speed must be compared with maximum permitted speed marked on the abrasive wheel, and in no way must the wheel be used if the spindle speed is greater than the permitted speed.

9.8 Fire regulations

New fire regulations for commercial and manufacturing premises came into force on 1 April 1989. The employer, manager or owner of a factory, office, shop or railway premises must decide whether or not the premises require a *fire certificate*. A certificate is needed if:

- more than 20 persons are at work in the premises at any one time;
- more than 10 persons are at work at any one time on a floor other than the ground floor;

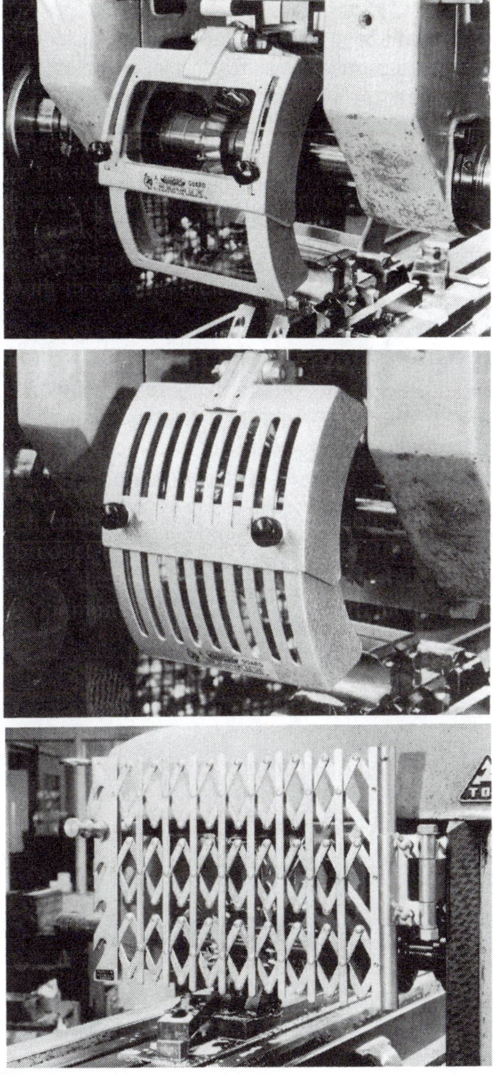

Fig. 9.2 Milling cutter guards

- the premises are used as a factory which has highly flammable or explosive materials stored in or under it.

Note that the persons 'at work' means everyone, including employees, self-employed persons, trainees and apprentices.

The decision as to whether a certificate is or is not required rests with the local fire service after a certificate has been applied for. From the time of application for such a certificate, all means of escape with which the premises are provided must continue to be kept safe and effective, all means for fighting fires must be maintained in efficient working order, and all employees must receive instruction what to do in the event of fire. The HMSO publication, *Guide to Fire Precautions in Places of Work that Require a Fire Certificate*, should be available on the premises and regularly consulted. Regular consultation with the local Fire Prevention Officer is time well spent, particularly if changes are to be made to the premises, plant, processes, working practices, and materials being handled and stored.

If the premises are exempted from a certificate by the local fire service, or if insufficient persons are working in the premises to make a certificate obligatory, it is still the responsibility of the employer, manager or owner of a factory, office, shop, or railway premises, for providing the means of escape and the means of fighting fires. Failure to provide proper means of escape or means of fighting a fire is a legal offence. In the case of premises not requiring certification, reference should be made to the HMSO publication: *Code of Practice for Fire Precautions in Factories, Offices, Shops and Railway Premises Not Required to Have a Fire Certificate*.

Assignments

1. With reference to the Health and Safety at Work Act (1974), compare the composition and responsibilities of the Health and Safety Commission with the composition and responsibilities of the Health and Safety Excecutive.
2. Describe the sanctions available to the Health and Safety Inspectorate for enforcing the provisions of the Act.
3. Discusss the responsibilities of Employers and Employees under the Health and Safety at Work Act.
4. Discuss how the European Safety Directive (1992) affects the manufacture, sale, leasing and importing of machinery in the U.K.
5. Summarise the requirements of the new fire regulations (1989) as set out in the following HMSO publications:
   (a) Guide to Fire Precautions in Places of Work that Require a Fire Certificate;
   (b) Code of Practice for Fire Precautions in Factories, Offices, Shops and Railway Premises not required to have a Fire Certificate.
6. Discuss the special guarding and safety requirements of:
   (a) grinding machines;
   (b) horizontal milling machines;
   (c) power presses.

10 Cost control

10.1 Cost control

Some basic principles of estimating and costing were introduced in section 11.5 of *Engineering Materials: volume 1*, where the processes of estimating and costing were used to arrive at the selling price of a product in advance of that product being manufactured. The purpose of the exercise was to arrive at a price that could form the basis of a 'quotation' or 'tender' when seeking a new order for a product yet to be made. It is relatively easy to tender for a batch of motor cars which are in regular production and for which all the costs are known and the selling price has been fixed. It is quite a different matter to tender for an oil rig of high complexity to a new design that has yet to be built.

Once the tender has been accepted and the order has been put into work, it is necessary to monitor the costs to see if they agree with the estimate. If the manufacturing costs are greater than estimated, then the firm will make a smaller profit or even a loss on the order. If the manufacturing costs are less than estimated, then the firm will make an increased profit. Unfortunately, tendering in a highly competitive market leaves little, if any, room for making allowances for cost over-run when estimating and tendering. The object of this chapter is to analyse the costs of manufacture and examine some methods of cost control.

Historical costing is the term used when the costs of manufacturing and marketing a product are arrived at after the product has been completed.

It is the true cost of making and selling a product and the various cost elements are determined from records kept of the cost of materials, of bought-in components, of the labour used, of the machines used, and of anything else that directly or indirectly is attributable to that product. Unfortunately, any discrepancy between the true cost and the estimated cost does not become apparent until the product has been completed and it is too late to retrieve the situation. If the product is the first of a batch, then it may be possible to review the manufacturing processes used and recoup any loss in the remainder of the batch. However, if the product is a 'one-off', no such corrective action can be taken.

Standard Costing is the term used when estimated cost elements are used to set standards that must not be exceeded at each stage of manufacture. The estimated cost elements are those arrived at during the process of estimating for tendering purposes. Standard costing allows greater management control and corrective action can be taken as soon as any cost element approaches or exceeds the standard set.

10.2 Cost elements

Figure 10.1 shows the basic cost elements that go to make up the selling price of a product. They consist of:

(i) *Prime costs*. These are the actual costs of raw materials, bought in parts, and labour used in the manufacture of a product. Since they refer directly to the product they are also referred to as 'direct costs'.

(ii) *Production overheads*. These are costs which result from the manufacture of the company's products but which cannot be directly charged against any particular product. Hence they are referred to as

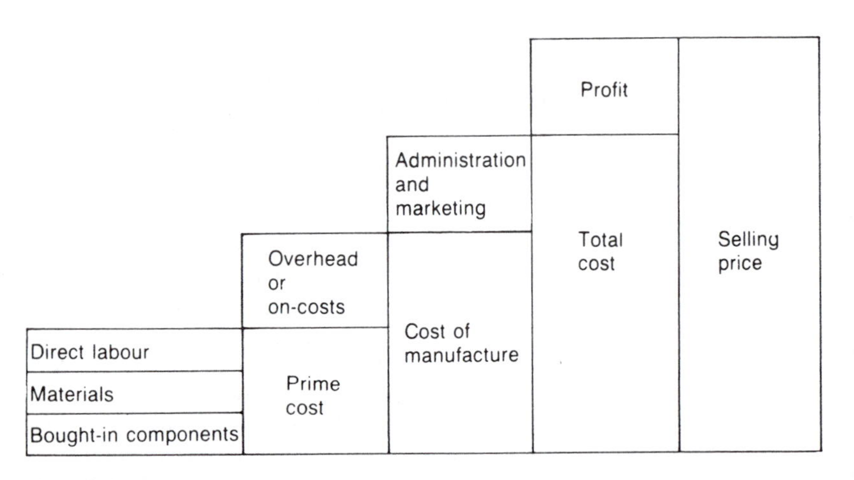

Fig. 10.1 Cost structure

Table 10.1 Typical production overhead costs

Workshop premises	**Plant**
Rent or capital loan charges Unified business rate Depreciation Insurance Maintenance and repairs Heating and lighting	Leasing or capital loan charges Maintenance and repairs Energy to operate plant Insurance Depreciation
Indirect labour	**Indirect materials**
Supervision Inspection Stores Tool room Clerical assistants Transport Time-office Drawing office	Any materials assoicated with the activities listed in this table

'indirect costs'. Table 10.1 lists some typical production overhead costs.

(iii) *Administration, selling and fixed overheads*. Items (i) and (ii) together represent the manufacturing costs of a particular product. In addition, allowance has to be made for the general management of the company together with the costs of marketing and selling. Some costs, such as rental for the premises, servicing of loan interest, and local taxation (uniform business rate), continue even when no production is taking place. These are 'fixed overheads' and are independent of the level of manufacturing activity within the company.

(iv) *Profit*. If items (i), (ii) and (iii) could be accurately assessed and allocated to a company's products, then those products could be sold without making a profit or a loss. However, a company is primarily in business to make a profit. The profit is needed to build up the company's reserves against any future down-turn in activity, to provide a fund for the replacement and updating of plant and equipment, to pay a dividend to the shareholders, and to pay any taxes due on the profits themselves. When setting the profit level, attention has to be paid to the competitiveness of the market and allowance has to be made for inflation. This last item is particularly important when the period of manufacture is protracted, for example, when building a large ship or an oil rig. Failure to allow for inflation, in such circumstances, can wipe out any profit that should have been made on an order.

The level of a company's overhead expenses is often taken as an indication of that company's efficiency, and that high overheads indicate low efficiency. This is not necessarily true. For example, a highly-automated company, which is working very efficiently, may have minimal direct labour costs so that, in proportion, its overheads appear to be high as a percentage of those labour costs. No matter whether the costing system adopted is simple or complex, the problem still remains as to how to allocate the various costs to any individual product so that it carries its fair share of the costs, yet remains competitive and saleable. Some alternative methods of allocating costs will now be considered.

10.3 Prime costs (direct labour)

The simplest method of determining the direct labour cost of a product is to multiply the time booked on each job element by the appropriate labour rate and then adding together all the job elements which make up the product. This is how the direct labour cost is calculated when a historical costing system is used. When a standard costing system is used the calculation of direct labour costs is a little more complicated.

$$\text{Standard labour cost} = \text{standard time} \times \text{operator rating} \times \text{labour rate for the job}$$

The standard time for the job can be defined as the time a qualified worker would take to complete the job when working to a preset level of performance. Standard time is arrived at by the use of 'work measurement' techniques. Operator rating is a measure of how any individual worker compares with the 'ideal' standard operator. The standard labour cost is not the actual cost, which can only be determined by historical costing, but a target that needs to be achieved or improved upon if a profit is going to be made. Actual work performance must be constantly compared with the standard so that remedial action can be taken early while there is still time to prevent a cost over-run.

10.4 Prime costs (direct material)

The materials used directly in the manufacture of any product must be accurately booked against the job and carefully priced. This applies equally to materials bought-in specifically for the job and materials that are drawn from the stores. The cost of materials kept in the stores may vary depending upon market conditions at the time of purchase. Allowance must also be made for any change in cost when the material is replaced. There are a number of ways of allocating cost to stored materials.

- *Market price*. The material cost is based upon the market price for the material at the time of issue, irrespective of the price originally paid.

- *Replacement pricing*. The material cost is based upon an estimated price that will have to be paid when the material is eventually replaced.
- *Standard cost*. This is a cost set for a material as a matter of company policy and need not be the same as the actual market price for the material.
- *Average cost*. This is where the material has been bought over a period of time from a variety of sources at various prices. The material is then charged to the job at the average of all the prices paid.
- *First in, first out (FIFO)*. Where the material has been purchased in batches over a period of time, it is assumed that it will be issued from the stores in the same date order. It will be charged against the job at the original purchase price of each batch.
- *Last in, first out (LIFO)*. Assuming the worst situation of the last batch of material being issued first, all the material is charged against the job at the price of the last batch of material received into stores.
- *Highest in, first out*. This assumes that the most expensively priced batch of material, irrespective of the date of purchase, will be issued first and all the material used on the job will be charged at this price.

10.5 Allocation of overhead costs

Since overhead costs are indirect costs, their fair allocation against any particular job or product is difficult and a number of systems have been devised to try and arrive at an equitable solution. At present, the most widely used system is *absorption costing*. The procedure is to divide the company into *cost centres*. These cost centres may be production cost centres which are concerned with actual manufacturing such as the foundry and the machine shop, and service cost centres such as the stores, the maintenance department, the sales department, the accounts department, and the design and development department.

Having set up the cost centres, the overhead costs then have to be apportioned in an equitable way between the centres. For example, the rent could be charged against a particular cost centre in proportion to its floor area as a percentage of the total site area. Lighting costs could be apportioned according to the number of light fittings and their power consumption, maintenance could be allocated on the basis of the man-hours spent in a particular cost centre over a set period of time. In the case of service departments, wages and salaries must also be included as they cannot be charged directly against a specific job or product. Thus the total indirect overhead costs of each centre can be built up. Next, the overheads of the service cost centres must be transferred to the producing cost centres according to how much use they make of the service cost centres. Finally, a system has to be devised so that the collective overheads of the production cost centres can be charged fairly against the goods produced by the company.

The methods of allocating overheads and the absorption of collective charges in common use are as follows:

- *Percentage of prime costs*. If the total overheads for operating a production cost centre for a year are known, and if the total prime costs generated within that cost centre for a year are known, then the overheads can be expressed as a percentage of the prime costs. Thus the cost of manufacturing a particular product within that cost centre is:

  the prime cost of the product
  + the overhead percentage of the prime cost of production.

  This method is simple and is used where there is little variation between the different products manufactured and where there is a reasonable balance between material and labour costs.
- *Percentage of direct productive labour*. This method is used where manufacturing of the product is labour intensive and the amount of material is negligible. Here, the total cost centre overheads are taken as a percentage of the total of the cost centre's productive direct labour costs over a set period such as one year. The material is charged at cost. Since the direct productive labour costs usually reflect the skills involved, the allocation of overheads reflects the sophistication of the process and the investment in the plant and equipment used. When this method is used, the cost of manufacturing a particular product is:

  material cost for the product + the direct labour cost for the product
  + the overhead percentage of the direct labour cost of the product
  for the cost centre.

- *Differential percentage*. With this method, the cost centre overheads are loaded onto the direct material costs and the direct labour costs at different rates. Usually the rate loaded onto the material is about one-quarter or one-third the rate loaded onto the direct labour costs. This system is frequently used in small companies as it is not only simple to apply but it also allows greater flexibility in pricing strategy.
- *Man-hour rate*. With this method, the total charges (prime costs and overheads) for the cost centre are divided by the total productive man-hours of the centre and allocated pro rata to the number of man-hours spent on each job. This takes into account various rates of pay but ignores the use of different types of equipment. Therefore it is more suitable for fitting and assembly shops than it is for machine shops where there is a wide diversity of equipment and processes.
- *Machine-hour rate*. The total charges (prime costs and overheads) for the cost centre are allocated to the individual machines in a workshop, taking into account such items as the initial cost of the

equipment, the power required, the size and type of product, and routine servicing and maintenance costs. The machine-hour rate is determined by dividing the individual machine costs by the number of productive hours for which that machine works.

10.6 Depreciation

Depreciation is the reduction in the intrinsic value of an item of plant or equipment due to normal wear and tear or due to the fact that it has become technologically out-dated. It must be remembered that each item of plant represents a capital asset in a company's accounts. Since the value of each item of plant is continually diminishing, this represents a reduction of the capital assets of the company and must be shown in the balance sheet if this is truly to represent the financial status of a company. Further, a provision must be made for the eventual replacement of the items of plant under consideration. To avoid repeatedly using the phrase 'item of plant' throughout this section, the term 'asset' will be used instead. The original cost of each asset together with all expenditure thereon, such as servicing and repairs and maintenance, but minus its residual 'trade-in' value or scrap value must be charged against revenue (the money earned by that asset) and this charge must be spread over the asset's economic service life as fairly as possible. Remember that an asset's competitive service life, is usually very much shorter than its actual usable life. To provide for the replacement of an asset, a sum of money must be set aside annually from revenue earned by the asset, so as to accumulate the required amount by the end of its service life. Allowance must also be made for inflation. It is essential to create this depreciation reserve fund so that new capital does not have to be raised when replacement becomes necessary. From time to time technological innovation may make it necessary to replace plant prior to the planned date and before adequate funds are available. Under such circumstances it may well be necessary to raise additional capital and the cost of servicing such capital must be included in the estimates of revenue the new plant will earn over its lifetime. There are various methods of writing down the value of an asset over its lifetime and some of these will now be examined. For simplicity, the fact that the depreciation funds can be invested to accrue interest (compound) will be ignored, but it can be helpful in offsetting the effects of inflation to some extent.

- *Fixed instalment method*. This is also known as the 'straight line' method since the value of the plant is written down in equal amounts over equal increments of time, thus producing a straight line graph. Let I equal the initial cost, R equal the remanent (final disposal) value, and n equal the number of years over which the value of the item of plant is being written down. Then, the reduction in asset value and the sum set aside each year should be $(I - R)/n$. This is shown in Fig. 10.2. This simple method makes no allowance for the

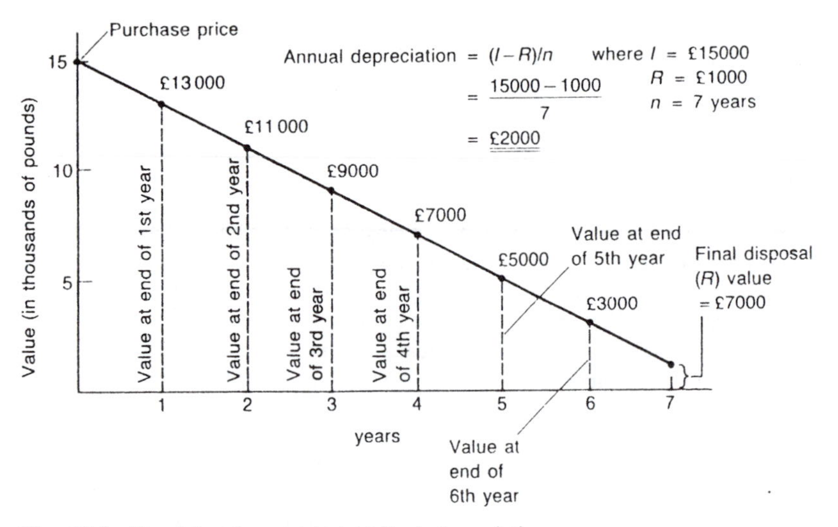

Fig. 10.2 Fixed instalment (straight line) depreciation

fact that the cost of repairs and maintenance is greatest at the end of the life of an asset.

- *Reducing balance method*. This allows for the fact that, financially, most assets depreciate more rapidly in the early years of their life and that the depreciation — as a percentage of the original purchase price — becomes less each year. Using the symbols from the previous example, and letting r equal the fixed percentage, then r can be determined from the expression $r = [1 - (R/I)^{1/n}] \times 100$. The application is shown in Fig. 10.3.
- *Interest law method*. This provides for depreciation by crediting the asset account with equal yearly amounts. The asset account is then debited with the interest which would otherwise be earned by the capital invested in the asset had that capital been invested in, say, a bank deposit account. The notional interest so deducted is then credited to the *profit-and-loss account*. The annual increments, less the notional interest on the yearly balances, indicates the amounts to be set aside for replacing the asset at the end of its forecasted service life. Let S equal the the sum set aside each year. Then, using the symbols from the previous examples, $S = [P(I - R)]/[(1 + P)^n - 1]$, where P = notional rate of deposit account interest. This method of calculating the depreciation allowance is frequently used in connection with asset leasing. Whichever of the three previous methods is used, the accumulated reserves at the end of the service life of the asset (nth year) will be $I - R$, as shown in Fig. 10.4.

There are other systems of allowing for depreciation. For instance, in the 'depreciation fund' system the asset is shown at its full value

Year	Start of year value (£)	Depreciation (£) (n = 32%)	End of year value (£)
1	15 000	32% of 15 000 = 4800	15 000 − 4800 = 10 200
2	10 200	32% of 10 200 = 3264	10 200 − 3264 = 6936
3	6936	32% of 6936 = 2221	6936 − 2221 = 4715
4	4715	32% of 4715 = 1509	4715 − 1509 = 3206
5	3206	32% of 3206 = 1026	3206 − 1026 = 2180
6	2180	32% of 2180 = 698	2180 − 698 = 1482
7	1482	32% of 1582 = 474	1482 − 474 = 1008

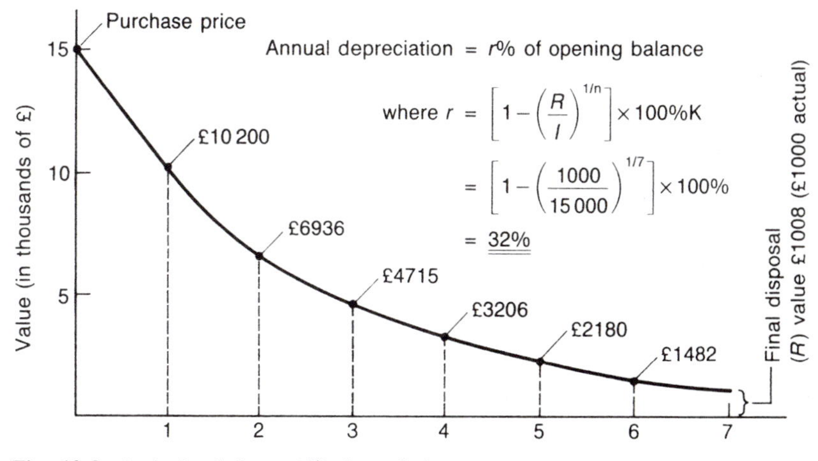

Fig. 10.3 Reducing balance (%) depreciation

throughout its life but the profit-and-loss account is debited with a fixed annual sum which is paid into a deposit account or into gilt-edged securities. The 'insurance policy system' is similar to the previous example except that the annual sum debited from the profit-and-loss account is credited to an insurance policy which secures a sum sufficient to replace the asset at the end of its service life. The 'revaluation' system requires each asset to be revalued each year and a charge equal to the difference between the book value and the assessed value is debited against revenue earned by the asset, and placed in a sinking fund ready for eventual replacement of the asset, the book value being duly written down. An independent professional valuer is required in the interest of accuracy and impartiality. Finally, in the 'single charge' system, a single charge is made annually against revenue earned by the asset to cover repairs, renewals and depreciation. The charge may vary and revaluation may also be required from time to time.

Year	Sum set aside each year (£)	Balance carried forward (£)	Accrued interest (£) at 8%	Balance at year end (£)
	(*S*)			
1	1569	—	—	1569
2	1569	1569	8% of 1569 = 126	3264
3	1569	3264	8% of 3264 = 261	5094
4	1569	5094	8% of 5094 = 408	7071
5	1569	7071	8% of 7071 = 655	9206
6	1569	9206	8% of 9206 = 737	11 512
7	1569	11 512	8% of 11 512 = 921	14 007

$$\begin{aligned} S &= [p(I-R)]/[(1+p)^7-1] \\ &= [0.08\,(15\,000-1000)]/[(1+0.08)^7-1] \\ &= \underline{\underline{£1569}} \end{aligned}$$

where: p = 8% interest

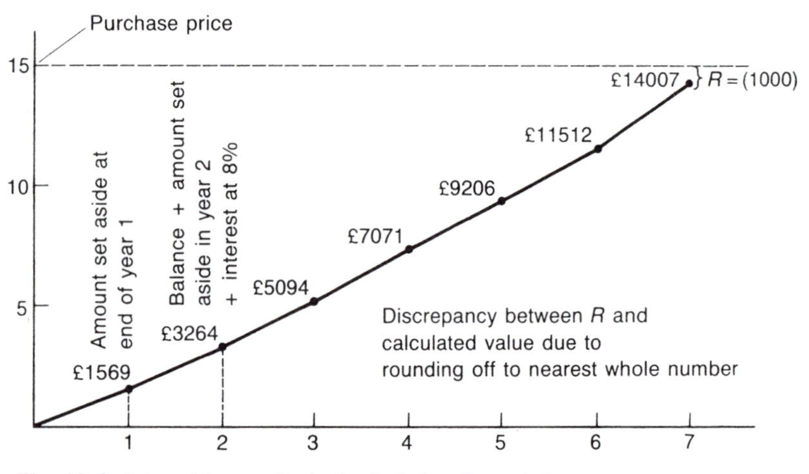

Fig. 10.4 Interest law method of calculating depreciation

10.7 Costing applications

Job costing

Job costing is used for costing the production of 'one-off' components or assemblies. It is essentially a historical costing operation with the prime costs of materials and production labour being charged daily against a specific works order and the overheads being allocated on completion. If standard costs have been ascertained then the actual costs can be compared with them on a daily basis and some degree of budgetary control maintained. However, if a cost over-run has occurred it is very difficult to recoup the situation except by seeking ways to reduce the costs of later stages in the process.

Batch costing

Batch costing is similar to job costing except that a number of identical components or assemblies are being made against the same works order. If the batch is large enough, then special tooling such as jigs, fixtures and press-tools may have to be made. The cost of such tools must be recovered. This cost can be charged against the order or, sometimes, the customer will pay 'part-cost' towards the tools to reduce the unit component cost. This is often done if a succession of batches is envisaged. Tools wear out and have to be maintained and replaced from time to time and allowance must be made for this in a similar way to asset depreciation, as described earlier in this chapter. Where batches of components or assemblies are being produced, and particularly if repeat batches are being manufactured, closer budgetary control can be exercised than for simple job costing. Standard costing techniques can be applied and any deviation between actual costs and standard costs can be investigated and corrections made to the manufacturing processes employed.

Process costing

Process costing is applied to continuous (mass) production and continuous processing (e.g. chemical manufacturing and metal extraction). By its very nature, continuous processing lends itself to standard costing since there is adequate time to compare actual costs with standard costs, to investigate and analyse any discrepancies, and to carry out any remedial action that may be necessary. The aim of process control is to determine the average cost of a product over a period of time. This is achieved by adding together the direct labour and material costs with the overheads and dividing the sum of these costs by the number of units produced over a prescribed period of time. Work in progress also needs to be taken into account at the beginning and end of the time period. Data are collected from the various cost centres involved and the more finely these are subdivided, the more accurate the analysis becomes. However, too much subdivision leads to too great a clerical load in collecting and processing the data and, even with computerisation, a balance has to be achieved between the costs saved by accurate cost control and the cost of implementing the system. Separate costings have to be carried out for the service cost centres and the production cost centres and an equitable system has to be found for allocating the service centre costs to the production cost centres so that all costs incurred in operating the company may be carried in an equitable manner by the goods manufactured and, ultimately, by the customer. Since all the costs are directly allocated to, and absorbed by, the goods being manufactured, this is known as *absorption costing*.

Marginal costing

Overhead costs can be fixed or variable. The fixed overheads are independent of the level of manufacturing activity within the company, i.e. all the costs that would remain if the company suddenly stopped manufacturing and dismissed its productive labour force, and closed down its workshops, but kept all its offices and non-productive staff intact. The variable overheads are those directly related to manufacturing and which increase or decrease with changes in manufacturing activity level.

There is a school of thought that only the variable overheads should be added to the prime costs to give what are referred to as *marginal costs*. The difference between the marginal costs and the selling price contributes to paying all the remaining operating costs and the fixed overheads and are referred to as *contribution costs*. Marginal costing allows for a differential pricing policy so that items which sell strongly may be loaded with higher contribution costs, while items that sell less strongly may be more lightly loaded with contribution costs so as to keep their price down in a competitive market.

10.8 Break-even analysis

Now that some various cost elements and some various techniques of costing have been considered, the use of break-even graphs, such as the example shown in Fig. 10.5, can be examined. The vertical axis represents *cash* both as revenue from sales and the costs of manufacture. The horizontal axis represents units of production.

First consider the costs of manufacture. The build-up of these costs is shown in Fig. 10.5(*a*). Since the fixed overheads have already been defined as being independent of the number of units manufactured, it is plotted as a straight line parallel to the horizontal axis. Next the variable overheads must be added and these increase as the level of manufacturing activity increases. As soon as the decision to manufacture is taken, some expenditure will have to be incurred to prepare for manufacture even before the first item is made. These are the production overheads and they will increase as manufacturing becomes more active and more electricity is used and greater use is made of the service departments. Finally, the prime costs have to be added. Direct labour and direct materials are wholly proportional to the number of items manufactured.

Revenue from sales is directly proportional to the number of items manufactured and sold (ignoring discounts and other customer incentives). Revenue is plotted as shown in Fig. 10.5(*b*). Finally the two graphs are superimposed as shown in Fig. 10.5(*c*). The area of the graph where the operating costs are greater than the sales revenue represents a loss. The area of the graph where the sales revenue is greater than the operating costs represents a profit. The break-even point is where the sales revenue and the operating costs are equal. This graph only represents the situation at a particular instant in time. Any active company will be constantly ajusting the slope of the various elements of the graph to maximise profits. An increase in overheads resulting from the purchase of new and, hopefully, more productive plant should be more than offset by a reduction in direct labour costs. The company should constantly be

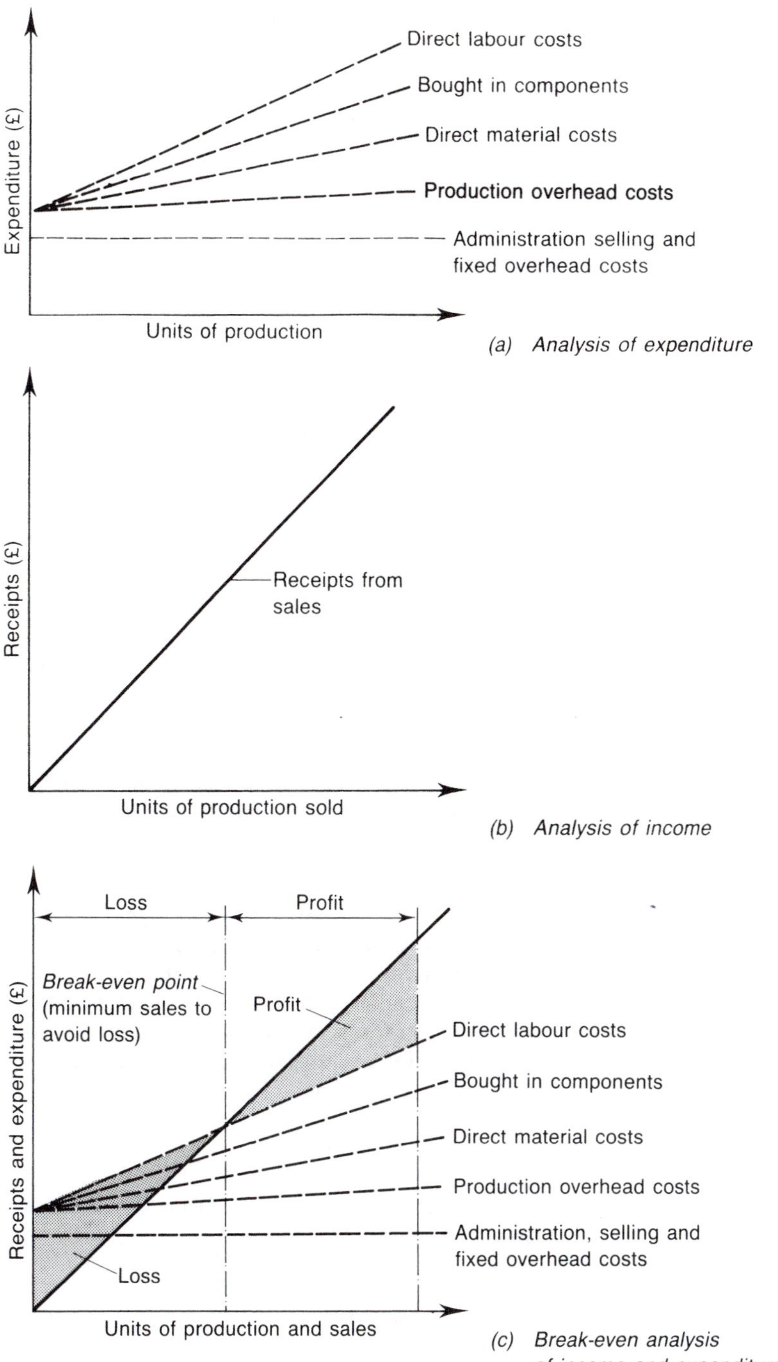

Fig. 10.5 Break-even analysis of profit and loss

looking for cheaper and better materials. The selling price will constantly need to be tuned to the maximum the market will stand.

10.9 Process selection

The first decision that has to be made concerning any new component is whether to *make* or *buy*. Standard components such as nuts and bolts should always be bought in since they are made by specialist firms in large quantities and the cost is less than making a few such items 'in-house'. To this end, designers should always make as much use as possible of standard components. Where these are made to BSI standards or to DIN standards, the quality is also guaranteed. For non-standard items a number of factors have to be taken into account.

- Is suitable plant and labour available 'in-house'? Is there sufficient capacity available on the plant in time to complete the components by the required delivery date? Can the plant and labour available operate to the required quality standards? Can the components be produced at an economic price? Will 'buying-in' result in plant and labour standing idle and thus representing a non-productive cost to be recovered? If the answers to any of the foregoing questions is 'yes', then the decision should be to 'make'.
- If the answer to any of the above questions is 'no', then the purchase of new plant and/or tooling must be considered. Is there sufficient demand for the component under consideration to warrant the outlay on new plant after allowing for depreciation or servicing any loan charges incurred? If not, can the expenditure on new plant be offset against further, similar work? Again, if not, then the decision must be to 'buy-in' the components from specialist sub-contractors.

The selection of manufacturing process must also be cost related. As stated in section 3.1, it might appear to be more costly to manufacture a component by powder metallurgy than by drop forging. However, provided the quantity of components required will justify the tooling, the components produced by powder metallurgy will be more accurate, have a better surface finish and will require less secondary machining and finishing compared with forging. Thus overall the powder metal process may result in a cheaper and better component.

Assignments

1. (a) Discuss the needs for cost control in manufacturing companies.
   (b) Compare the advantages and limitations of:
      (i) historical costing;
      (ii) standard costing.
2. Discuss the cost elements of a typical manufacturing company with which you are familiar.

3. Compare the advantages and limitations of any TWO methods of allocating direct material costs from the following list:
   (a) market price;
   (b) replacement pricing;
   (c) standard cost;
   (d) average cost.
4. Compare the advantages and limitations of any TWO methods of allocating direct material cost from the following list:
   (a) first in, first out (FIFO);
   (b) last in, first out (LIFO);
   (c) highest in, first out (HIFO).
5. Describe TWO methods of allocating direct labour costs.
6. Describe what is meant by overhead costs and compare and contrast the advantages and limitations of any TWO methods of allocating overhead costs from the following list:
   (a) percentage of prime costs;
   (b) percentage of direct productive labour;
   (c) differential percentage;
   (d) man-hour rate;
   (e) machine-hour rate.
7. Describe what is meant by depreciation and compare and contrast the advantages and limitations of any TWO methods of calculating depreciation from the following list:
   (a) fixed instalment method;
   (b) reducing balance method;
   (c) interest law method.
8. Compare and contrast the advantages and limitations of any TWO of the following costing techniques and state under what circumstances you would use them:
   (a) job costing;
   (b) batch costing;
   (c) process costing;
   (d) marginal costing.
9. Describe how 'Break-even Analysis' can be used to determine the point at which a manufacturing process ceases to make a loss and commences to make a profit.

11 Control of manufacture

11.1 Introduction to management of manufacture

The earliest organisations had owner-manager supervision and, even today, many small businesses are operated successfully in a similar manner. One of the first decisions that has to be made in any organisation is which of the many duties must the owner-manager undertake personally. Even in a 'one-man' business, materials have to be purchased, products manufactured, invoices prepared, jobs costed and tenders submitted in addition to the raising of capital, and tax considerations. The problems become magnified as the organisation expands and labour has to be hired, paid for, and employment legislation complied with. The high failure rate of small businesses gives testimony to the fact that many persons set up in business on the basis of their technical prowess and craft skills but without the necessary organisational skills or without giving sufficient thought to the problems of being one's 'own boss'. Managing an enterprise either alone or, as is more usual, as part of a team requires training and skill in just the same way as operating a machine tool. Small to medium sized companies engaged in engineering activities did, and still do, appoint supervisors who are more skilled than the most skilled workers under their control. Such supervisors also have the responsibilty for production planning, control of quantity and quality, and the despatch of finished goods. Increasingly, however, even small and medium sized firms are having to adopt the principles and practices of scientific management in order to survive.

The scientific approach to management involves the application of the following principles:

- *Forecasting and planning*: this requires an assessment of the future and decision-making for future action.
- *Organisation*: the division of labour, the allocation of duties, and the lines of authority and responsibility.
- *Command*: the issuing of instructions to ensure that decisions are activated.
- *Control*: the setting of standards, the comparison of physical events against the set standards, and the taking of any necessary corrective action.
- *Coordination*: the unification of effort to ensure that all activities of the business are pursuing the same objectives.
- *Communication*: the transfer of information between different people and/or sections of the organisation, in particular between management and the workforce.
- *Motivation*: this is the driving force behind all actions. Psychological considerations make it important to recognise the motivation behind:

  — the customer buying the goods or services supplied by the organisation;
  — the people working in the organisation.

In addition to the principles of management postulated above, is the introduction of *behavioural science*. This assists the modern manager in the understanding of human behaviour, for example, job satisfaction, attitude to work, and putting the right person in the right job. Such skills are essential to all managers since: 'management is usually about people'.

11.2 Methods of manufacture

There are four basic methods of manufacture and these are summarised in Table 11.1.

Table 11.1 Methods of manufacture

Method	Description	Examples
Job or unit	Single or very small quantity production of articles to individual customer requirements can involve a single operator or large groups of operators.	Bridges, ships, and special components.
Batch production	Involves the multiple production of articles from, say, five units to many hundreds of units; either to customer requirements or in any quantity in anticipation of orders.	Gear boxes, pumps, electronic assemblies.

Table 11.1 Methods of manufacture (*cont'd*)

Method	Description	Examples
Mass, flow or line production	The manufacture of very large quantities of products, usually consumer goods made in anticipation of orders.	Cars, household appliances.
Continuous or process production	The plant resembles one huge machine with materials taken in at one end and the finished articles despatched at the other.	Plastic and glass sheet, plaster board.

Note:

- To differentiate between 'job' and 'batch' production, it is not the number of components which is the deciding factor but the manner in which the production is organised. Consider the manufacture of, say, four components. These could be made by four operators, with each operator making a component outright. This is what normally happens when 'job' or 'unit' production is employed. Alternatively, the components could be passed from operator to operator, with each operator specialising in and completing a particular feature of a component. The manufacturing method would then be classified as 'batch' production.
- Job and batch production have the following characteristics in common: the flow of production will be intermittent; some parts will be for customers' orders, others for stock; schedule control of orders will be necessary to ensure that delivery times are met; there will be a large variety of products.
- 'Flow' and 'continuous' production have the following characteristics in common: the flow of production is, ideally, continuous; the production is usually in anticipation of sales; profit margins will be smaller than for batch or unit production; close control of costs is required at all stages of manufacture; the rate of flow at each stage of production is strictly controlled to avoid over-production or shortages; there will be a small range of standard products.
- Increased production does not necessarily result in increased profits. It could result in a reduction in manufacturing costs but the selling price might also have to be reduced to increase sales volume in line with the increased rate of production.

11.3 Plant layout

Plant layout usually depends upon the method of manufacture. For job and batch production the machines are grouped according to type. For example, all the lathes would be in the *turning section* of the machine shop. For flow and continuous production the machines would be arranged to ensure a smooth flow of work from one operation to the next.

The laying out or rearrangement of machines and/or processes is usually costly in time and labour. Assuming that planning consent can be obtained and environmental constraints overcome, the ideal situation is to layout a '*green-field*' site. Prior to plant considerations, such factors as the availability of services and labour, attitudes of the local community, access for transport, proximity of markets, terrain of the new site and, more recently, the availability of local and/or government funding must be taken into consideration.

More often, engineers are called upon to rearrange the layout of an existing plant to accommodate a new product or method of manufacture. Since non-productive *down-time* must be kept to a minimum, the changeover usually has to be achieved during weekends and holiday shutdowns. Specialist firms are available with the necessary equipment and skilled manpower to affect a speedy changeover.

In either of the above cases, the new layout must be carefully planned in advance, in full consultation with the key personal concerned with its operation. The main factors affecting layout can be summarised as:

- volume of production
- sequence of the process
- type of building and work area
- transport area
- the type of production and its handling
- monitoring of work flow
- power sources
- temporary stocking area requirements
- permanent stocking area requirements
- service areas
- pedestrian areas
- inspection requirements

11.4 Types of layout

Process layout

In process layout the requirements of various manufacturing activities are grouped together. For example, all the lathes in the turning section, all the milling machines in the milling section and all the grinding machines in the grinding section. While such a layout is traditional and still widely used for jobbing and small batch production, it leads to problems of storage and handling of work in progress. An example of this layout is shown in Fig. 6.19.

Product layout

Product layout is also known as 'line' layout. In this type of layout, the needs of the process take precedence. That is, the machines and processes are laid out in the sequence of operations specified for the production process. An example of this type of layout is shown in Fig. 6.20. The advantages of this arrangement are as follows:

(i) greater specialisation and speed of production is possible,
(ii) high-capacity dedicated machines can be used,
(iii) the layout lends itself to automation including mechanical handling and robotics,
(iv) because of (iii), it is less labour intensive.

Group layout

This is also known as 'cell' layout. In this type of layout, similar machines and processes are grouped together to manufacture small batches of a variety of products having family resemblences. An example of such a layout is shown in Fig. 6.21. The advantages of this arrangement are as follows:

- Machine utilisation is high as work can be switched to the first available machine.
- The layout does not have to be changed to suit changes in the product.
- Material handling is reduced.
- In-process stocking areas are reduced due to the balanced flow of production.
- Production control is simple; delays are quickly apparent.
- There is greater job satisfaction for personnel.
- The manufacturing 'cell' can be automated to accept FMS techniques, as previously discussed in section 6.14.

11.5 Methods of plant layout

The basic procedures for laying out equipment in a plant are similar to those used for 'method study'. As the processes of layout are very much dictated by the unique circumstances in each case, there is no hard and fast set of rules that can be applied. However, the application of method study principles will provide a starting point:

(1) *Select* the type of layout required, i.e. product layout, process layout or group layout.
(2) *Record* all the facts concerning the manufacturing method; either the existing method or a new approach.
(3) *Examine* the facts critically and in an ordered sequence.
(4) *Develop* the most effective layout.
(5) *Critically examine* the proposed layout.
(6) *Install* the layout.
(7) *Supervise* the installation of the layout.

Since it is much easier to move models around than large pieces of equipment, simulation of the process will enable a critique and adjustments to be made easily. The final layout should clearly indicate that the flow of work and information can be monitored. There should be no 'black holes' down which materials, workpieces and information documents can disappear, only to reappear at suitably embarrassing moments. People, materials and information documents should be on view at all times. It is surprising how often cupboards and small offices can suddenly appear in the layout. Most of the time they are just status symbols and make no useful contribution to the processes and management of manufacture.

Simulation makes use of *flow diagrams* which are detailed scale drawings of the progress of either material or components in relation to the physical environment, i.e. the path such materials or components follow through machines, benches, stores and departments. Using this method the most efficient flow path can be determined. Wasteful journeys and back-tracking can be eliminated and the sequence to provide the minimum delay can be established.

Matrix diagrams

Matrix diagrams are scaled work areas marked with a grid to a convenient scale. Scaled cardboard pieces are cut to represent machinery and equipment. These are moved around on the grid until the most suitable layout is obtained. The pieces can then be glued in position or drawn around to produce the final layout drawing. The grid ensures that the width of gangways and the working areas around items of plant are adequate for safety and comfortable working. The maintenance team should also be involved at the planning stage so that adequate room is left for removing and/or replacing plant components. Because of bad planning and lack of foresight, it is not unknown for bricks having to be knocked out of walls so that drive shafts can be removed. The main disadvantage of this system is that it is only two-dimensional.

Scale models

Scale models provide a three-dimensional layout enabling any overhead obstructions and lifts to be taken into account. The departure from paper plans permits greater use of colour and enables layouts to be visualised from the point of view of appearance as well as convenience. Colour coding can be used to emphasise any particular storage areas and mechanical handling equipment.

Scale models are available for a large range of standard machine tools, equipment, benches, storage cupboards, etc., and their bases are prepared to plug into a 'Lego' type base plate which forms a scale grid of the working area. Figure 11.1 shows an example of a typical layout using scale models. Although the models are expensive, a complete layout for a large machine shop running into several thousand pounds, it can be claimed that the initial outlay is insignificant compared with the effected savings achieved by getting the layout right first time. The main advantages of using scale models can be summarised as follows:

- They enable dimensions, including overhead clearances, to be accurately measured.

Fig. 11.1 Plant layout using models

- They enable visual obstructions to be avoided.
- They reduce explanation since the layout is obvious to all concerned, including non-technical staff such as accountants.
- They stimulate the imagination and increase the accuracy and efficiency of the layout.
- They encourage aesthetics and the use of colour.

In some instances, such as large-scale chemical plants and oil refineries, standard equipment models are not available and highly-detailed scale models have to be hand-crafted by skilled model-makers. They are used not only to prove the feasibilty of the design layout but can also be used for management presentations to such bodies as shareholders and planning authorities. Such models are very costly and invariably end up in glass cases at company headquarters.

Computer programs

CAD software is available for planning works layouts. The floor area is drawn out to scale and representations of the machines and equipment can be called up from the system database and scaled to match the layout of the working area. One widely-used example is 'CRAFT' (Computerised Relative Allocation of Facilities Technique). This uses workpiece handling cost as its computational basis. Other programs are based on the need for similar and associated plant to be grouped together.

Once the layout has been finalised, it should be submitted to the following critique before being put into practice:

- Is the flow of production logical, with minimum back-tracking?

- Is the layout compatible with plant services (e.g. electricity, gas, water, waste disposal) and any mechanical handling requirements?
- Are there any possible overhead obstructions to clear?
- Are the gangways adequate for the movement of materials and personnel?
- Can maintenance be carried out on every piece of equipment?
- Are storage areas adequate?
- Have all aspects of safety, fire regulations, and personnel comfort been taken into account?
- Has provision been made for monitoring the installation to ensure it conforms with the requirements of the planned layout?

11.6 Process charting in plant layout

In order to understand what is required when either re-laying out an existing plant or starting from scratch with a 'green-field' layout, it is important to chart the existing or proposed manufacturing process. For this purpose a form of shorthand is available from method study techniques which enables a pathway to be established to scale, indicating the route of a product through the various operations. This *process chart* can then be used as the basis of a critique to establish both the optimum method and the optimum layout to meet the production criteria. Figure

Activity	*Result*	*Symbol*
Operation	Produces, accomplishes, furthers the process	○
Inspection	Verification — quantity and/or quality	□
Transportation	Movement	⇨
Delay	Interferes or delays	D
Storage	Holds, retains or returns	▽

Fig. 11.2 Process charting symbols

11.2 shows the code of five symbols that is used to signify the operations undergone by a product in the course of either manufacture or handling. Example 11.1, which is concerned with the inspection, stencilling and filling of 250 kg drums, shows a typical process chart using these symbols. Figure 11.3 shows the plant layout derived from this process chart.

11.7 Group technology

In recent years, manufacturing has had to come to terms with an everchanging scenario in which the requirement for purpose-specific, or

Example 11.1

Consider a process chart for the inspection, stencilling and filling of 250 kilogram drums.

Chart begins: empty 250kg drums in stock.
Chart ends: filled 250kg drums in stock.

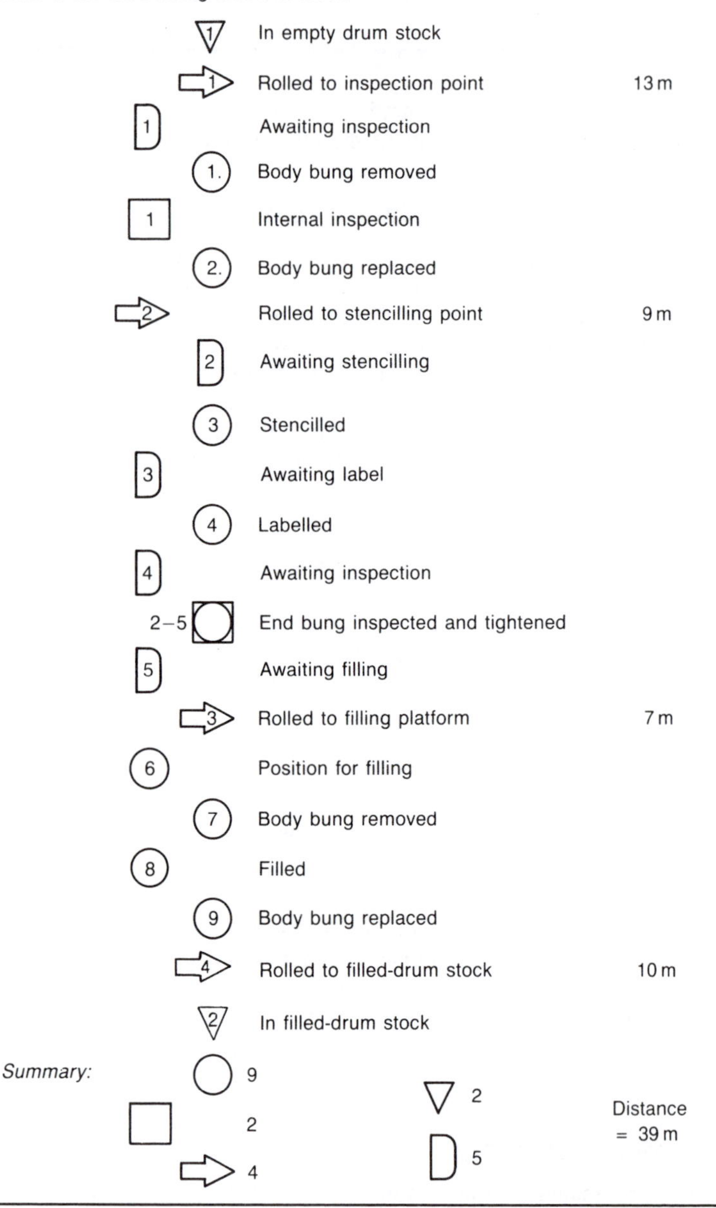

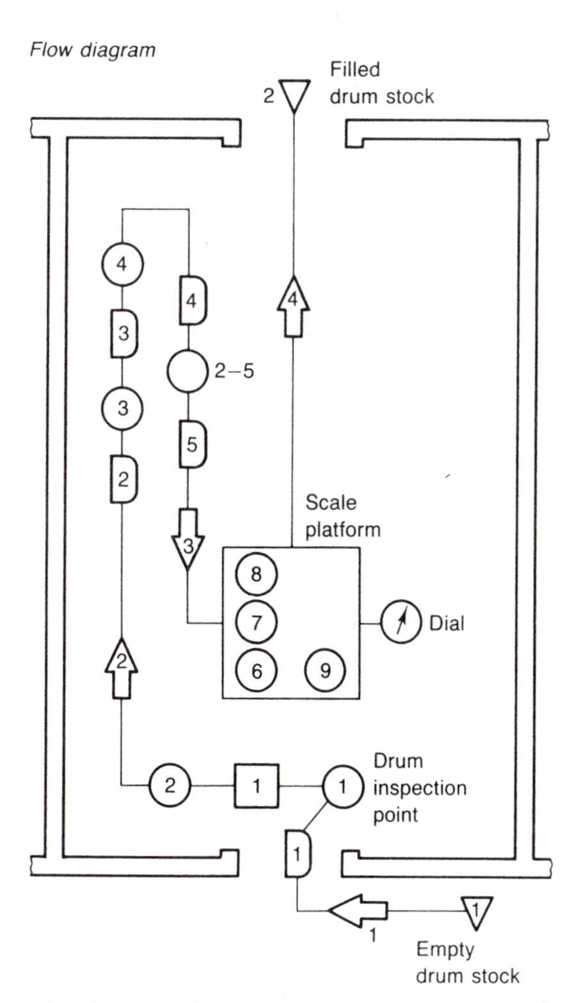

Fig. 11.3 Plant layout developed from the process chart in Example 11.1

customised, products has replaced the production of large numbers of identical products. In fact, about 90 per cent of the manufacturing base of the UK is involved with batch production rather than with continuous production.

Economically-successful batch manufacturing, incorporating modern technology, presents problems to companies whose plant layouts were probably designed many years ago. Although these layouts made sense at the time of their inception, they may no longer satisfy the requirements of modern computerised and robotised manufacturing techniques, i.e. techniques forced upon companies by fierce competition from abroad to reduce unit production costs and the need to reduce the time taken to get a new product or component into production to satisfy customer requirements. Further, there is a constant demand for better and better quality with no corresponding increase in cost.

In response to the need for greater flexibility and a more rapid response to the demand for frequent changes in design and manufacturing methods, the *group technology* approach was first adopted in America and Europe during the early 1970s and more recently in the UK.

Group technology (GT) sets out to achieve the benefits of mass production in a batch production environment. It identifies and groups together similar components in order that the manufacturing processes involved can take advantage of these similarities by arranging (or grouping) the processes of production according to 'families' of components. These grouped production facilities are referred to as *cells*. This allows different 'families' of components and/or assemblies to be manufactured using the cells almost like a number of mini-factories under one roof. These work (or manufacturing) cells have separate planning, supervision, control and even production and profitability targets. With small batch production in traditional workshops based on process layouts, the non-productive costs can be high for the following reasons:

- difficulties in scheduling,
- the transfer of work between stations which are widely separated; excessive handling,
- setting up production processes.

These non-productive costs increase as a proportion of the total cost as the batch size decreases. However, small batches are more easily handled by a cell because of its more closely knit and flexible organisation and the fact that, specialising in a limited range of similar components, set-up times are reduced.

11.8 Group technology (design)

New designs should make use of standard components wherever possible as this reduces costs, lead time, and ensures uniformity of quality. Where a new, non-standard component has to be used the designer will often work from scratch rather than spend time searching a retrieval system for a similar, previous design. However, in many cases, the new drawing may turn out to be merely a variant on some previous design. There may only be a change in a single variable such as a different diameter or a change of material specification. To change the design of an existing part would be more cost-effective than starting from scratch every time.

Such an approach would require a retrieval system to be established in which all the components produced in a company would be classified and coded according to various features. For example, geometric form or any manufacturing similarities. Obviously such a system would lend itself to computerisation. Thus the concept of a parts 'family' can be applied to the design process in addition to the manufacturing process. In fact it should, ideally, start in the design process.

11.9 Group technology (parts-families)

As previously stated, group technology seeks similarities not differences. Therefore, all similar parts can be collected into families. A 'family' or 'parts-family' is a collection of components that can have similarities in geometric form or can be manufactured by similar manufacturing methods. Once the family has been identified, then a composite component can be drawn up which contains all the features of the parts-family or components-family as shown in Fig. 11.4(*a*). The available process machines can then be sorted out into the best mix to produce the composite component. The group of machines selected is then physically moved together to create a *work cell*. The machines in the cell are then set up to produce the composite component. The actual production components are manufactured by omitting those operations not needed on any particular component. For example, consider the manufacture of moulding equipment for making glass bottles. It can be seen that these are mostly turned components that can be derived from the parts-family. Figure 11.4(*b*) shows a traditional, sectionalised (functional) manufacturing layout. The similarity between the components enables them, ideally, to be designed so that they can be brought together in a group technology cell as shown in Fig. 11.4(*b*).

11.10 Group technology (management)

With conventional process layouts for batch production, production control demands more and more up-to-date information. This has led to some very complex control systems which attempt to provide management data on a day-to-day basis and even, in some extreme cases, on an hourly basis. This results in progress chasing becoming a continuous function. Many of these paper systems have collapsed under the sheer volume of information and failed to provide the desired information from the shop floor.

Managing a group tecchnology system can benefit the information gathering process since each group runs as a separate entity and it is only necessary to plan the work in to and out of the cell. There is no necessity to plan work through each machine since each cell is self-monitoring. Group self-control is the biggest and most difficult change for conventional management to accept, since responsibility, and therefore authority, is placed firmly at the point of production. External interference by the detailed planning of each operation would effectively destroy the concept of group technology.

11.11 Group technology (personnel)

The job-satisfaction level, for many people, has considerably diminished with the advent of flow-line production and the introduction of high technology machinery. The work becomes de-humanising as operators find themselves repeating the same task day in, day out. The Volvo company in Sweden attempted to overcome the de-humanising effect of

(a) Component family

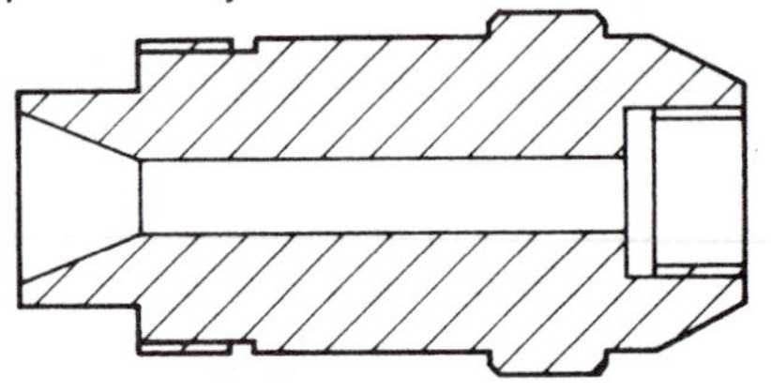

Composite component

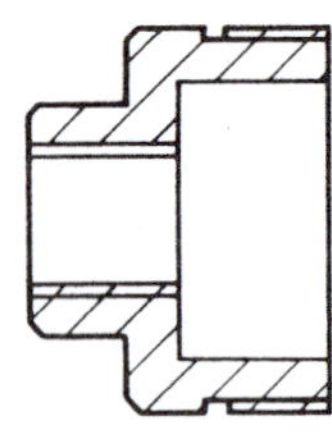

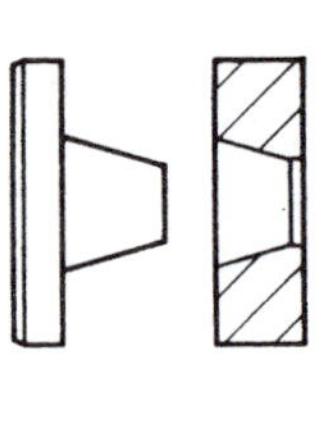

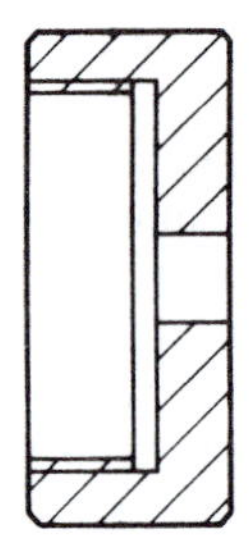

Derivations

(b) Sectionalised (functional)

Ring mould and plug (turning)	Mould face preparation (milling)	Baffles and blowheads (turning)	Blank mould forming (turning)	Finished mould forming (turning)	Mould bottoms (turning)	Engraving (engraving)

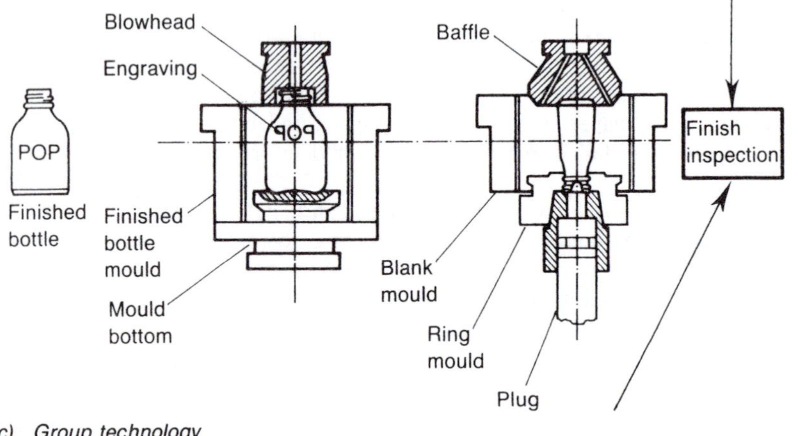

(c) Group technology

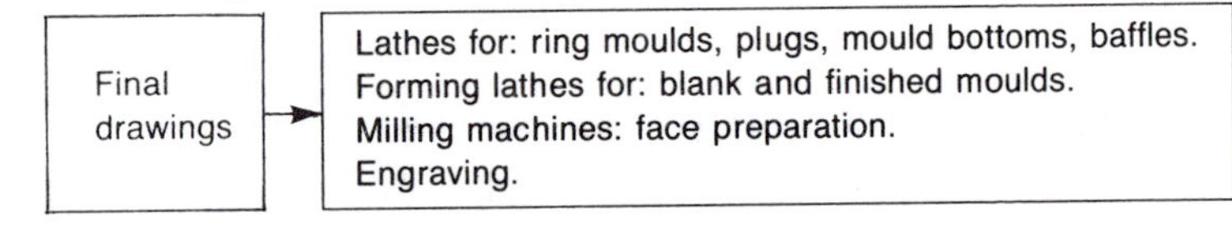

Fig. 11.4 Group technology (parts-families)

the mass-production of cars by creating a group effect. Workers were banded together in teams, with each team producing a complete vehicle rather than simply completing one task repetitively on many cars. Each car carries the group insignia and, in the case of problems, can be traced back to a particular group. All the component parts are delivered to the group rather than to individuals. This job enrichment provides a positive involvement in the company's products and hence a good influence on the quality of the finished product. The concept of small group working coincides with the workers' natural desire for maximum job satisfaction.

11.12 Group technology (group characteristics)

John Burbidge, in his book *The Introduction of Group Technology*, Mechanical Engineering Publication, 1979, suggests that seven characteristics are shown by an effective group:

- *The team*. Groups contain a specified team of workers who work solely or generally for the group.
- *Products*. Groups produce a specified 'family' or set of products. In an assembly department these products will be assemblies. In a machine shop the products will be machined parts. In a foundry the products will be castings.
- *Facilities*. Groups are equipped with a specified set of machines and/or other production equipment, which is used solely or generally in the group.
- *Group layout*. The facilities are laid out together in one area reserved for the group.
- *Target*. The workers in the group share a common product output target. This target output or 'list order' is given to the group at the beginning of each production period for completion by the end of period. How this is achieved is for the group to decide.
- *Independence*. The groups should, as far as possible, be independent of each other. They should be able to vary their work pace if they so wish during a period. Once they have received materials, their achievement should not depend upon the services of other production groups.
- *Size*. The groups should be limited to restrict the numbers of workers per group. Groups of 6 to 15 workers have been widely recommended. Larger groups up to 35 workers may be necessary for technological reasons in some cases. Such large groups have been found to work efficiently in practice.

11.13 Group technology (coding and classification)

A system of classification and coding is the foundation to the application of group technology. The grouping of components into part-families is the largest problem when considering the change to group technology. One method is to conduct a visual inspection to code components. It is the least expensive and least sophisticated method available. It is also the least accurate. The part-families are established by a visual examination of actual components, component drawings or from photographs. Difficulties occur when there are many variations of small details which are not easy to identify.

Another approach to the problem of classification is to use *production flow analysis* (PFA) which is based upon the method of manufacture. This system uses route cards as the foundation for sorting components into classes. The main disadvantage of this method is the acceptance of route cards as being the definitive method of manufacture without a critique being carried out. The first sorting is based upon the first operation (set 1, 2, 3, 4, etc.) Then set 1 is sorted into sub-sets by the second operation (set 12, 13, 14, 15, etc.). Set 12 are those components having their first operation performed on machine type 1 and their second operation on machine type 2. The components are then sorted again by the third operation (set 123, 124, 125, etc.). The procedure is continued until all the route cards have been dealt with. From this sorting, some common characteristics will emerge which will enable the machine loading to be equalised. Where a large number of route cards have to be examined, it is permissible to take random samples of 2000 to 10 000 cards. The PFA system has never found favour in the USA.

The most commonly used method is by parts classification and coding. Though it is the most complex and time-consuming of the available methods, it has the advantage that it lends itself to a computer database. A coding system can be alphabetic, numeric or alpha-numeric.

It can be *dependent* (monocode), in which each succeeding symbol is dependent upon the one preceding it. For example, the code 12 is a two-digit code where the first digit 1 might indicate a circular workpiece and the second digit 2 would indicate a dimension within a certain range. Because digit 2 is preceded by and dependent upon digit 1, the dimension refers to a circular component and will be a diameter.

Alternatively, it can be *independent* (polycode), in which each succeeding symbol in the sequence stands upon its own and does not depend upon any preceding symbol. The position of the symbol in the sequence determines its function. For example, the first digit may indicate geometric form and the second digit may indicate overall length within a given range. Thus the second digit of the code will always indicate overall length in this system. Two common systems are available and these will now be considered.

Opitz

The Opitz system was developed by H. Opitz in Germany. It uses an alpha-numeric code 12345 6789 ABC, in which the first sequence of five digits refers to the design attributes of the parts, the sequence of four digits refers to supplementary information such as material, dimensions, form of supply of raw material and accuracy, and the alphabet code is available for each company for its own unique purpose.

Miclass

The Miclass system is the Metal Institute Classification System developed in the Netherlands. The main advantages of this system are that it can be readily computer based and that it can classify a very large number of parts into their respective family groups. It is an interactive system in which the operator responds to a series of questions posed by the computer. The complexity of the part will determine the number of questions. The code can have from 12 to 30 numbers, with the first 12 digits being a universal code applied to any component. The remaining 18 digits are for use by individual companies for their own unique purposes. The universal code is set out as follows:

1	Main shape	2 and 3	Shape elements
4	Position of shape elements	5 and 6	Main dimensions
7	Dimension ratio	8	Auxiliary dimensions
9 and 10	Tolerances	11 and 12	Material

Advantages of group technology

The case for group technology can be summarised as follows:

- production can be 'customer led', i.e. production is responsive to customer demands in terms of quantity, quality, availability and design variants;
- the use of a coding classification system leads to improved product design and reduced design lead time;
- standardised machine set-ups and tooling can be used;
- more effective use can be made of machines and equipment;
- handling times and costs are lower due to less transportation between operations;
- planning procedures are simplified;
- buffer stocks are lower;
- management procedures are simpler;
- employee job satisfaction is greater.

11.14 Flexible manufacturing systems (FMS)

The natural extension of group technology is to move to manufacturing cells in which all the functions are externally controlled by computers, i.e. a people-less manufacturing cell. In group technology the human beings are part of the process of manufacture but in flexible manufacturing systems the human effort is confined to component and tooling design, setting up the machines and equipment in the cell, and providing very high level maintenance expertise. It is even possible to install machines that can change their own tools and set themselves.

Again there is a need for part-families but, in the case of flexible manufacturing systems, the cell can handle much larger families of parts and deal with them in a random order. Other differences are:

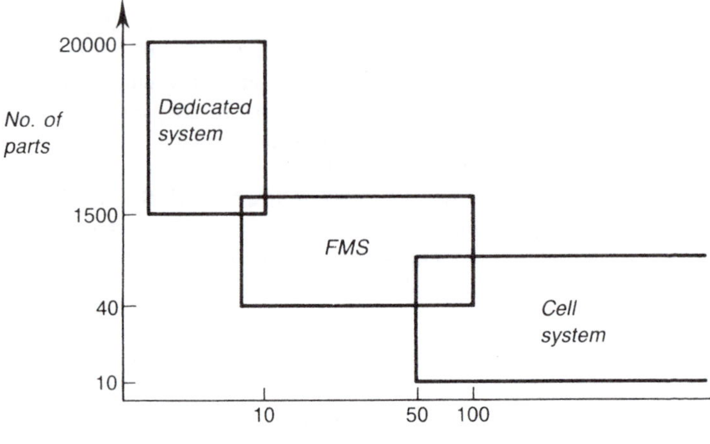

Fig. 11.5 Position of FMS in the production hierarchy

- the need for transport (mechanical handling) equipment that can move components from machine to machine in the correct sequence and maintain maximum loading for each machine.
- a computer that can control all the operations and materials handling at the same time.

The outstanding difference between a GT cell and an FMS cell is that the latter involves a very much higher level of capital investment. Before investing in an FMS cell, comprehensive feasibility studies have to be carried out and these must include a five-year market forecast, an estimate of the return on the capital invested, and the level of flexibility required in the system.

Basically, FMS is a group of computer numerically-controlled (CNC) workstations connected by a materials handling system. The individual computers of the workstations and the handling system are themselves under the control of a master computer (see also section 6.14). In terms of production volume, FMS lies between the high volume production of dedicated machines used for mass production and individual CNC machines which are best suited for batch production in a group technology cell where there is a large variety of products. This is shown in Fig. 11.5.

11.15 Flexible manufacturing systems (concepts)

Where the parts manufactured on flexible manufacturing systems are prismatic in shape, they can be loaded into pallets carrying the necessary fixtures and transported on conveyors or by automated vehicles whose path is determined by guideways in the floor. Such systems are referred to as 'primary conveying systems'. The pallets are coded with a bar code carrying the machining instructions. The bar code is 'read' by sensing devices which route the part to the particular machining destination or to a buffer store conveyor until the required machine is available. Thus, while the part is within the system, it is either in transit or being machined. The system has secondary conveyors whose function it is to receive the parts from the primary system and transport them to the individual machines or buffer stores.

The need for buffer stores is due to the different machining times for each operation. Thus buffer stores allow short operations to be accommodated with those requiring longer machining times. The part remains within the system until all the operations on it are complete. Remembering that this is a flexible system, then different parts requiring different operations can be accommodated within the system at the same time provided that the tooling is available. Since the development of *industrial robots*, the range of components that can be handled has grown beyond those with simple prismatic geometry and the overall flexibility has been enhanced. An example of a typical FMS cell is shown in Fig. 6.21.

11.16 Flexible manufacturing systems (parts/machine tools relationships)

The need for a parts-family is at the root of FMS. Thus there has to be a close liaison between the manufacturing engineers and the designers in order that the machining datums can be set and agreed. The location points for the fixtures are of equal importance, together with the state of the raw material entering the system. All these facts have to be agreed and established at the planning stage.

The part size will influence the choice of machine within the system (large part: large machine) and the part geometry determines the type of machine tool, as it does in general manufacturing. Where the variety of parts to be manufactured is large, then standard CNC machines should be considered. However, the larger the volume and the smaller the range of components to be manufactured, the greater the tendency to move towards dedicated machines and equipment. This will reduce the flexibility but increase the rate of production. Care has to be taken when considering the part/machine tool relationship to keep future production requirements in mind. If the market is fluid, then flexibility is the key factor but if the future demand is more stable, then less flexibility and greater production rates will be required.

The variety and complexity of the workholding fixtures in FMS make great demand on the ingenuity of production engineers. As in all machining operations, the accuracy of the final product is only as good as the accuracy of presentation of the work to the cutting tools. The concept of the parts-family leads to the identification of common datum and common location features. The object, as in all jig and fixture design, is to reduce loading time. It is particularly important in FMS, since gains in production times can easily be offset by loading costs. Inspection time and costs must also be reduced if the full benefits of FMS are to be achieved. Automatic inspection stations or probes set within the machine tools themselves must be employed. Actual dimensions are compared with standard dimensions by the computer, and adjustments are automatically made to the machine tools. Swarf will be produced in large quantities, so automatic swarf-handling equipment also has to be considered. All too often, high investment systems can be brought to a halt as tools become broken and movements clogged, simply because adequate swarf-removal facilities have been overlooked.

11.17 Flexible manufacturing systems (materials handling)

Materials handling is the element that builds the flexibility into the system. Its functions are to move the work between machines, buffer stores, inspection stations, etc., and also to present the work to the machines correctly orientated for cutting. The handling system must have the facility for the independent movement of workpieces between machines. That is, such workpieces must be able to flow from one station to another as the loading and routeing demands. There must also be buffer storage facilities for work awaiting machining.

At the machine tool, the secondary handling sytems must permit workpiece orientation and location for clamping with the workholding fixture ready for machining. For geometrically simple, prismatic parts-families, palletisation on conveyor systems or automatic guided vehicles (carts) is the most common primary transport system. Pusher bars, guideways or robots transfer the parts between primary and secondary sytems and load the workstations.

The development of industrial robotics has increased the flexibility and versatility of flexible manufacturing systems by enabling rotary parts to be included. The robots can transfer parts between systems and load them into machine fixtures under the control of the master computer. A single robot can be installed in the centre of the system and can load each machine with parts at random as the routeing requires (refer back to Fig. 6.21). It will, of course, also unload the parts when the process is complete. The machine tools must be laid out so that they are within the operating radius (reach) of the robot's arm. The *end effector* which is mounted on the 'wrist' of the robot must be able to hold all the components within the part-family of the cell.

11.18 Flexible manufacturing systems (people/system relationships)

While the system itself is automatic, people still have a role to play. This role is one of system management rather than machine operating. The supply of raw material to the FMS cell and the removal of the finished components is still a manual operation. In the long term, completely automated factories could become feasible but not necessarily desirable.

The major change is in the content of the work available. The need for trade skills are reduced, but more emphasis is placed upon office-based skills and upon design and production engineering skills. Sociologically the implications are profound, with a lessening in demand for skilled and semi-skilled operators, a shortening in the working week and the virtual elimination of the need to work 'unsocial' hours. Robots and computers do not have to see, so round-the-clock 'lights out' operation of the factory can achieve substantial savings in operating costs. The manning needs for such a system is for technicians and engineers of the highest skills in the fields of product and tool design using CAD/CAM techniques, tool setting and changing, and for multi-skilled maintenance engineers with expertise in electrical, electronic, mechanical, hydraulic and pneumatic systems. The computer system will make its own demand on programmers and system analysts.

11.19 Flexible manufacturing systems (benefits)

The introduction of FMS to a manufacturing company must be market driven. It must not be introduced just because the system is available. Properly introduced FMS can bring commercial benefits by providing:

- a faster response to market changes, demanding changes in design or new designs;
- improved product quality;
- reduced direct and indirect labour costs;
- better production planning;
- increased machine utilisation;
- minimum work in progress inventory since the time spent in the system is reduced.

It is not possible to buy a ready-made flexible manufacturing system; each installation has to be tailored to suit specific company needs. Therefore, once a company has decided to follow the FMS route, it has to look at every aspect of its production practices, both good and bad. It is in this analysis and during pre-planning that the greatest benefit to the company occurs. Even if the decision is made not to invest in FMS, the feasibilty study will have revealed flaws within the present methods of production which can be corrected and the company will benefit. Also, the level of management commitment, so essential in the implementation of FMS, will have been explored.

11.20 Materials requirement planning (stock control)

Most organisations 'buy in' materials in one of the following ways: in the raw state for processing into a finished product; in a part-finished state (e.g. castings or forgings) for processing into a finished product; as standard components (e.g. nuts and bolts) ready for assembly; or as a mixture of one or more of these foundation materials. The cost of foundation materials can form a large part of the final cost of the finished product. If the elapsed time between the delivery of the foundation materials and the sale of the finished product is small, then the cost of stockholding is low. However, if the elapsed time between the delivery of the foundation materials and the sale of the finished product is high, then the cost of stockholding is significant. Since there is no return on the working capital invested in foundation material, this can adversely affect the profitability of the company. At best it can affect the 'cash flow' of the company since funds tied up in the material represent 'dead money' which cannot be used for more profitable purposes. At worst, if bank borrowing was necessary for the purchase of the material, the interest charges can turn a paper profit into an actual loss.

The object of the department charged with the function of materials control is to maintain a supply of foundation materials at the lowest possible cost. A term that is frequently used in *stock control* is 'lead time'. This is the time taken from the initiation of an order to the delivery of the material into the company stores. If the foundation material is a stock item at the supplier, the lead time will be short. However, if the material has to be manufactured by the supplier, the lead time can be very significant. Thus the problem is that of raising orders at the correct time, both from outside suppliers and from internal production departments. *Material requirements planning* (MRP) has been in industry for a very long time but, with the advances made in recent years in computer technology and software, a computer-based system for material requirements planning — MRP 1 — has been developed. Before examining MRP 1 in detail, it is advisable to review the more common methods of organising stock control in a company.

The term 'stock control' describes a system of purchasing in which the level of stocks being held in store is used to regulate the raising of buying or production orders. Its function is to ensure that an adequate supply of foundation materials is available at the correct time and in the correct quantities to provide continuous production. Items can be bought in via the *purchasing office* or manufactured 'in-house'. For the reasons previously mentioned, stocks should be kept to the minimum necessary to maintain production. The amount of foundation material held in stock is governed by the following factors:

- operational needs,
- lead-time,
- cost of storage,
- deterioration of materials while in stock,
- the cash-flow position of the company — stock ties up working capital.

Stocks should only be held if:

- delivery cannot be exactly matched to supply;
- delivery is uncertain;
- substantial price reductions provide a cost advantage;
- bulk buying attracts discounts which offset the cost of storage;
- they provide a buffer of finished products providing a service to customers;
- in-house production is prone to operational risks (process break-downs).

11.21 Material requirements planning (stock control systems)

The basic feature of these systems is that the stock level initiates any new orders. When the stock level falls to some pre-determined level ('re-order level' or 'order point') then a new order is issued for a further supply. Figure 11.6 shows the basic features of a stock-control ordering system based upon variation of stock over a time period. *Note:*

- the *order point* (OP) is the level of stock at which a new order is issued;
- the *order quantity* (OQ) is the quantity ordered to restore stocks to the agreed level — also known as the 'batch quantity' or 'batch size';
- the *lead time* is the time required either to buy in or to manufacture, in-house, a replacement batch;
- if the batch quantity is known, then the lead time can be calculated;
- the order point can be determined by setting off the lead time and batch quantity.

The steps required to set up an elementary stock control system are as follows:

(1) Choose the order quantity (OQ).

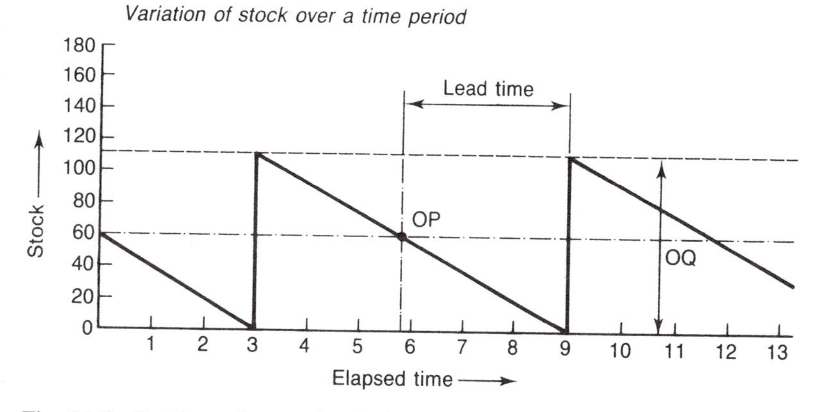

Fig. 11.6 Simple stock-control ordering system

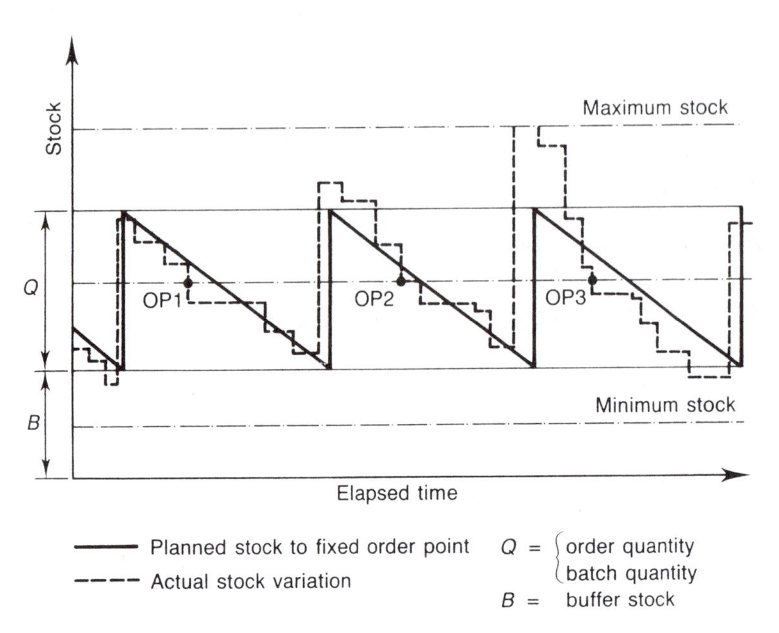

Fig. 11.7 Minimum and maximum stock levels

(2) Estimate the lead time to obtain a replacement batch of the required size.
(3) Calculate the number of parts which will last, at the normal consumption rate, during the lead time and thus find the order point (OP).
(4) Maintain a permanent record of the stocks of each item.
(5) Arrange for a new order to be issued each time that the stock level falls to the order point.

Buffer stock

The simple system shown in Fig. 11.6 is impracticable because of the imprecision of actual production completion times. If a delay in production occurs and the lead time becomes increased or there is an increase in demand, then stocks would run out, causing an interruption in production at some stage. An improved model is shown in Fig. 11.7. The buffer stock represents an insurance against the risk of the exhaustion of stock. Its value is obtained by balancing the cost of a shortage, causing interrupted production, against the working capital locked up in holding additional stock.

Stock levels

The *minimum stock* is a control level to reduce the risk of exhausting stocks, as shown in Fig. 11.7. When stocks reach the *minimum stock level*, the fact is reported so that an investigation can take place and corrective action taken. Buffer stock (B) and minimum stock need not be

Batch quantity:	100	Description:
Order point:	56	Bearing Block
Buffer stock:	40	Part No.: 3890 A
Maximum stock:	160	
Minimum stock:	30	

Date	*In*	*Out*	*Balance*	*Date*	*In*	*Out*	*Balance*
3.6.91	98	—	98	2.9.91	—	12	36
10.6.91	—	12	86	9.9.91	—	2	34
17.6.91	—	16	70	16.9.91	104	—	138
24.6.91	—	14	56	23.9.91	—	12	126
1.7.91	—	14	42	30.9.91			
8.7.91	—	4	38				
15.7.91	—	14	24				
22.7.91	102	—	126				
29.7.91	—	16	110				
5.8.91	—	16	94				
12.8.91	—	14	80				
19.8.91	—	16	64				
26.8.91	—	16	48				

Significant Dates
24.6.91 — Re-order.
15.7.91 — Minimum breached; investigate.
26.8.91 — Re-order.

Fig. 11.8 Stock-control record card

the same amount because the former is a fixed amount of stock maintained as an insurance against high demand or delays in delivery, while the latter is a control level.

The *maximum stock* is also a control level and indicates when stock levels are unacceptably high in economic terms (see Fig. 11.7). When this point is reached, production must be reduced or even ceased until stocks have reached a more manageable level. The maximum stock level is set so that it is in excess of buffer stock by an amount sufficient to allow for small variations in the production plan. Only when something is wrong in the system are the maximum and minimum stock levels normally reached. A typical stock-control record card is shown in Fig. 11.8.

Batch quantity

Although it was thought possible to calculate the *economic batch quantity* (EBQ) using mathematical formulae, EBQ is more commonly selected for each case on merit and previous experience. That is,

$$\text{Batch quantity} = (\text{total requirement per year})/(\text{batch frequency})$$

This method avoids batch quantities being set to suit production convenience without regard to stock costing.

11.22 Materials requirement planning (demand patterns)

Consider an everyday item such as a bicycle. Figure 11.9 shows the relationship between the component parts of the bicycle, with the completed assembly at the highest level which is classified as level 0. The items making up the constituent parts are classified as the lower levels 1, 2, 3 and 4. Therefore, if the items classified at these lower levels are not being sold as spare parts, they can be classified as *dependent* items, i.e. the demand is solely dependent upon the number of complete bicycles assembled.

The completed bicycle is an end in itself and it is not used in, or as part of, a higher level item. It is in fact the finished product which the customer purchases from the local cycle shop. The volume of production

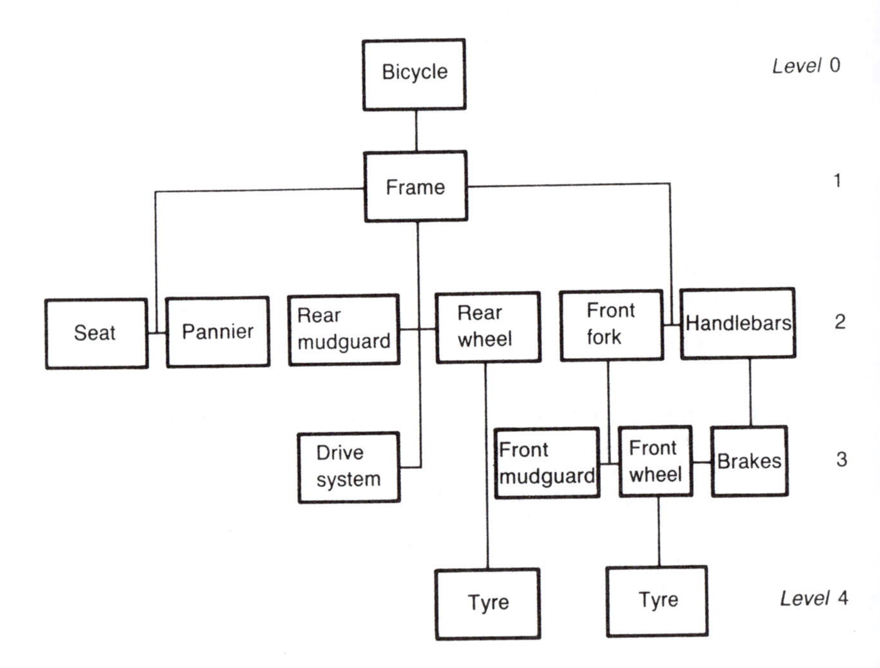

Fig. 11.9 Material requirement planning

is based upon sales forecasts for that particular model and is *independent* of the demand for any constituent parts. Thus the demand for a company's products can be classified into two distinct categories:

- *Dependent demand* is demand generated from some higher level item. It is the demand for a component that is dependent upon the number of assemblies being manufactured. As such, it can be accurately calculated and inserted into the master schedule.
- *Independent demand* is the demand or market for the finished product. The magnitude of the independent demand is determined from firm customer orders or by market forecasting. For MRP purposes it is independent of other items. It is not used in scheduling any further assemblies.

A stock-control system could keep all the parts shown in Fig. 11.9 in stock using either order point methods or sales forecasts for each item. To do this each separate item has to be treated as independent. MRP 1, however, can calculate the amount of stock of dependent items required to meet the projected demand for the independent higher level item which, in this example, is a bicycle.

It is claimed that MRP 1 is superior to the standard methods of stock control previously discussed in this chapter. This is because, while the demand for the independent items (level 0) cannot be accurately predicted — and are therefore unsuitable for stock control methods — lower level items (1, 2, 3, 4) are dependent on a higher level item and can thus be calculated with accuracy. Therefore, although it is only possible to forecast next year's sales of complete bicycles approximately, it is possible to predict that exactly as many tyres will be required as there are wheels.

Further, stock control methods assume that usage is at a gradual and continuous rate. In practice, however, this is not the case. The call on parts will occur intermittently, with the quantity depending upon the batch size of the final product. This often results in the holding of excessive stocks. Material requirement planning assists in the planning of orders so that subsequent deliveries arrive just prior to manufacturing requirements, as shown in Fig. 11.10.

11.23 Master production schedule

The master production schedule is the schedule or list of the final products (independent or higher level items) that have to be produced. It shows these products against a time or period base so that the delivery requirements can be assessed. It is, in fact, the definitive production plan which has been derived from firm orders and market forecasts. In order to minimise the number of components specified, care has to be exercised in deciding how many of the basic level 0 assemblies are to be completed. The total lengths of the time periods (the planning horizon) should be sufficient to meet the longest lead times.

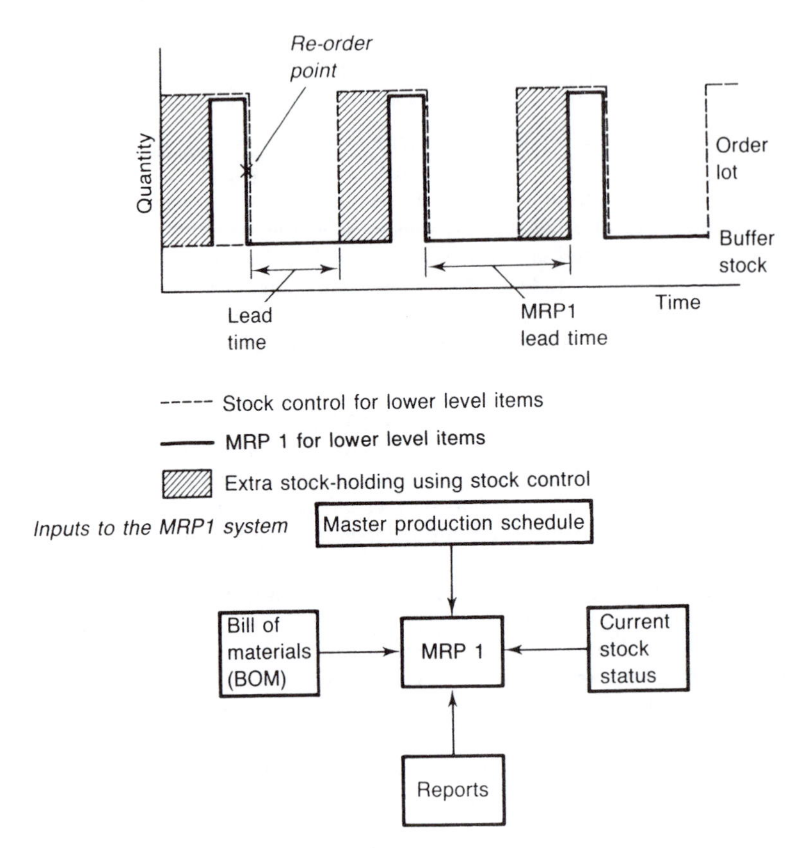

Fig. 11.10 Comparison: stock control versus MRP 1

11.24 Bill of materials

The bill of materials (BOM) contains all the information regarding the build-up or structure of the final product. This will originate from the design office and will contain such things as the assembly of components and their quantities, together with the order of assembly. Figure 11.11 shows how the BOM is structured and presented. The number of parts should also be shown at each level.

Current stock status

This is the inventory or latest position of the stock in hand, pending deliveries, planned orders, lead time, and order batch quantity policy. All components, whether in stores, off-site, or in progress need to be accurately recorded in the inventory record file.

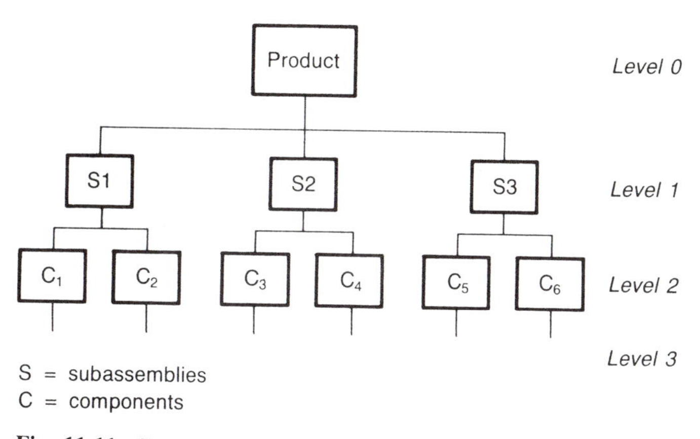

Fig. 11.11 Structure for bill of materials

11.25 MRP 1 in action

Upon receiving the input data, the system has to compute the size and timing of the orders for the components required to complete the designated number of the final product (the independent or highest level item). It performs this operation by completing a level-by-level calculation of the requirements for each component. Since the latest stock information is available to the system, these calculations will be the net value of the orders to be raised.

Requirements = net requirements + stock currently available
Stock currently available = current stockholding + expected deliveries

Inherent in the calculations is the factor for lead time, i.e. ordering and manufacturing times. The system must determine the start date for the various subassemblies required by taking into account the lead times, i.e. it offsets these lead times to arrive at the start dates. A complication that the system has to overcome is the use of some items which may be common to several subassemblies. The requirements of these common use items must then be collected by the system to provide one single total for each of these items. Figure 11.12 shows a standard planning sheet for one of the outputs of MRP 1. Other reports required in addition to the order release notices are any:

- re-scheduling notices indicating changes;
- cancellation notices where changes to the master schedule have been made;
- future planned order release dates;
- performance indicators of such things as costs, stock usage, comparison of lead times.

	1	2	3	4	5	6	7	8
Gross requirements				140				
Scheduled receipts			50					
On-hand	100		150					
Net requirements				100				
Planned order releases				100				

Fig. 11.12 Standard planning sheet for one of the outputs of MRP 1

A number of advantages are claimed for MRP 1 software. These include:

- a reduction in stocks held in such things as raw materials, work in progress, and 'bought in' components;
- better service to the customer (delivery dates are met);
- reductions in costs and improved cash flow;
- better productivity and responses to changes in demand.

11.26 Manufacturing Resource Planning (MRP 2)

MRP 1 is a large step forward, compared with manual systems, in the procurement of materials and parts in as much as it provides a statement of exactly what is required and when. Unfortunately, it takes no account of the production capacity of a particular manufacturing facility. Most factories have a production output that can only be changed over a period of time either by improved productivity or the injection of capital to purchase more productive equipment.

Clearly, by themselves, the outputs from MRP 1 have limited value to the master production schedule except to detail the material requirements. Without the information regarding production capacity, it is difficult to match MRP 1 to actual production schedules. Remember that *capacity planning* is concerned with matching the production requirements to available resources (personnel and plant). It would be impractical to have a master production schedule that exceeded the capacity of the plant to meet its requirements. This could lead to decisions to extend the plant capacity by increasing the labour force, shift working, or sub-contracting, when what is required is more sensible scheduling.

The shortcomings of MRP 1 led to the development of MRP 2, which brought into account the whole of the company's resources, including finance. Thus the initials MRP also stand for Manufacturing Resource Planning. This is quite a different and more difficult philosophy when one considers the many variables that can affect shop-floor production. For example, machines may breakdown, staff may become ill or leave, and rejects may have to be reworked. In addition, MRP 2 takes into account the financial side of the company's business by incorporating the *business plan*. MRP 2 is used to produce reports detailing materials planning together with the detailed capacity plans. This enables control to be effected at both the shop-floor and the procurement levels. Note that MRP 2 does not replace MRP 1 but is complementary to it and *the two systems are used together*.

As in all computer systems, the information produced is only as good as the information fed in, hence it is dependent upon the feedback of information relating to those things under its control, i.e. such data as the state of the manufacturing process and the position of orders. MRP 2 contains enough information, and the power to organise it, to be able to run the whole factory. Figure 11.13 shows a schematic layout of production incorporating MRP 1 and MRP 2 organised around the master production schedule.

MRP 2

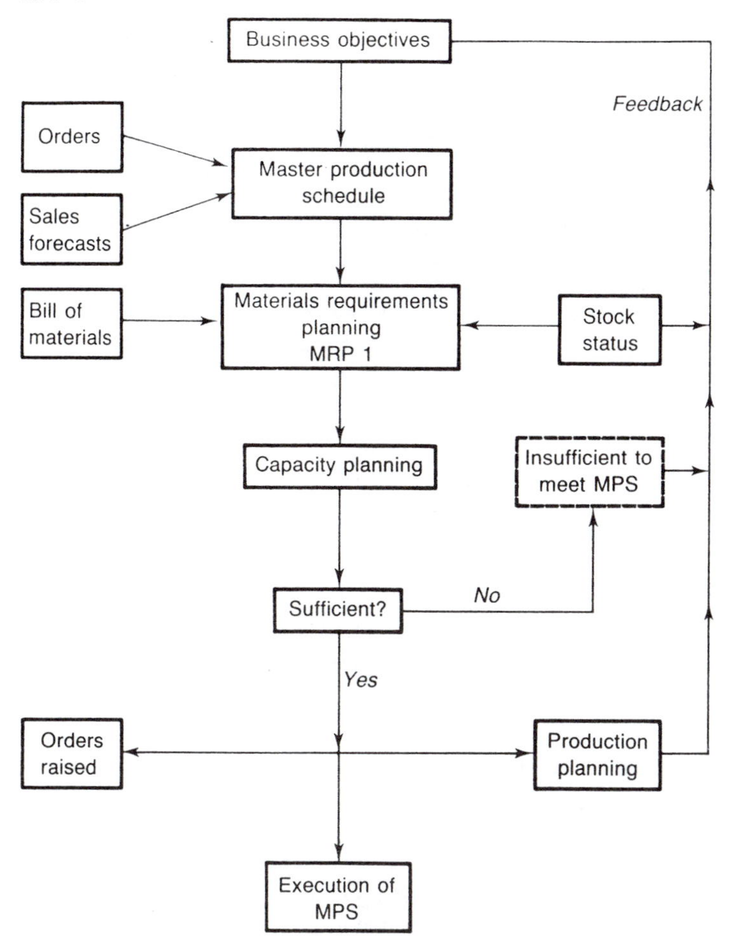

Fig. 11.13 Schematic layout of MRP 1 and MRP 2 organised around the master production schudule

MRP 2 is a management technique for highlighting a company's objectives and breaking them down into detailed areas of responsibility for their implementation. It involves the whole plant and not just the materials provision. The key to MRP 2 lies in the master production schedule (also called the *mission statement*) which is a statement of what the company is planning to manufacture and, because it is the master, then all other schedules should be derived from it. The master schedule should *not* be a statement of what the company would like to produce, as its own derivation lies in the company objectives. It cannot in itself reduce the company's lead times, but it will provide the incentive for the plant personnel to take action in that area if required. It is important, as in all planning, to monitor adherence to the master production schedule. MRP 2 will not stop over-ambitious planning but it will show up the consequences.

11.27 'Just-in-time' (JIT)

The phrase 'just-in-time' (JIT) originated in Japan, with the production plant of Toyota as the most frequently quoted application. This led to the incorrect conclusion that since Toyota is involved with mass (repetitive) production of cars, then JIT only applies to this mode of manufacture. In fact JIT applies to any mode of manufacture. In the USA the term 'zero inventories' is used but the philosophy is exactly the same as JIT. Perhaps the most apt definition of JIT belongs to R. Schonberger (*Japanese Manufacturing Techniques*, Free Press, New York, 1982):

> Produce and deliver finished goods *Just-in Time* to be sold, sub-assemblies just-in-time to be assembled into finished goods, fabricated parts just-in-time to go into sub-assemblies, and purchased materials just-in-time to be transformed into fabricated parts.

A second approach is propounded by D. Potts (*Engineering Computers*, September 1986):

> A philosophy directed towards the elimination of waste, where waste is anything which adds to the cost but not the value to the product.

A third approach which appears to combine the other two is put forward by C. Voss (*Just-in-Time Manufacture*, IFS, London, 1987):

> An approach that ensures that the right quantities are purchased and made at the right time and in the right quantity and there is no waste.

So there we have it; JIT philosophy is about minimum stocks being available at the point of manufacture, reducing costs by eliminating those things which add nothing to the value of the product (e.g. avoiding interest charges on working capital tied up in servicing unnecessarily large stocks), while maintaining the availability and quality of the product. However, any manufacturing system should have the foregoing objectives, so it is in the application that JIT gets nearer to the

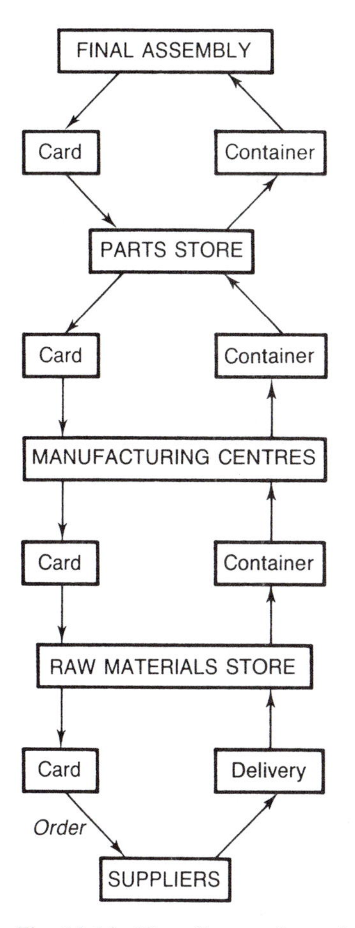

Fig. 11.14 Flow diagram for a simplified kanban system

manufacturing ideal than many traditional systems.

Toyota uses a system known as *kanban* (kanban = 'card' in Japan) so, in this form, their JIT system is a card system. In the kanban system, assembly schedules are derived from a master plan which is drawn up to meet a specific 'planning horizon'. The 'planning horizon' is the time period that has previously been agreed as realistic and trustworthy, for example, six weeks may be appropriate. Using the master plan and the time scale, the daily production requirements can be set. Referring to the dependency theory of MRP 1 and the bicycle example considered earlier, then a card (kanban) would be issued to the store holding the parts for the final assembly and a container of parts despatched to the assembly point. A card would then be issued to the manufacturing centres for replacements to be made for the final parts store. The manufacturing centres would, in turn, issue a card for any replacements which they require. Cards would be passed down through the system to raw materials purchase and sub-contractors. Thus, ideally, at each level the replacements arrive 'just-in-time' to maintain production and no stocks are held.

The issue of a card from the master plan or schedule 'pulls' the work through the system, since no production can take place without its authorisation. Figure 11.14 illustrates this point. In such a 'pull-through' system in which everything is dependent upon the arrival of an authorisation card, there must be times when no production takes place. This is because of the overriding principle that nothing is produced until it is required except, of course, for the minimum (small) inventories kept in the appropriate stores.

The impression might be given that the labour force sits around doing nothing until a card arrives. This is not the case. The major advantage gained from any inactivity is that the reason for that stoppage can be immediately investigated. Everyone — management and production workers — can be engaged upon tackling the problem. In the meantime, those members of the production team not engaged directly in solving the problem can be engaged in quality circle work, training or in maintenance activities.

The Toyota system is just one example of the philosophy of JIT because, at the end of the day, that is what JIT is — simply a philosophy. It covers a range of production control techniques which have the objectives of eliminating every facet of the business which fails to make a contribution. It focuses minds on achieving production which is not only 'just-in-time' but which produces 'just enough' with no surplus and costly excess stocks. It streamlines the production process by removing the cushion of safety stocks, excess stocks, and/or inflated lead times which can cover up many real problems in a company. It concentrates the minds of the entire workforce on the business of the company. By making every action important, there has to be involvement and commitment from everyone for the successful implementation of JIT. There has to be a team approach to the problem-solving inherent in JIT.

Without a JIT philosophy, 'waste' can be overlooked in a company because it is not recognised and because of a 'we have always done it this way' mentality. When waste is considered to be any activity failing to contribute by adding value to the product, it can cover such things as the production of scrap and reworking, excessive transporting by poor routeing and poor plant layout, overproduction to compensate for scrap, machine breakdowns and absenteeism. All these things can all too easily become part of the production scene by the acceptance of their inevitability. The analysis of management and working practices necessary when introducing the JIT philosophy immediately shows up these faults in the system.

It has to be borne in mind that each element in a company is dependent on all other elements and therefore JIT can become a very vulnerable system. The employees and suppliers are all part of the system so that any conflict will halt production very quickly. In a large organisation, for example, an industrial dispute in one link in the supply chain can quickly bring the rest of the chain to a halt.

11.28 Just-in-time and MRP 2

The kanban (card) system just described is only one way of applying the JIT philosophy. As the production becomes more varied and involved, then the amount of data required to run the system increases significantly and computer support is required. To this end MRP 2, which was introduced in section 11.26, is a computer-based system that has the objective of unifying all the associated functions in a company from initial planning through to delivery of the finished products. It is applicable to any form of company, irrespective of size and diversity of production. It breaks down the company's Business Plan into detailed operational tasks for each part of the organisation. The effective planning element removes any confusion which could exist by concentrating the minds of the workforce — management and productive labour — on what has to be achieved.

Since JIT has the major objective of eliminating waste in all its forms and, as confusion leads to wasteful excesses, then the adoption of MRP 2 promotes the concept of JIT. The key to successful implementation is correct planning and scheduling. Both have the main aim of producing only what is required and when it is required, and are therefore compatible. Thus, the benefits of JIT can be summarised as:

- better quality by closer control;
- reduction of lead-times;
- reduction of stocks and work in progress;
- more involvement of the workforce leads to greater job satisfaction;
- increased efficiency by the process of continually challenging the existence of everything connected with the production in a never-ending search for greater efficiency.

Assignments

1. Produce a critique of a factory layout with which you are familiar.
   (a) Discuss the criteria upon which the critique is based.
   (b) List the necessary information which would be required before planning the layout of a new workshop.
2. The design function of a product lies between the marketing and operation functions. Discuss the influence of the marketing and operation functions on the design of the product.
3. (a) List and explain SIX control arrangements which make up a manufacturing control system.
   (b) How are they implemented in your own place of work?

4. With reference to component grouping, define, in your own words:
   (a) (i) classification;
       (ii) coding.
   (b) For any group of components with which you are familiar, design a simple classification and coding system.
5. Explain the difference between MRP 1 and MRP 2, and discuss the circumstances in which a company would wish to develop from MRP 1 to MRP 2.
6. Prepare a report for senior management outlining the philosophy of JIT and its possible implications for the production process at your own plant.

12 Quality in manufacture

12.1 Quality in manufacture

The word 'quality' is bandied about quite loosely these days, especially in publicity material. As applied to manufacturing it is defined in BS 4778 (1987) as '*the totality of all features and characteristics of a product or service that bear upon its ability to satisfy stated or implied needs*'. Put more simply, it can be defined as 'fitness for purpose'. The survival of manufacturing companies in today's international markets depends upon the achievement of acceptable levels of quality at minimum cost to the customer. There is no such thing as absolute quality. A customer's idea of quality will change with time and with competition in choice. Therefore a company's attitude to quality must be constantly reviewed in the light of these changes in customer perceptions so that, at all times, the company's products represent 'fitness for purpose', 'value for money' and 'state-of-the-art technology'.

The price of any commodity is negotiable: quality is definitely not negotiable. This is widely appreciated in industry because most manufacturers are, themselves, customers in one way or another. Any apparent savings made by the purchase of inferior materials or components are quickly dissipated in loss of customer confidence. The quality of goods or services goes hand in hand with reputation. This goodwill is a company's greatest asset. It does not just happen; it has to be part of a company's philosophy, i.e. it has to be managed so that quality becomes the responsibility of every member of a company. This is the concept of *total quality control* (TQC) and its implementation is *total quality management* (TQM). The concept of TQC has replaced the former system where quality was the sole preserve of the quality control

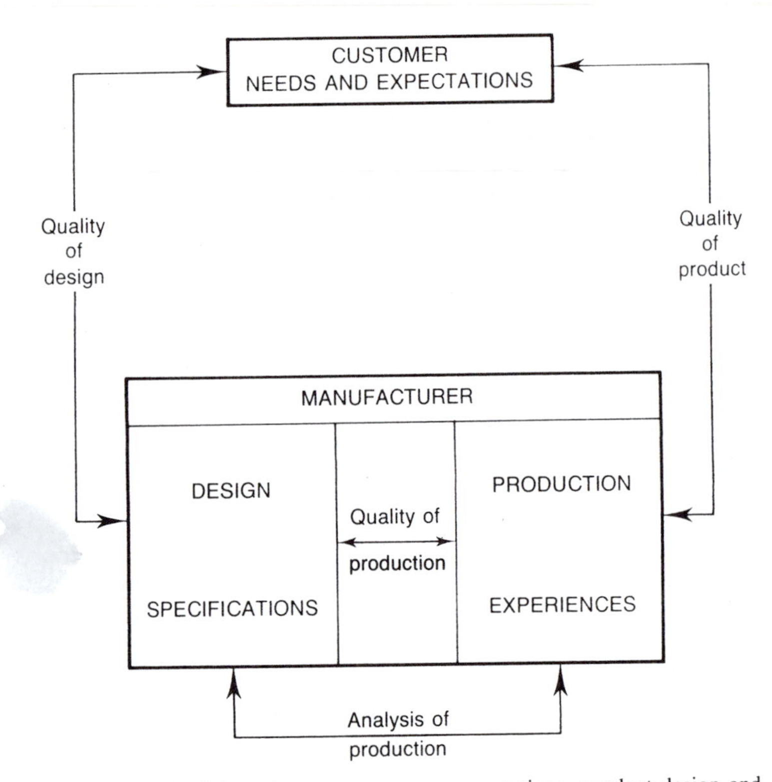

Fig. 12.1 The linkage between customer expectations, product design and manufacture

department. Such departments still have an important role to play, but that role has changed with the advent of TQC and TQM.

Figure 12.1 shows the importance of the customer in manufacture. The concept of supplier and customer exists within an organisation to provide a *quality chain* between the external supplier to the company and the delivery of goods to the customer. Each department is the customer of the departments supplying it with goods and services and, in turn, each department becomes the supplier of those departments making use of its goods and services. Similarly, within each department, there is a supplier/customer relationship between the individual members of the personnel of the department. The typist is a supplier of documentation to the manager and the manager, as a customer, has the right to expect a quality, error-free typing service.

Throughout and beyond all organisations, be they manufacturing concerns, banks, retail stores, universities or hotels, there exists a series of quality chains as shown in Fig. 12.2. This may be broken by one person or one piece of equipment not meeting the requirements of the customer, internal or external.

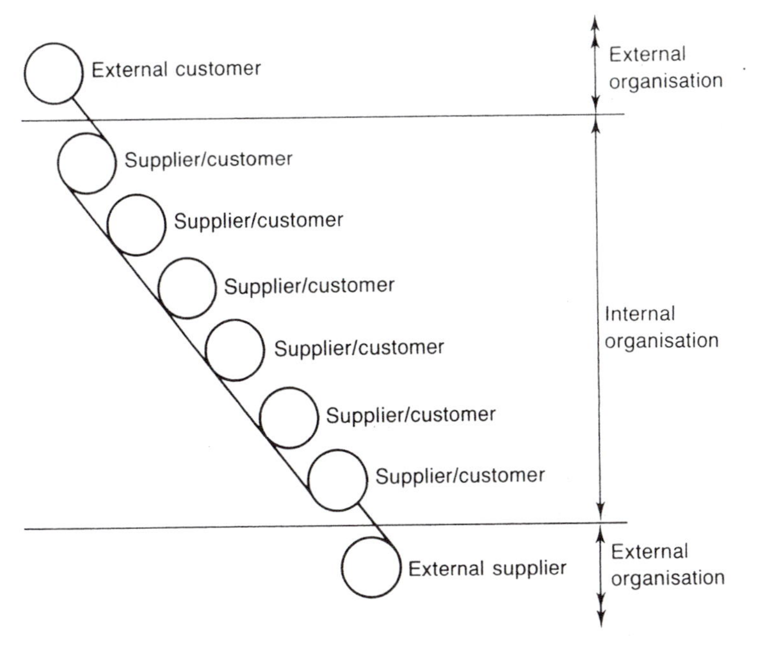

(courtesy of D.T.I. *Total Quality Management — A Practical Approach*)

Fig. 12.2 Quality chain

12.2 Reliability

This is the ability to provide 'fitness for purpose' over a period of time. Obviously any component will fail at some stage of its life, although the time span may be such as to give the impression of being virtually everlasting. Failure is considered as being the point at which a product or service no longer meets its fitness for purpose. Increasingly, reliability is a major factor when a customer is making a purchasing decision.

Therefore there is an increasing need to design reliability into a product from the start. A difficulty exists in the testing of a new design for its reliability due to the time factor involved as well as the widely varying environments and conditions under which the product has to operate during its working life. There are many tests in existence which are specifically designed with reliability in mind. For example, in the aircraft industry, there are rigs set up to test the fatigue resistance of wing assemblies. Hydraulically-actuated pistons are positioned along the underside of a wing and are programmed to move up and down to simulate the forces acting on the wing during flight conditions. By this method, many thousands of flying hours can be compressed into days as the assemblies are safely tested to destruction. Again, rolling roads and specially-designed test tracks are widely used in the automobile industry.

In addition, prototype vehicles are test driven under varying climatic conditions over very large distances throughout the world to test their reliability before entering large-scale production and being sold to the motoring public.

From the outset, a designer can take some practical steps to prevent premature failure. For example:

- the use only of components with a proven reliability in assemblies. This can prove to be difficult when a designer is being innovative and is introducing new materials and manufacturing methods;
- the use of uncomplicated designs incorporating as many previously-proven features as possible;
- the use, within the cost parameters, of back-up sytems and duplicated components where total failure cannot be tolerated, as in the controls of automobiles or aircraft. Duplication decreases the probability of total failure by the square of the chance of a single component failing. Triplication of the number of components in parallel decreases the probability of failure by the cube of the chance of a single component failing. Wherever possible, components should be designed to be *fail-safe*. That is, failure of the component shuts down the system of which it is a part or a back-up system is automatically brought into operation. The pilot light on a modern gas appliance must be so designed that, should it fail, the gas supply to the appliance is automatically cut off;
- only tried and tested methods of manufacture should be employed until new methods have been proved to be equally reliable or better. This may sound dull to a thrusting young designer, but the cost of failure can be prohibitive both financially and in the loss of human life.

12.3 Specification and conformance

Accepting that the customers' needs are paramount when considering the definition of quality as 'fitness for purpose', then it becomes necessary to establish at the outset just exactly what those needs are. Figure 12.1 shows how the customer has to be involved at every stage from design through to delivery. In the production of goods or the delivery of services, there are always two separate but interconnected factors relating to quality:

- quality in design;
- quality of the conformance with the design.

12.4 Quality in design

The first stage in the development of a product or a service is 'customer needs'. This is the province of the marketing department working in conjunction with the design engineers. This first stage is the most difficult and yet the most crucial as it sets the scene for the whole process of manufacture in the case of goods, and delivery in the case of services. The term '*quality of design*' is a measure of how the design meets with customer requirements, i.e. the stated purpose.

The translation of customer needs, as set out in the initial design, into manufactured goods or delivered services is achieved by the drawing up of specifications. These specifications can best be described as detailed statements of the requirements with which the goods or services must conform. Correct specifications control the purchase of goods or services as well as directing the efforts of the manufacturing processes. Specifications can be expressed in terms of such things as mechanical properties of materials, tolerances on machine parts, and surface finish requirements, all of which are easily recognisable in engineering terms. Specifications may also refer to more abstract concepts such as the *aesthetics* of design. Further, specifications can also apply to the internal supplier/customer relationship. For example, a designer may request the company legal department to draw up a specification for a product such that it does not infringe any patent or copyright and that it does not infringe any current legislation such as that concerned with product liability.

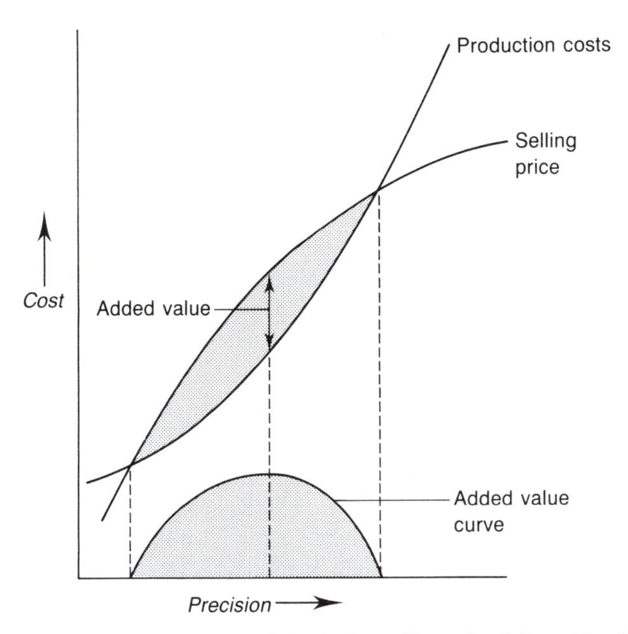

Increased precision = an initial rise in the selling price followed by decreasing returns.

Fig. 12.3 The influence of added value on design specifications

The translation of customer needs (the demand specification) into a design specification and then into a feasible manufacturing specification is very time consuming. This is because each demand need has to be broken down into sub-needs which are then issued either as purchasing orders or manufacturing requirements. The design specification must contain information concerning the minimum functional, reliability and safety requirements together with the information required to achieve minimum cost factors. Thus the concept of minimum functional requirements is closely related to the achievement of minimum cost factors. Any designer must realise that demanding higher specifications, such as greater precision, leads to an increase in manufacturing costs. Figure 12.3 shows the influence of the design specifications on the value added to a product. Increased precision often results in increased production costs, leading to an initial rise in the selling price followed by decreasing sales and returns as indicated by the added value curve. To demonstrate that the 'quality of design' meets customer demand, the design engineers can test materials and components, then build and test prototypes. Customer involvement at this stage is essential in order to secure their acceptance before quantity production commences.

12.5 Quality of conformance to design

The ability to manufacture goods or deliver a service to agreed specifications at the point of acceptance is referred to as the *quality of conformance*. Many companies now employ conformance engineers whose function is to ensure that goods and services match the design specifications and hence the demand specifications.

Quality cannot be inspected into goods or services. The traditional concept of quality control consisted of inspecting samples of components to ensure that they had been made 'according to the drawings' and rejecting any components that were out of tolerance. Unfortunately this concept that quality is solely the province of the inspection department still persists in many companies.

If the philosophy of customer satisfaction and customer demand led design has been adopted from the start, this philosophy must be followed throughout the whole system from the basic idea to the finished product or service. This is achieved by regular conformance checks during production to ensure that the specifications are being achieved. Thus conformance testing is about checking the quality of the whole process and not just about inspection of the finished product. This avoids the waste of making rejects and the costs of reclamation and rectification. A high level of end inspection indicates that a company is trying to 'inspect-in' quality. The retrieval and analysis of production data plays a very significant part in achieving specified levels of quality, so that when the product or service reaches the customer and *fully satisfies* all the specified requirements, then *quality of conformance* will have been achieved. Figure 12.4 shows the relationships between inputs and outcomes.

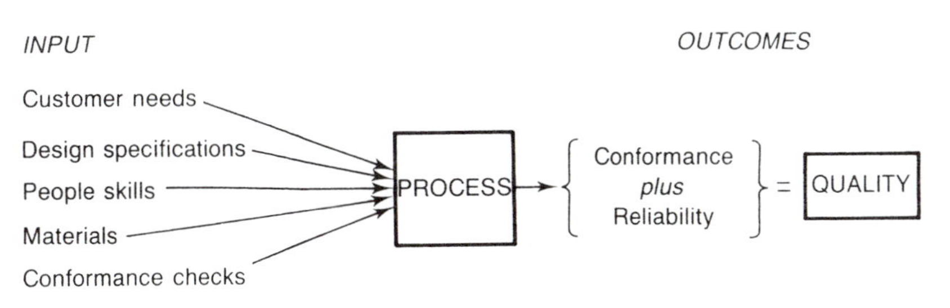

Fig. 12.4 Quality: inputs and outcomes

12.6 Accreditation of conformance

In today's demanding and competitive markets, good product design and efficient manufacturing must be underpinned by properly authenticated measurement and testing. For a company to have absolute confidence when issuing a certificate of testing, its own inspection instruments must, themselves, have been certificated by an independent accreditation agency. One such agency is the *National Measurement Accreditation Service* (NAMAS).

NAMAS is a service of the National Physical Laboratory (NPL), one of the Research Establishments of the Department of Trade and Industry. NAMAS assesses, accredits and monitors calibration and testing laboratories. Subject to its stringent requirements, these laboratories are then authorised to issue formal certificates and reports for specific types of measurements and tests. NAMAS accreditation is voluntary and open to any UK laboratory performing objective calibrations or tests. This includes independent commercial calibration laboratories and test houses, and also laboratories which form part of larger organisations such as a manufacturing company, educational establishment or a government department. It includes cases where the laboratory concerned provides a service solely for its parent body.

Every industrialised country requires a sound metrological infra-structure (metrology is the science of fine measurement) so that government, manufacturers, commerce, health and safety and other sectors can have access to a wide range of measurement, calibration and testing services in which they can have complete confidence. Figure 12.5 shows the structure of the National Measurement System which is in place in the UK. It shows that the National Physical Laboratory (NPL) is at the focus of the system and maintains the national primary standards both for the SI base units (mass, length and time) and the SI derived units such as force and electrical potential. Commerce, industry and other users do not have direct access to the primary standards, but they have access to secondary standards which have, themselves, been calibrated against the primary standards. NAMAS provides an essential part of this hierarchical structure through accredited laboratories and test houses which give their customers access to authenticated measurements,

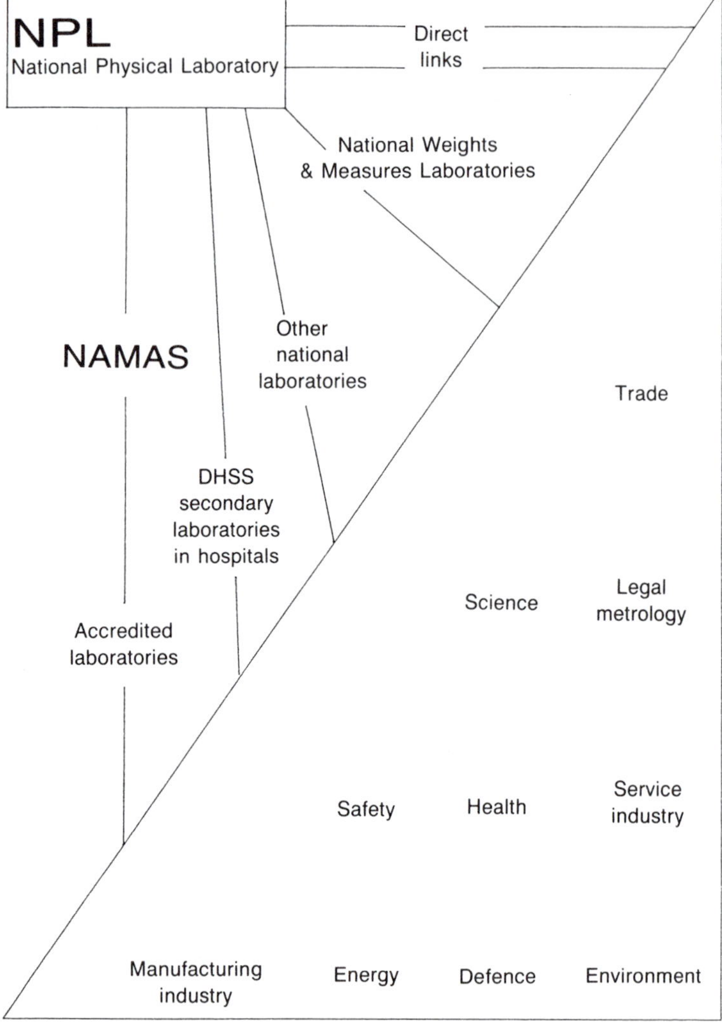

Fig. 12.5 National Measurement System

calibrations and tests of all kinds. These have formal and certificated traceability to the national primary standards at the NPL. Confidence in the accuracy of measurement and test data is a vital part of ensuring product quality. The manufacturing industry is constantly being encouraged to improve in every aspect of its quality procedures, and NAMAS plays its part by emphasising the importance of trustworthy measurements and tests.

On an international scale, laboratory, test centres and inspection departments accredited by a body which is trustworthy and whose competence is beyond question, makes the products and services of a country more readily acceptable overseas. This is particularly important with the advent of the Single European Market. For this reason, NAMAS cooperates fully with other international agencies concerned with laboratory accreditation. Mutual agreements recognising each other's competence are negotiated where appropriate.

12.7 Total quality management (TQM)

Quality is not only about 'good design' and 'conformance to design specifications'. Quality should permeate every aspect of a company's business, always focusing upon the needs and hence the care of the customer. The traditional approach to quality has been that 'production makes it' and 'quality control inpects it'. This approach leads to financial loss being incurred every time defective products are scrapped or time is involved in reworking and reclaiming defective products. This also leads to late delivery and dissatisfied customers who have to be pacified, and who have the same problems with their own customers. It has been estimated that as much as two-thirds of all the effort utilised in business is wasted by this domino effect. Thus the objectives of any company can be defined as:

- optimum quality;
- minimum cost;
- shortest delivery time accurately adhered to.

To achieve and, more importantly, maintain these objectives will require something better than the traditional approach to quality. It will require more than an attempt to 'inspect-in' quality. It requires the efforts of every resource; the total commitment of every member of the company to satisfying the customers' requirements at all times. The slogan 'Quality is everyone's business' just remains a slogan if not backed by positive action throughout the whole of a company's activities. It can soon change to 'Quality is nobody's business' without managerial determination from the Board of Directors downwards. From this approach of total involvement has developed the concept of *total quality management* (TQM). Total quality management can be defined as:

> an effective system of co-ordinating the development of quality, its maintenance and continuous improvement, with the overall objective of the achievement of customer satisfaction, at the most economical cost to the producer and, hence, at the most competitive price.

In theory, since quality affects everyone in the factory, it can be truly stated that 'Quality is everyone's business'.

12.8 Background to TQM

The philosophy of TQM was introduced to Japan by Dr W. E. Deming of the USA shortly after the Second World War. He found a receptive audience for ideas that were based upon the principle of continuous improvement of quality, i.e. quality as a dynamic and not a static process. Western industrialists — particularly in the car industry — began to become aware of the high levels of quality being achieved in Japan and realised that they had to set their own houses in order if they were to survive.

The Deming philosophy of continuous improvement was set out in a 14-point plan. The first point states:

> Create constancy of purpose towards the improvement of products and service, with the aims of becoming competitive, staying in business and providing jobs.

This first point is addressed to management, urging them to create an environment in which constant review and improvement can flourish. Quality is not a definitive programme with a finite end, but a continuous process. Philip B. Crosby, also of the USA, finishes his 14-point plan for total quality management with the advice: 'Do it over again'. Thus the clear message is that quality is as much about organisation as it is about production. Total quality management must involve the whole of the organisation starting at the very top with the Senior Directors and the Chief Excecutive.

Total quality management includes not only the entire personnel within an organisation but the suppliers as well for, at the heart of TQM, lies the internal and external customer–supplier relationships introduced originally in Fig. 12.2. By the introduction of TQM, everyone becomes accountable for their own contribution to quality: everyone is motivated to attain the highest quality by being alert to that which is taking place. This cannot be achieved by coercion; TQM has to be 'user-driven' and the challenge to top management is to obtain a change in the culture of the company towards quality without issuing edicts. They must create an environment of job satisfaction where most people want to perform well and be able to take a pride both in their own achievements, and in those of the company, in addition to feeling a sense of involvement.

Ideas for improvement must come from people with the know-how and experience at all levels of the organisation. Who knows better how to make a product than the person who actually makes it? The objective is to move the pride of workmanship possessed by a single craftsman into the arena where products or services are the result of the contributions of many people of differing skills and abilities. Thus TQM is about changing attitudes and utilising everyone's knowledge and experience in a never-ending quest for the three objectives of highest quality, minimum cost and shortest delivery time. The culture of the company becomes that of identifying failures as early as possible and striving to get things right the first time. Thus TQM is much more sensible than the traditional approach of trying to 'inspect-in' quality by coercion and the efforts of a small band of inspectors; an approach where everyone else's opt-out was 'It's not my job to detect faults'.

The advantages of TQM can be summarised as follows. It enables a company to:

- identify clearly the needs of their market and hence satisfy the needs of their customers;
- devise and operate procedures for achieving a quality performance resulting from a clearly defined quality policy;
- develop good lines of communication throughout the whole of its operation, both internally and externally;
- critically examine every process that contributes to the final product or service, in order to remove all those activities that are wasteful;
- measure its own performance and be in a position to continually update and improve its performance;
- be competitive by understanding all the competition, both at home and abroad;
- provide high quality goods and services which represent maximum 'value for money' by providing increasing job satisfaction for the workforce through a commitment to the philosophy of teamwork which, in turn, harnesses all the talents of all the people in the organisation;
- understand *itself* as never before.

12.9 Quality assurance: BS 5750 (ISO 9000: 1987)

Although, under TQM, everyone from the Boardroom down is responsible for the achievement of quality standards, the 'buck stops' at the desk of the company officer who signs the quality assurance certificate. Quality assurance (QA) is defined in BS 4778: 1987 as: 'all activities and functions concerned with the attainment of quality'. In practice, it is some formal system in which all the operations are set out in written form as procedures and work instructions. The documents are subject to regular review and up-date and are usually under the authority of one person, the quality manager.

Quality assurance is not a new concept. For many years in certain areas of industry customers have demanded certification of quality from suppliers. These certificates of quality were used to verify that products or services met the specifications demanded by the customer. Some examples are:

- material specifications in which the results of various prescribed tests are recorded on the acceptance certificate which bears an authorised signature;
- lifting equipment including slings and chains;
- pressure vessels such as boilers, compressed air receivers and compressed gas bottles.

What is new, is that products and services not usually associated with certification have been brought into line by having written evidence that the company has achieved Nationally Accredited Standards of quality.

These standards are such that they are also accepted internationally by keeping the standards of accreditation common amongst all the industrial communities.

In the UK the British Standard for Quality Assurance is BS 5750 (ISO 9000: 1987). The first standard for quality assurance was published in the UK as BS 5750: 1979. However. the interlinking of world trade both within and outside of the European Community led to other countries copying the British approach but with their own particular amendments. This international interest in quality standards led to the introduction of ISO 9000: 1987. This combines international expertise with eight years of British experience in operating quality assurance. It has been adopted without alteration by the British Standards Institution, hence the dual numbering used. The new ISO standard also incorporates the European Standard EN29000.

12.10 The basis for BS 5750/ISO 9000

The definition of quality upon which the Standard is based is in the sense of '*fitness for purpose*' and '*safe in use*', and that the product or service has been designed to meet the customer's needs. The constituent parts of this Standard set out the requirements of a *quality-based* system. No special requirements are set out which might only be needed or even be achievable by a few companies, for BS 5750 is essentially a practical basis for a quality system or systems which can be utilised by any company offering products or services both within the UK or abroad. The principles of the Standard are intended to be applied to any company of any size. The basic disciplines are identified with the procedures and criteria specified in order that the product or service meets with the requirements of the customer. The benefits of applying BS 5750 are very real. They include:

- cost effectiveness, because a company's procedures become more soundly based and criteria more clearly specified;
- reduction of waste and the necessity for reworking to meet the design specifications;
- customer satisfaction, because quality has been built-in and monitored at every stage before delivery;
- a complete record of production at every stage being available to assist in product or process improvement.

By the adoption of BS 5750, a company can demonstrate its level of commitment to quality and also the ability of that company to supply goods or services to the defined quality needs of customers. For many companies it is merely the formalising and setting down of an existing and effective system in documented form so that the validity of the system can be guaranteed by obtaining external accreditation.

By having an agreed standard as the basis of supply there can be major benefits to all the parties concerned. A customer can specify detailed and precise requirements, knowing that the supplier's conformance can be accepted since the quality system has been scrutinised by a third party, i.e. it has been accredited. However, successful implementation will be dependent upon the total commitment of the whole management team and, in particular, the person at the very top of the organisation.

12.11 Using BS 5750/ISO 9000

Customers may specify that the quality of goods and services which they are purchasing shall be under the control of a management system complying with BS 5750/ISO 9000. Together with independent third parties, customers may use the Standard as an assessment of a supplier's quality management system and, therefore, the ability of a supplier to produce goods and services of a satistfactory quality. The Standard is already used in this way by many major public sector purchasing organisations and accredited third-party certification bodies. Thus it is in the interests of suppliers to adopt BS 5750/ISO 9000 in setting up their own quality systems.

The direct benefits to companies who have been assessed in relation to the Standard and who appear in the DTI *Register of Quality Assessed United Kingdom Companies* consist of:

- reduced inspection costs;
- improved quality;
- better use of manpower and equipment;

Where a company is engaged in exporting goods or services, then there is a direct advantage in possessing mutually recognised certificates which could be required by overseas regulatory bodies.

12.12 The quality manager

Responsibility for the implementation of BS 5750/ISO 9000 is best placed in the hands of a professionally-qualified quality manager. Such a position must carry all the authority and responsibility for the operation of the agreed quality system. Although such a person may initially have a line management position, it is better that the quality manager is seen to operate across departmental boundaries and reports directly to the chief executive, since 'Quality is everyone's business'.

The function of the quality manager is to coordinate and monitor the quality system, ensuring that effective action is taken at the earliest moment so that the requirements of BS 5750/ISO 9000 are met. However, the first task of the quality manager will be to convince the other managers in the organisation that quality must become the vital part of all the company's operation. Any narrow view of quality is totally alien to an environment aspiring to total quality management (TQM). This means that the education and training of all personnel will be at the top of the agenda for any effective quality manager.

12.13 Elements of a quality system

Setting up

The quality system must be planned and developed to encompass every function within the organisation and, in particular, the functions of customer liaison, design, manufacturing, purchasing, contracting and installation. The infrastructure of the installation will have to be fully understood in respect of responsibilities, procedures, and processes in order that their functions can be presented in a documented form. Existing quality control techniques will need to be examined and updated in terms of equipment and trained personnel in order that the capability exists for the fulfilment and recording of the quality plans.

Control of design quality

The objective of any design is to meet the customer's requirements in every respect. Therefore, it is vital that the input to the design function is established and documented. This input being the result of consultation by the marketing and the design staff with the customer. There should be adequate trained staff and resources available to ensure that the design output meets the customer's requirements, and that any changes or modifications have been controlled and documented.

Traceability

All the products needed in the fulfilment of the customer requirements should be clearly identified in order that they can be traced throughout the organisation. This is necessary in order that a capability exists for the tracing of any part which could be delivered to a customer. The need for this traceability might arise in the case of a dispute regarding non-conformity or for safety or even statutory reasons. Identification is also important where slight differences exist between the requirements of different customers.

Control of 'bought-in' parts

All purchased products or services should be subjected to verification of conformance to previously agreed specifications. Remember that, usually, all organisations are both customers and suppliers. Control is exercised by documentation of the purchasing requirements, and the inspection (and hence verification) of the purchased product. Many companies insist on details of the quality system used by subcontractors and suppliers. Hence the value of all companies having accreditation under BS 5750/ISO 9000. Even where a customer supplies a product to be processed, the customer must be assured that the receiving company is, itself, 'fit for purpose' to carry out the process. While the product is in the possession of the processing company, they are responsible for its well-being, i.e., that it remains free from defects.

Control of manufacturing

Clear work instructions are at the heart of any manufacturing process. They eliminate any confusion by showing, in a simple manner, the work to be done or the service to be provided and also indicate where the responsibility and authority lies. If the customer's requirements, via the design function, are clearly specified on the work documentation together with quality criteria, then the task of the manufacturing function becomes more easily controlled. BS 5750/ISO 9000 spells out the items that work instructions should include and emphasises the control of any additional processes, such as heat-treatment, thus avoiding expensive errors.

Control of quality of manufacture

Prompt and effective action is essential to the maintenance of quality throughout the system. Not only must defective parts be identified but the causes must be quickly rooted out. This could lead to changes in the design specifications or the working methods. The control of the manufacturing process starts with the inspection of the incoming goods for conformance to agreed specifications. Documentary evidence of such conformance should be provided either by the external supplier or by the company's own in-coming inspection. The accreditation of every company's measuring and testing equipment by an external body such as NAMAS provides consistency and avoids disputes between supplier and user.

A means of indicating the status of goods with regard to their inspection, that is 'Inspected and approved', 'Inspected and rejected', or 'Not inspected' should be part of the quality system. Any part which does not conform to the specifications must be clearly identified to prevent unauthorised use or despatch. Documentation must be raised to indicate the nature of the non-conformance and of either its disposal or any remedial work performed upon it.

Summary

BS 5750/ISO 9000 is about documentation procedures. It is not a panacea of quality but it is a mechanism by which a company can formalise its operation in a standardised way. The written procedure must be unambiguous and clearly state that which is to happen, NOT what the company thinks ought to be happening! The system chosen should be compatible with the business and should be capable of being maintained. Communication throughout the organisation is essential. The workforce must know what is happening and how it will affect them. This is particularly important during the assessment period, when the accreditation is being applied for, as the assessors may talk to anyone in the company. It is important, subsequently, to maintain and improve upon the standards achieved during the accreditation assessment. Successful companies who have achieved accreditation under BS 5750/ISO 9000 can display one or more of the marks of quality shown in Fig. 12.6 on their products and advertising media.

BSI leads the world in quality assurance and product certification schemes. We are continually assessing companies and testing products against the most stringent standards.

In meeting the requirements of our achemes, companies can look forward to more streamlined production and substantial internal cost savings. Moreover, they are entitled to display one of our marks of quality.

THE REGISTERED FIRM SYMBOL indicates that the company operates a quality system in line with BS 5750 — the British Standard for quality systems — providing confidence that goods or services consistently meet the agreed specification.

THE KITEMARK indicates that the product which displays it complies with the appropriate published specification.

THE SAFETY MARK indicates that the product meets the required safety specification.

THE REGISTERED STOCKIST SYMBOL indicates that the stockist operates a quality system which ensures the continuity of the quality chain through to the purchaser.

Small wonder, then, that our marks are recognised by purchasers the world over as top marks.

Fig. 12.6 BSI marks of quality

12.14 Prevention-oriented quality assurance

The attainment of quality in general terms means some form of inspection at some stage in the production process. The crucial decisions are:

- WHAT type of inspection procedures to use;
- WHERE in the process the inspection will take place;
- HOW process variations will be detected before they become defects.

The need for inspection is rooted in the fact that any process is subject at any given moment to disturbances which lead to variations during a production run. No two parts are produced exactly the same; there are always some differences resulting in deviations occurring either side of the objective. *Variability* can be defined as the 'amount by which the process deviates from a pre-determined norm' and, therefore, some variation must be tolerated.

This inherent *production variability* is a characteristic of the process of production and can be the result of:

Random causes. Certain aspects of production variability occur by chance and it is difficult to ascribe them to any one factor. Examples of random causes are:

- variations in material properties;
- process machine conditions such as wear in bearings and slides;
- temperature fluctuations;
- differences in the condition and supply of coolant;
- operator performance.

Assignable causes. These are causes of variation which can be identified. Examples of assignable causes are:

- tool wear;
- obvious material differences (e.g. different sources of supply);
- changes in equipment.

Assignable causes usually result in large process variability and must be kept under control.

Quality can be classified into two characteristics that will determine the techniques to be used in their inspection.

Variables. These are characteristics where a specific value can be measured and recorded and which can vary within a prescribed tolerance range between prescribed limits and still remain acceptable. For example:

- length, height and diameter inspected by measurement;
- mass;
- electrical potential (voltage);
- pressure and temperature.

Attributes. These are characteristics that can only be acceptable or unacceptable. For example:

- colour (if you want and have ordered a blue car, a red one will not be acceptable);
- size (a toleranced dimension inspected by gauging will be accepted or rejected on a 'go' or 'not go' basis).

12.15 Types of Inspection

There are basically two types of inspection which can be carried out at any stage of a process. These are:

100% inspection

With 100 per cent inspection every part will be examined. It may be thought that this will guarantee that all defectives will be identified. However, this is not the case where human beings are the inspectors and arbiters of conformance. This type of monotonous, repetitive inspection procedure leads to loss of concentration and the missing of some defectives. This can be as high as 15 per cent at the end of a shift. Some products themselves preclude 100 per cent inspection. For example, the ultimate test of a ballistic missile lies in the actual firing! Therefore, some other means must be found.

Sampling

With sampling, a decision on quality is based upon the close inspection of a randomly-selected batch of materials or products. The decision as to whether or not the remainder of the material or products are acceptable is based upon the outcome of mathematically-based statistical procedures. The process involves the collection of data and making decisions about conformance as a result of plotting the information on various control charts.

Whichever system of inspection is adopted, it should be clear that to employ only 'end-inspection' is wasteful and uneconomical. In many cases it is far too late, resulting in delayed deliveries and either total rejection of a batch of work or, at best, considerable reworking to reclaim the batch at considerable extra cost. Inspection should be used as a check on a process while it is in progress and adding value to a job. It is most effective when it is in the hands of the production operatives, with the quality control section only acting in a validating role. This is the principle that *prevention is better than cure*.

12.16 Position of inspection relative to production

Consideration must be given to the position of any in-process inspection points in order that costly time is not wasted on processing parts that are already defective. Such inspection points could be:

- prior to a costly operation (e.g. there is no point in gold-plating a defective watch case);
- prior to a component entering into a series of operations where it would be difficult to inspect between stages (e.g. within a FMS cell);
- prior to a station that could be subjected to costly damage or enforced shut-down by the failure of defective parts being fed into it (e.g. automatic bottling plants, where the breakage of a bottle could result in machine damage or failure, in addition to the cost of cleaning the workstation);
- prior to an operation in which defects would effectively be masked (e.g. painting over a defective weld);
- at a point of no return where, following the operation, rectification becomes impossible (e.g. final assembly of a 'sealed for life' device).

12.17 Areas of quality control

Input control

It would be foolish and wasteful to utilise resources on the processing of material or parts which are already defective when received from the supplier. The function of in-coming inspection is to detect any faults that could affect the finished product. An example of this is the machining of rotor and turbine shafts for the Electricity Generating Industry. The very large forgings for the shafts are inspected internally and externally for cracks before being subjected to extensive and costly machining processes.

One way of effecting input control is to rely upon the supplier's outgoing inspection. This could mean the production of documentary evidence in the form of certificates, hence the value of BS 5750/ISO 9000 in standardising quality. Alternatively, the customer has two other courses of action. One is for the customer to provide their own inspection at the supplier's plant. The other is for the customer to carry out in-company inspection at their own plant before accepting a delivery. This could be 100 per cent inspection or sampling to arrive at a decision regarding acceptance or rejection (see section 12.19).

Output control

Output control is the last link in the quality control chain. It is necessary, not because the previous inspection has failed, but out of the fallibility of any system of inspection and the resulting consequences of the customer receiving products or services which do not conform to the agreed specification. Again, the choice is between 100 per cent inspection, or sampling.

Process control

Process control is the 'let's see how it is doing' approach and, if it is successful, it reduces the output control effort to a minimum. Considering that any process has an inherent amount of variablity, it is important that the amount of this variation is known. Thus a capability study has to be performed and the process has to be kept within acceptable limits. The problem for *Process quality control* is, therefore, how to detect unacceptable process variation before it leads to non-conformance with the agreed specification.

12.18 Quality control techniques

Quality assurance (QA) can be defined as:

> 'the provision of documentary evidence that an agreed specification has been met and the contractual obligations have been fulfilled'.

It is a powerful definition indicating that all the legal implications associated with the law of contracts have been covered. Quality control (QC) can be defined as:

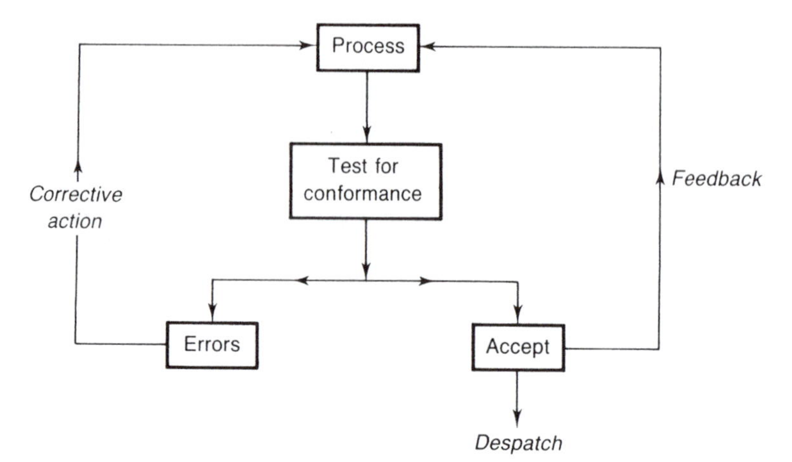

Fig. 12.7 The quality control (QC) loop

> 'the application of a system for programming and coordinating the efforts of the various groups in an organisation to maintain or improve quality at an economical level which allows for customer satisfaction'.

The application of these techniques and procedures is necessary to ensure that any goods or services actually meet the design specifications and hence the customer's contractual requirements.

Thus, QA is the objective to be attained and QC is the means of achieving that objective. Before any type of control can be exerted in any sphere, ground rules have to be established. In the case of QC, in common with any type of control, those ground rules consist of:

- the setting of standards against which performance can be compared;
- the mechanics for assessing performance against the prescribed standards;
- the means of taking any necessary corrective action where an error exists between actual performance and the prescribed standards as shown in Fig. 12.7 (the QC loop).

12.19 Acceptance sampling

There is a need for some sort of verification that goods being delivered for use in production, and finished products leaving production, have conformed to the agreed specifications. Where 100 per cent inspection has to be ruled out on the basis of cost and/or impracticality, the decision either to accept or reject a batch of work is based upon a sample taken at random from the batch. In this context, a sample can be defined as:

> 'a group of items, taken from a larger collection, which serves to provide the information needed for assessing the characteristics of the

population, or as a basis for action upon the population or the process which produced it.'

Further, a random sample can be defined as:

'one in which each item or collection of items has an equal chance of being included in the sample'.

The constant factor in the sampling of goods is that a clear definition of non-conformance must be available. In addition, there has to be prior decision on both the sample size and the number of defects found in the sample which would constitute either acceptance or rejection of a batch. There are risks involved in basing a decision on a sample. There is the risk of accepting a batch which, overall, contains an unacceptable level of defectives. This is known as *consumers' risk*. There is also the risk of rejecting a batch which, overall, is acceptable. This known as *producers' risk*.

The risk involved is a calculated one, based upon the laws of mathematics contained in a statistical analysis. Where risks are concerned, the *probability* of an event occurring has to be taken into account. This probability is measured on a scale between 0 (no possibility) and 1 (a certainty). The symbols used are:

p = the probability of an event occurring;
q = the probability of an event NOT occurring.
Therefore: $p + q = 1$
where: p = (possible outcomes)/(total outcomes)

It is often convenient to convert probabilities into percentages. For example, consider the probability of drawing a king from a full pack of cards. There are 52 cards in the pack and only 4 kings, so the probability is 4/52 = 0.077 or 7.7%. It must be remembered that if a decision is taken to accept a batch on the statistical evidence deduced from a sample, there is *no* absolute guarantee of the actual quality of the whole batch! Sampling only provides a basis for making a decision that takes into account all the foreseeable probabilities involved.

12.20 Operating characteristics of a sampling plan

To operate a sampling plan, a sample of n items is selected at random from a total batch of N items. The sample is then inspected against the conformance criteria. Usually the inspection is by attributes (i.e. accept or reject). This is because inspection by variables involves measurement, which is usually too slow and costly. Variables such as dimensions can be converted to attributes by assigning them limits of size so that the associated component feature can be accepted or rejected by a simple GO and NOT GO gauge. Dimensions within limits are acceptable, but dimensions that have drifted outside the limits are rejected.

The simplest type of acceptance sampling scheme is one in which a

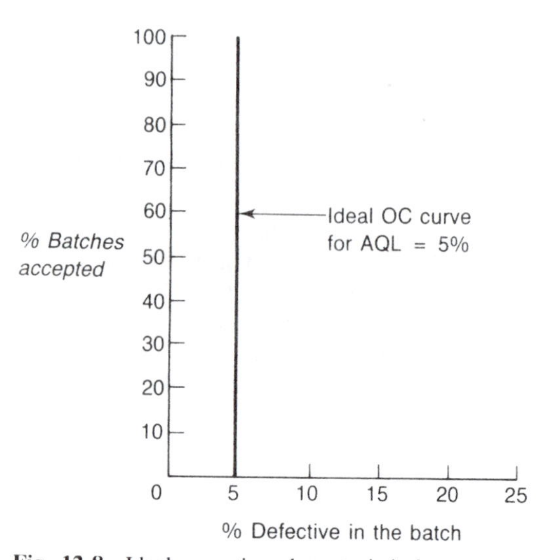

Fig. 12.8 Ideal operating characteristic loop curve

single sample (size *n*) is taken from a batch (size *N*) and the decision to accept or reject the whole batch is dependent upon the number of defectives found in the sample. If the number of defectives found in the sample is equal to or less than an agreed acceptance number *c*, then the whole batch is accepted. But if the number of defectives exceeds *c* then the whole batch is rejected.

The effectiveness of any sampling plan in detecting either acceptable or unacceptable batches can be shown graphically by plotting its *operational characteristic* (OC) curve. It can be seen from Fig. 12.8 that the vertical axis represents the percentage of batches accepted and the horizontal axis represents the percentage of defectives in the batch. The ideal sampling plan is one in which any batch having a non-conformance equal to or less than a predetermined percentage is accepted on the basis of a single sample. Similarly this sample will reject all batches with a non-conformance greater than the predetermined level. This level is known as the *acceptance quality level* (AQL). Figure 12.8 shows the operating characteristic curve for an ideal sampling plan; in this example, 100 per cent of all the batches with 5 per cent or less defectives are accepted and there is total rejection of any batches with a greater percentage of defectives.

In practice, however, such ideal performance is never encountered since all processes have some inherent variability. A more typical OC curve is shown in Fig. 12.9, where:

PAPD is the *process average percentage defective*. This usually coincides with the AQL but is, in fact, the average percentage of defectives being produced when the process is considered to be operating at a level of quality that is acceptable to the consumer.

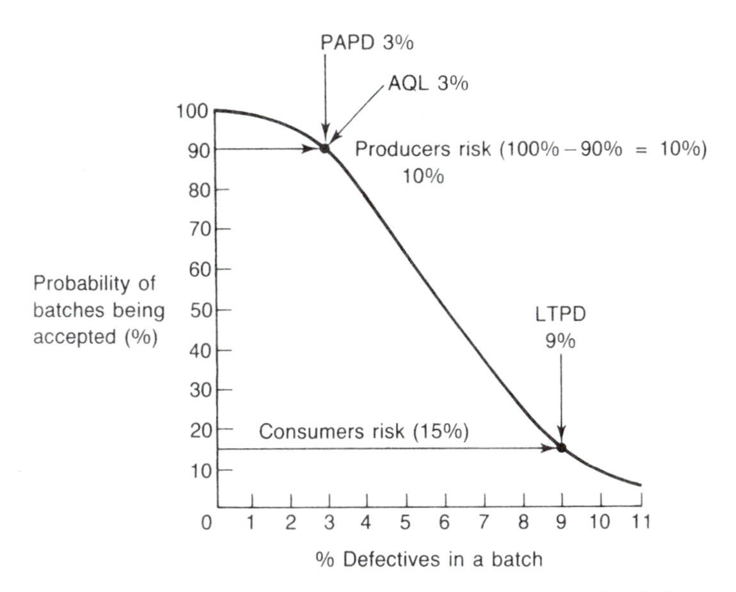

Note: Each individual sampling plan has its own unique operational characteristic curve

Fig. 12.9 Typical operating characteristic curve

LTPD is the *lot tolerance percentage defective*. This is also known as the *consumer* or *customer risk*, i.e. the percentage of defectives that the consumer would find unacceptable.

It is usual to locate the two important values of AQL and LTPD on the curve. The calculations for the probability are based on either a binomial distribution or, where the lot total is very large, on a Poisson distribution. It is not necessary to perform these calculations, as tables of the values are readily available in BS 6001 and BS 6002. The OC curves are also plotted in these Standards to asssist in the choice of the most appropriate sampling plan for both attributes and variables.

If the sample size is increased, then the sampling plan becomes more sensitive, i.e. it discriminates more between good and bad batches as the sampling approaches 100 per cent inspection. However, as previously stated, increasing the sample size increases the inspection costs proportionately. Similarly, if the number acceptable (c) in a sample is lowered, say, to zero, so that batches are rejected if a single defective is found in a batch, then the sampling plan becomes less discriminatory. It may be thought that a plan with $c = 0$ would be a desirable method of sampling. However, while many bad batches would be rejected, many batches with economically viable, and normally acceptable, defective rates would also be rejected. Further, $c = 0$ does not guarantee that all batches that are accepted are free from defectives.

The size of the batch has little effect on the OC curve provided that the batch size is at least five times the sample size. However, the batch size can become important if it is small enough to be affected by withdrawing a sample from it. The smaller the batch, the nearer the sample approaches 100 per cent inspection so that the cost saving of sampling becomes too small to be viable. As previously stated, any sampling plan only gives a basis for making a decision to accept or reject a batch: it does not offer any guarantees; it is a way of weighing up the probabilities. When selecting a sampling plan, consideration must be given to the cost of inspecting an individual part, the cost of rejecting a batch containing a high level of acceptable parts, and the cost of accepting a batch containing too many defectives.

12.21 Classification of defects

In many manufactured goods, more than one type of defect may be present on which the decision for acceptance or rejection can be based. Some defects are more significant than others and, therefore, some classification is required.

- *Critical*. This class of defect is based upon experience that a failure in service brought about by such a defect could result in situations where individuals would be at risk.
- *Major*. This class of defect could result in failure which could seriously impair the intended function of the product.
- *Minor*. This class of defect is not likely to reduce the effectiveness of the product, although non-conformance with the specification is present.

Critical defects may require 100 per cent inspection, while the sampling plan for major defects would require a small AQL, with a larger AQL being used for minor defects.

Some products have so many quality characteristics that to attempt to grade them by normal means would be difficult. For example, some complex assemblies such as furniture, vehicles, caravans and houses could contain several defects which could have differing effects upon the acceptance of the product. In these cases decisions are based upon a system of points and grading. The more important the defect, the higher the score and the poorer the product quality. For example:

Critical defects:	20 points
Major defects:	8 points
Minor defects:	2 points

The record of such 'quality scores' indicates trends, so that a constant score shows consistent quality at the level of the score, i.e. consistently good or consistently bad. Scores drifting higher shows that the quality is deteriorating. Usually the ceiling for scores is set at the score of the critical defects.

12.22 Vendor assessment and rating

One of the spin-offs from acceptance sampling is the fact that when applied to in-coming deliveries, a check can be kept upon a company's suppliers (the *vendors*). Their effectiveness can have a marked influence on the quality of the user's products. It is in the customer's interests to know the capability of all suppliers.

This is achieved by investigating all possible vendors and assessing them in terms of price, quality of deliveries, lead time, dependability, strike history and quality control procedures. Once the vendors have been selected then a continuous assessment is made by rating each delivery. A typical method would be to have 100 points divided between quality, price and delivery. For example:

Maximum points for quality = 40
Maximum points for price = 35
Maximum points for delivery = 25
Total = 100

Rating of a supplier (vendor):

- 98% of batches are accepted on sampling inspection;
- Price £100 per batch; lowest market price £90;
- 75% of deliveries are on time.

Quality rating = (40 × 98)/100 = 39
Price rating = (35 × 90)/100 = 32
Delivery = (25 × 75)/100 = 19
Total = 90 points.

All suppliers are rated in a similar fashion and a rating 'league table' is drawn up for use by the purchasing department.

12.23 Statistical process control (SPC)

Despite the advanced technology used in many current production processes, customer quality requirements have also advanced. Therefore, although process variation is of a lower order, it is still a problem and the output from a manufacturing process still needs to be monitored.

Achieving the necessary quality must lie within the capability of the production department as they have the responsibility for output. Yet there are financial constraints which make measuring and testing for quality an overhead cost since inspection does not add to the value of the product, it does not 'transform' anything. It must be remembered that quality is never free, and any company using the TQM route must provide adequate resources to deliver the quality demanded.

The problem of any production process is the detection of unacceptable process variations before they become defects. *Statistical process control* (SPC) helps to overcome this problem by measuring samples of items during various stages of the manufacturing process. It is then possible, using statistical means, to make reasoned judgements regarding the level of probability that all the items at that stage of production are being manufactured within the required specification. Further, the drift in variation can be seen at any time and this enables adjustments to be made to the process before the variation results in defects.

Traditionally, SPC was — and still is to some extent — carried out manually, with inspectors taking measurements from random samples and plotting the results on statistical charts known as *control charts*. From these charts, trends could be detected and the timing of any remedial adjustments could be determined. However, the introduction of computerised SPC has speeded up data collection and processing, thereby reducing the time lag associated with manual SPC during which the process could go out of statistical control.

SPC uses the results from individual samples to focus attention on the whole process, indicating overall performance. It also provides the information needed to make changes and adjustments to the processing system before defective parts are produced as a result of tool wear or other process variables. Figure 12.10 shows a summary of the flow of information when SPC is being used.

Backward control

Backward control is the system shown in Fig. 12.10(*a*) and gets its name from the fact that the components are inspected following an operation and the results are *fed back* to adjust the process so that it provides maximum quality.

Forward control

In a forward control system the components are again inspected following an operation but the results are *fed forward* to the next operation where defectives can be reworked to produce conformance. This is shown diagrammatically in Fig. 12.10(*b*).

Accept–reject control

In an accept–reject control system the components are subjected to 100 per cent inspection after a given operation. Those components which conform are moved to the next production process, while those components which do not conform are rejected. This is shown diagrammatically in Fig. 12.10(*c*). Because 100 per cent inspection is involved, this system is only economical where fully-automated inspection is possible, with no bottlenecks to hold up the flow of production.

12.24 Some principles of statistics

Before progressing to the preparation and use of control charts, it is necessary to review some of the basic principles of statistical probability. Statistics is the collection, classification and analysis of numerical information. In terms of manufacturing production, the application of

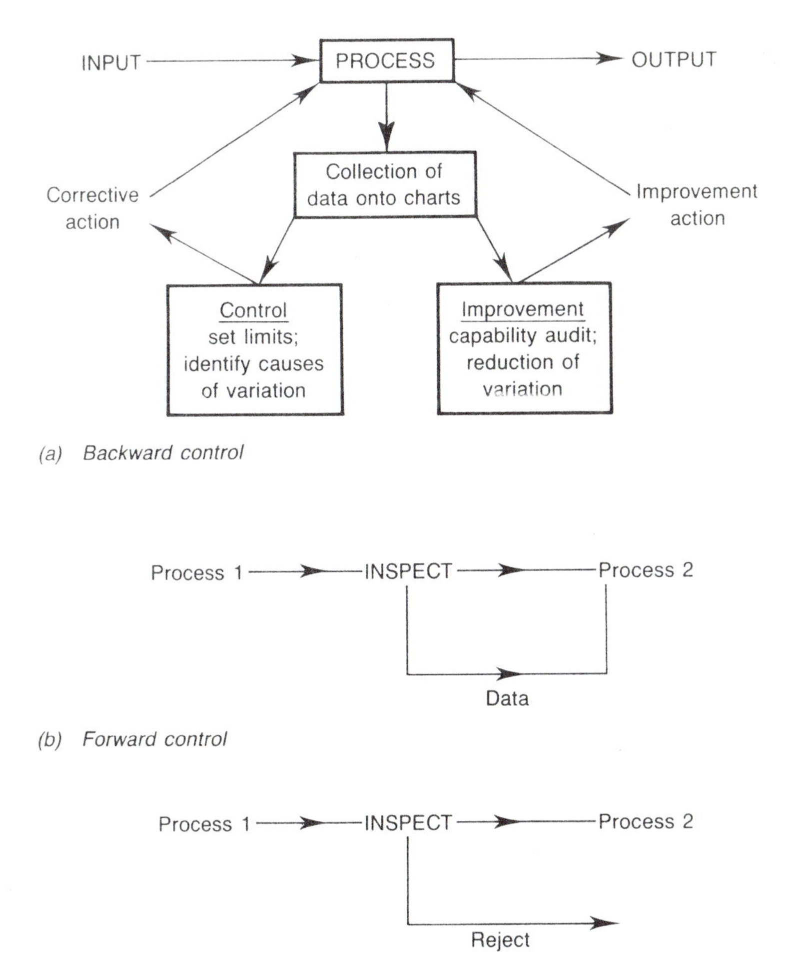

Fig. 12.10 Statistical process control (SPC)

statistics provides a very useful tool for predicting the outcomes from a particular process.

The numerical data obtained by sampling spreads over a range of values. This spread is referred to as the *dispersion* of the data and, for control purposes, production engineers are interested in the pattern or *distribution* of these variations about a mean value. The larger the sample the more nearly it represents the characteristics of the total population. In this context, *population* means the total batch from which the sample is taken.

The data obtained from the sample can be shown graphically as a *histogram* as shown in Fig. 12.11(*a*). The numerical values of the items

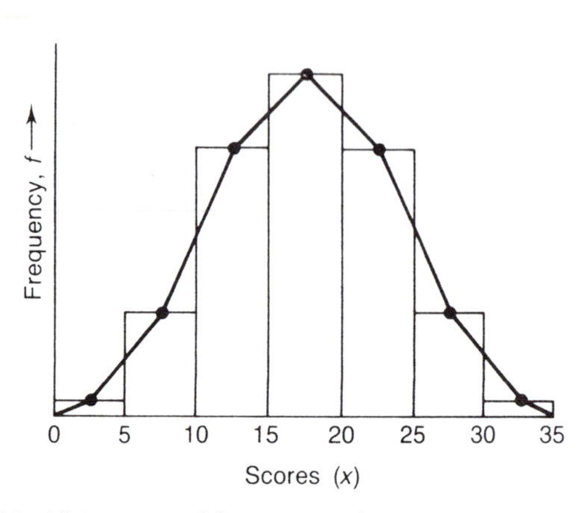

(a) Histogram and frequency polygon

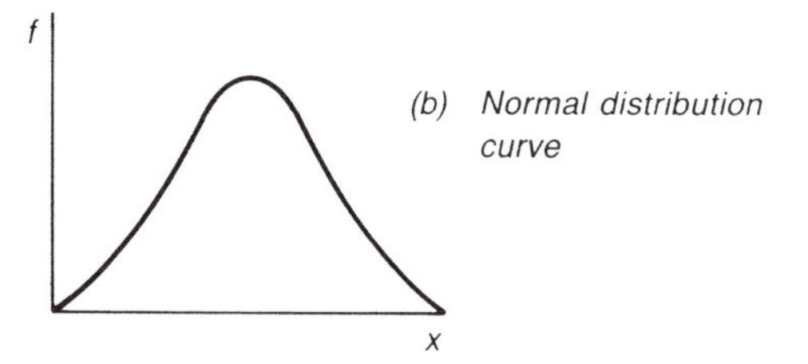

(b) Normal distribution curve

Fig. 12.11 Introduction to distribution curves

being plotted on the horizontal axis and the frequency with which they occur is plotted on the vertical axis. The width of each vertical column of the histogram (as measured on the horizontal axis) is called the *class interval*. The mid-points of the tops of the vertical columns (arithmetic means of the class intervals) can be connected by straight lines to form a *frequency polygon* as shown in Fig. 12.11(*a*). However, it is more convenient to convert the frequency polygon into a smooth curve as shown in Fig. 12.11(*b*). This curve is the basis of all the predictions derived from the sample and is the *curve of normal distribution* (also called a *Gaussian curve* or a *bell curve*).

It can be seen from the distribution curve that most of the sample data readings lie on or near to the centre of the variables plotted on the horizontal axis, with progressively fewer points further away from the central point. A single value, the *arithmetic mean* ($\bar{x}$) can be calculated to represent the whole of the distribution.

$$\bar{x} = (\text{sum of all readings } (x))/(\text{number of readings } (n)) = \Sigma x/n$$

For example, if a machine is set up to cut off lengths of tubing, then the mean value would coincide with the nominal length, and actual lengths of the cut tubes would vary slightly about this size. The distribution of the variations in size about the mean value represents the process variability and, if plotted graphically, would follow a *normal curve of distribution*, as discussed earlier. Control chart theory is based upon this assumption that the measurements follow a normal curve of distribution.

There are a number of measures of variability (dispersion) such as 'range', 'mean deviation', and 'standard deviation'. These all have their own particular advantages and limitations but the most satisfactory for control purposes is *standard deviation* and this will now be considered in some detail.

Standard deviation is, in fact, the distance from the mid-point of a normal distribution curve to the point on the curve where the direction of curvature changes (stops curving downwards and starts curving outwards). There are several variations of the formula for calculating standard deviation depending upon how the data have been collected and collated. The symbol for standard deviation is σ (sigma), and:

$\sigma = \sqrt{[1/n\ \Sigma(x - \bar{x})^2]}$ if individual values of x are given;
$\sigma = \sqrt{[1/n\ \Sigma f(x - \bar{x})^2]}$ if frequencies (f) are given for various stated exact values for the variable;
$\sigma = \sqrt{[1/n\ \Sigma f(m - \bar{x})^2]}$ if frequencies are given for the values falling in various stated ranges or class intervals.

12.25 Properties of normal distribution

One of the most important properties of normal distribution is that the proportion of a normal population which lies between the mean ($\bar{x}$) and a specified number of standard deviations (σ) away from the mean is always the same, irrespective of the shape of the curve. The total area under a curve of normal distribution = 1 (i.e. 0.5 either side of the mean), when the mean = 0, and the standard deviation (σ) = 1. The values of the areas under the curve between the mean and various multiples of the standard deviation are:

Between 0 and 1 standard deviation: area = 0.341 35
Between 0 and 2 standard deviations: area = 0.477 25
Between 0 and 3 standard deviations: area = 0.498 65 (approximately 0.5)

Figure 12.12 shows two examples of normal distribution curves with different shapes. The areas are those bounded by the mean, 2 deviations, the horizontal axis and the curve. The relationship between the mean and the normal distribution curve is shown in Fig. 12.13, and it can be seen that:

- 68 per cent of the total population is contained under the curve between +1 and −1 standard deviations about the mean value (see Fig. 12.13(*a*)). *Note*: for a symmetrical curve the mean, median and mode are all the same value;
- 95 per cent of the total population is contained under the curve

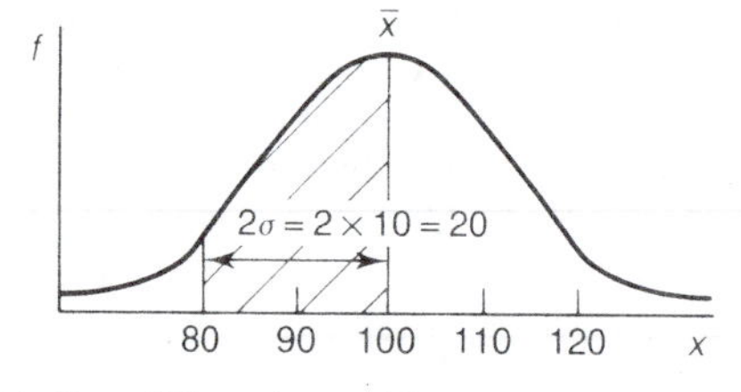

(a) $\bar{x} = 100$ *and* $\sigma = 10$

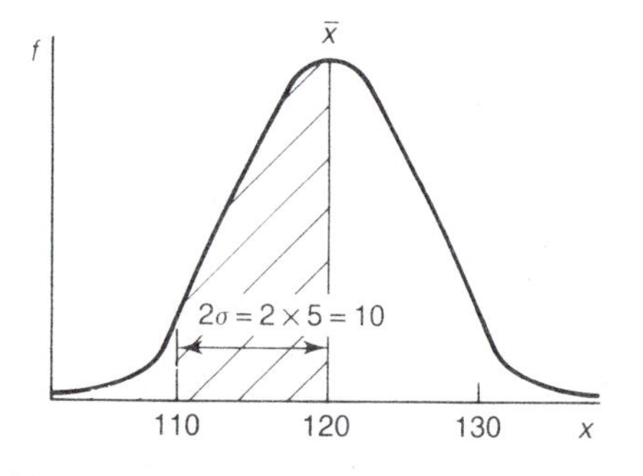

(b) $x = 120$ *and* $\sigma = 5$

Fig. 12.12 Typical curves of normal distribution

between +2 and −2 standard deviations (see Fig. 12.13(*b*));

- 99.73 per cent of the total population is contained between +3 and −3 standard deviations, i.e. for all practical purposes, +3 deviations determines the upper limit of the distribution and −3 deviations will determine the lower level of the distribution.

A process with a distribution showing a mean value that is equal to the nominal or target value is classed as being *accurate*, while a process with a distribution showing only a small spread, or scatter, is said to be *precise*. High quality is associated with high accuracy and high precision.

While the usual objective of any process is to produce items that are the same, some variation has to be tolerated. However, providing that limits to the variation are set, then items within the limits will conform.

The relationship between the variability of a process and the design tolerances is known as the *process capability* (C_p). For a process to produce items that conform to design tolerances, the spread of the distribution must be equal to or less than the prescribed tolerances.

Where only random variability exists (when the process is under statistical control), the capability index can be calculated from:

$$C_p = (\text{upper limit} - \text{lower limit})/6\sigma$$

Note: for values greater than 1, variability is contained within the limits; for values less than 1, variability is outside the limits.

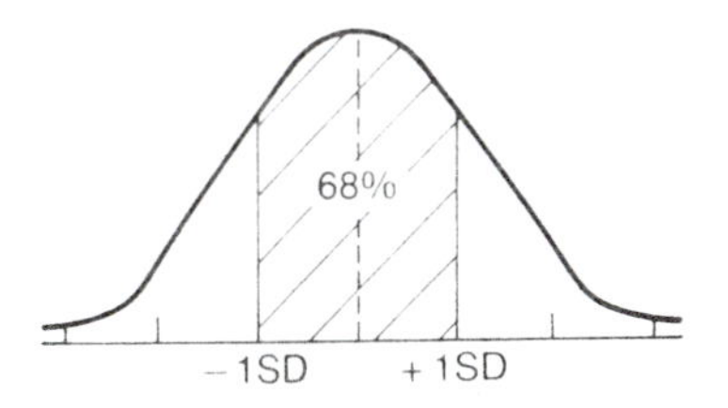

(a) 68% of all the total population will be contained under the curve between +1 and −1 standard deviations about the mean value

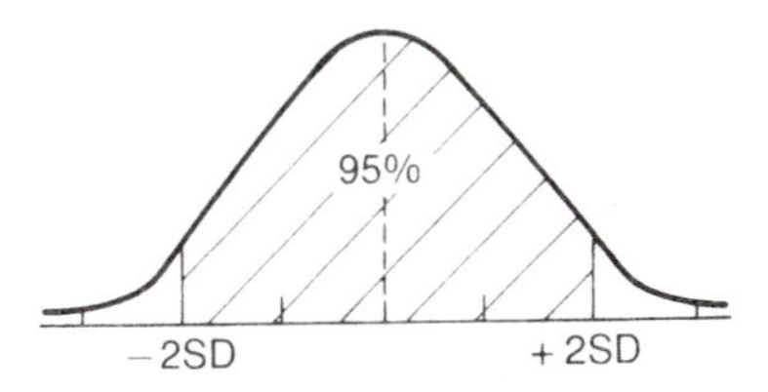

(b) 95% will be contained under the curve between +2 and −2 standard deviations about the mean value

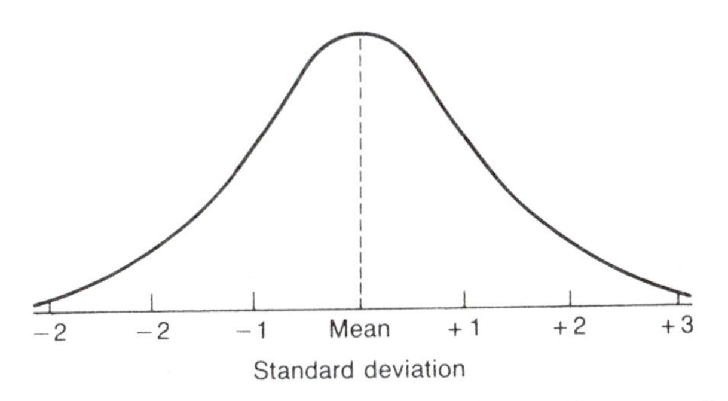

(c) Approximately 100% (99.73% actual) will be contained between +3 and −3 standard deviations about the mean value, i.e. +3SD gives approximately the upper limit of the distribution and −3SD gives the lower limit of the distribution

Fig. 12.13 Relationship between mean, standard deviation and normal distribution

12.26 Control charts for variables

Rather than plot a distribution for each sample as it is taken from the process, it is more expedient to plot the distribution of the mean values of the samples to obtain information about the process as samples are taken from a process.

Consider repeated samples of size n drawn from a process, giving a normal distribution about a mean $\bar{x}$ and a standard deviation σ, then the calculated means $\bar{x}_1$, $\bar{x}_2$, $\bar{x}_3$, etc., can themselves be arranged into a

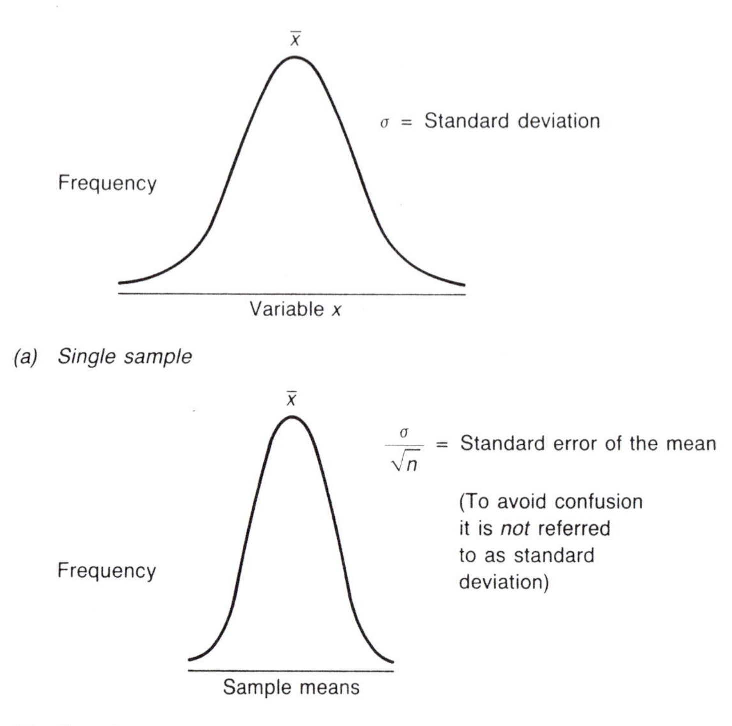

The spread is narrower for the plot of sample means

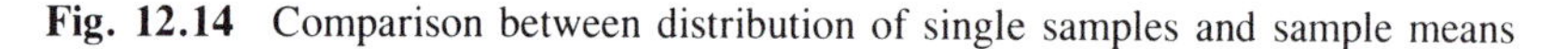

Fig. 12.14 Comparison between distribution of single samples and sample means

frequency distribution known as the *sampling distribution of the mean*. This distribution will also be normal and have its own mean value and its own standard deviation. To avoid confusion, the standard deviation of the sample means is referred to as the *standard error of the mean*, where:

Process mean $= \bar{x}$

Standard error $= \sigma_n = \sigma/\sqrt{n}$

Figure 12.14 shows a comparison between the distribution of a single sample and the distribution of the sample mean. It can be seen the spread of the latter is much less than for the single sample, i.e. the variation is less for the sample mean.

Control charts are graphs on to which some measure of the quality of a product is plotted as manufacture is proceeding. The measure of quality is taken from small samples and recorded on the control chart. Trends can be observed from the plots and corrective action taken before the process comes out of 'statistical control', i.e. before items are produced that do not conform with the agreed specification. Control charts are designed to give two signals:

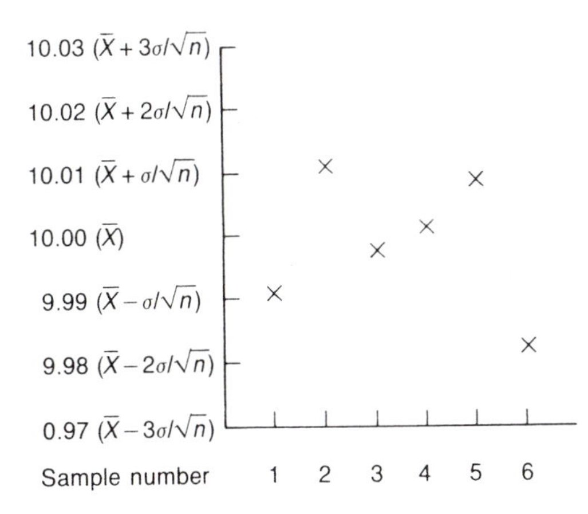

Fig. 12.15 Simple control chart

- a *warning signal*, which alerts the production controllers that a drift is taking place and to start planning the appropriate corrective action;
- an *action signal*, which indicates that the process is about to get out of statistical control and the pre-planned corrective action must be implemented immediately.

Consider the control of production of 10 mm diameter steel rollers. Sampling shows that the diameters achieved during production are normally distributed about a mean of 10 mm with a standard error from the mean of 0.01 mm. Periodic samples are measured and the sampling means of their diameters are plotted onto a chart as shown in Fig. 12.15.

From what has already been discussed about normal distribution, it can be expected that 68 per cent of all the readings will fall between $\bar{x} \leq \sigma/\sqrt{n}$, that 95 per cent can be expected to fall between $\bar{x} \leq 2\sigma/\sqrt{n}$, and that almost all the readings will fall between $\bar{x} \leq 3\sigma/\sqrt{n}$.

As it is necessary to control the measured diameter of the rollers during a long production run, then the control chart provides an easily understood, pictorial representation of the process, and it will show clearly any process variation. In Fig. 12.16(*a*) the last point to be plotted falls outside the $3\sigma/\sqrt{n}$ limit. This indicates a sharp deviation from the standard due to either a random variation or, more likely, a technical fault such as tool failure. Figure 12.16(*b*) shows a slower 'drift' from the specification. A general drifting of the plot above or below the mean indicates a gradual change in the process settings. The time for resetting can be estimated from the control chart.

12.27 Warning and action limits

It is important that indications are available so that it can be seen when the plot, and therefore the process, is drifting out of control in order that corrective action can be planned and implemented. For this purpose,

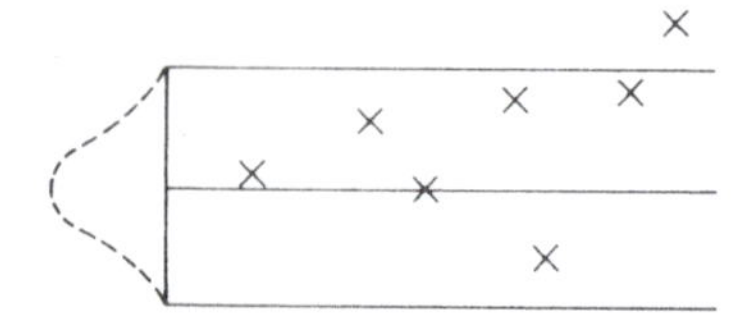

The plot falling outside the $3\sigma/\sqrt{n}$ limit would indicate a sharp deviation from the standard due to either a random variation or a technical fault.

(a) A sharp deviation from the specification

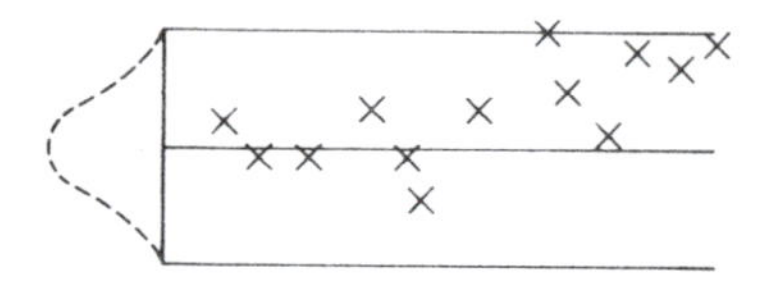

A general drifting of the plot above or below the mean would indicate a gradual change in the process settings. The time for resetting can be estimated from the control chart.

(b) A slower drift from the specification

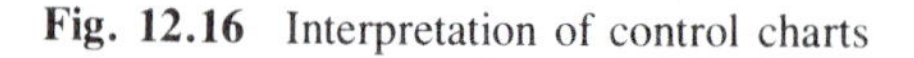

Fig. 12.16 Interpretation of control charts

further use can be made of the probabilities offered by the relationship between the curve of normal distribution, the mean and the standard deviation. It can be seen from Fig. 12.17(*a*) that the probabilities are as follows: 2.5 per cent (1 in 40) of all readings will lie above $\bar{x} + 2\sigma$; 2.5 per cent will lie below $\bar{x} - 2\sigma$, and 0.1 per cent (1 in 1000) of all readings will lie above $\bar{x} + 3.09\sigma$ and that 0.1 per cent of all readings will lie below $\bar{x} - 3.09\sigma$. It is conventional to use these values as:

upper warning limit (UWL) at $\bar{x}$ + 2 standard errors;
upper action limit (UAL) at $\bar{x}$ + 3 standard errors;
lower warning limit (LWL) at $\bar{x}$ − 2 standard errors;
lower action limit (LAL) at $\bar{x}$ − 3 standard errors;

where one standard error = $\sigma/\sqrt{n}$.

A typical chart layout is shown in Fig. 12.14(*b*), and the charts can be roughly interpreted as follows:

- If a point falls on or outside the *warning limits*, take an immediate second sample. If the second point also falls on or outside the limits, immediately adjust the process.
- If a point falls outside the *action limits*, assume the process is out of control; shut it down and take immediate corrective action.
- If points fall within the warning limits, assume the process is in control and take no action.
- If points fall within limits but are drifting, either above or below the mean, towards one or other of the warning limits, determine the time for action to be taken and the type of action to be taken.

The same principles apply to control charts which indicate the spread of the distribution, i.e. *range charts* (also known as '*w*' charts). The

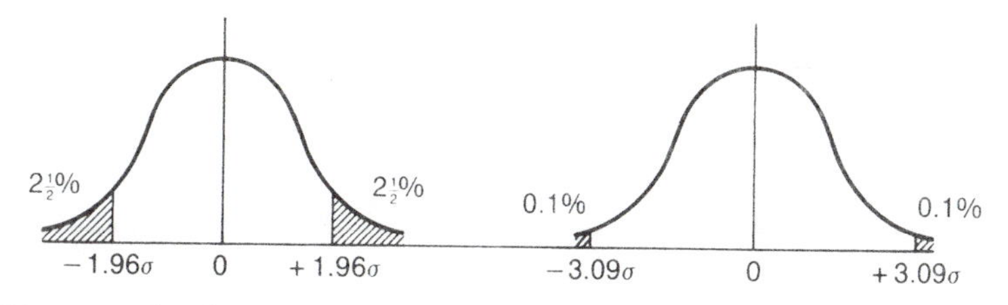

(a) It is conventional to use these values as follows:
upper warning limit (UWL) at X + 2 standard errors,
upper action limit (UAL) at X + 3 standard errors,
lower warning limit (LWL at X − 2 standard errors,
lower action limit (LAL) at X − 3 standard errors.
Note: standard error is the standard deviation from the distribution of sample means: $\sigma_n = \sigma/\sqrt{n}$

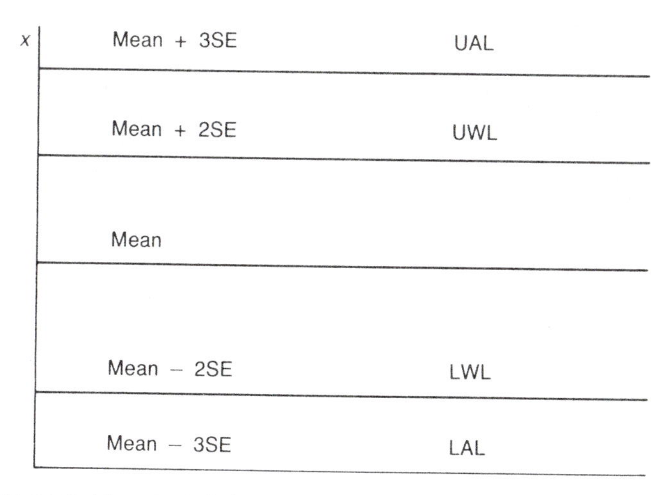

(b) Limits applied to a control chart

Fig. 12.17 Application of standard deviation to control chart limits

difference between the highest reading and the lowest reading in the sample is taken and plotted as before on a control chart similar to that shown in Fig. 12.18 except that the mean is $\overline{w}$ instead of $\overline{x}$. The lower limits are less important on this chart, since a trend towards the lower limit indicates less spread and, therefore, improved quality.

For both types of control charts the position of the limits is not calculated from first principles. To save much unnecessary calculation, tabulated data are available for establishing the standard deviation and the standard error of the distribution of means (see BS 2564).

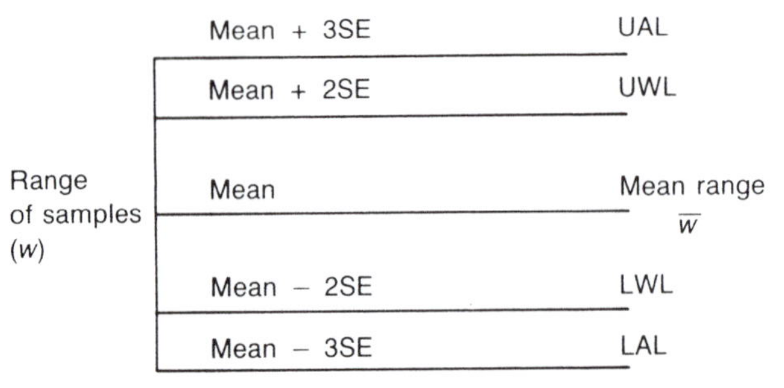

Fig. 12.18 Range chart (*w* chart)

12.28 Control charts for attributes

When inspecting for attributes, the decision making is reduced to 'good' or 'bad', 'right' or 'wrong', and the decision is either to allow the process to continue or to stop it for adjustments. There are two types of control chart which are based upon defectives, i.e. any item that does not conform to the agreed specification. There are also two types which are based upon defects, i.e. any feature on an item that is unwanted; for example, blemishes or other surface marks which detract from the appearance of an item but do not render it defective. However, a predetermined number of such minor defects on an item could, in fact, render it defective overall.

Control of defectives

Number defective charts (np *charts*). The sample size is a constant value and there is no requirement for a range chart. As with control charts for variables, the control limits are set at $\pm 2\sigma$ and $\pm 3\sigma$ from the mean. Then:

$$\text{the standard deviation } (\sigma) = \sqrt{[np(1 - p)]}$$

where: p = (total defectives/total output) or 'fraction defective', and
n = sample size.

Consider the control limits that can be derived from the following process information:

Total items rejected as non-conformants = 142
Number of batches = 12
Constant batch size = 400
Then $p = 142/(12 \times 400)$ = 0.0296
n = batch size = 400
and $np = 400 \times 0.0296$ = 11.84
therefore $\sigma = \sqrt{11.84(1 - 0.0296)}$ = 3.390
Warning control limits = $np \pm 2\sigma = 11.84 \pm 2 \times 3.39 = 18.62$ and 5.06
Action control limits = $np \pm 3\sigma = 11.84 \pm 3 \times 3.39 = 22.01$ or 1.67

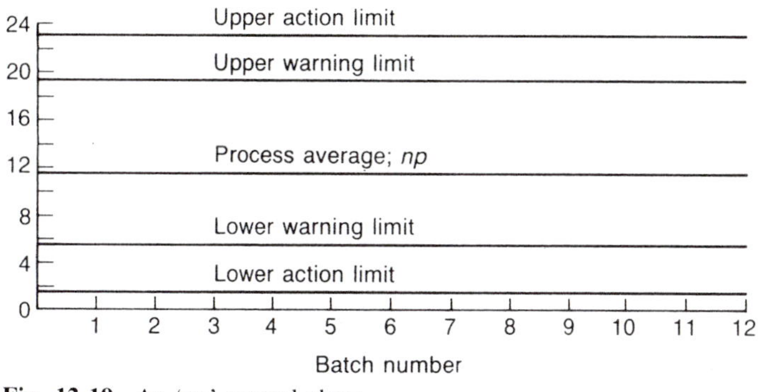

Fig. 12.19 An '*np*' control chart

The control limits would be entered on the control chart as shown in Fig. 12.19 and the number of non-conformants in each of the 12 batches would then be plotted on the chart to give a graphical representation of the production run.

Percentage defective charts (p charts). The sample size for *p* charts can be variable and this allows samples to be taken on a time basis rather than on a batch-size basis. However, even *p* charts cannot cater for *wide* variations in sample size. Such charts are only satisfactory where the sample size does not vary by more than 25 per cent either side of the average sample size. The procedure is similar to that for the previous (*np*) charts.

Process average $\bar{p}$ = (total number of non-conforming items)/(total number inspected)

Standard deviation $\sigma = \sqrt{\bar{p}(100-\bar{p})/\bar{n}}$ where: $\bar{n}$ = average sample size.

Consider the control limits for the following derived process information:

Total non-conformants in a day's production = 127
Total production for the day = 5050
Number of samples taken = 12

Control limits:

$$\bar{p} = 127/5050 = 0.025 \text{ or } 2.5\%$$
$$\bar{n} = 5050/12 = 421$$

$\sigma = \sqrt{2.5(100 - 2.5)/421} = \sqrt{0.579} = 0.761$

Action control limits $= \bar{p} \pm 3\sigma = 2.5 \pm (3 \times 0.761) = 4.783\%$ and 0.217%

Warning control limits $= \bar{p} \pm 2\sigma = 2.5 \pm (2 \times 0.761) = 4.022\%$ and 0.978%

These control limits can now be entered onto the chart as shown in Fig. 12.20.

Sometimes this type of chart is referred to as a *proportion defective chart*, the difference being that the numbers are not expressed as percentages but are left as proportions. For example, $\bar{c}$ would be left as 0.025 instead of 2.5 per cent. The standard deviation would then read: $\sigma = \sqrt{[\bar{p}(1 - \bar{p})/\bar{n}]}$. However, the percentage chart is more easily understood on the shop floor.

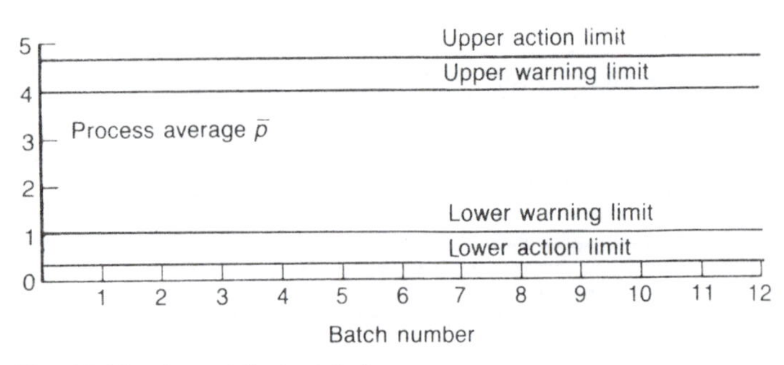

Fig. 12.20 Control limits '*p*' chart

Control of defects

C-Chart. This is used with a fixed sample to count the number of defects contained within that sample. Each defect in itself will not render the item defective but the cumulative effect may result in the item being deemed to be non-conforming. The total number of defects present in each sample is plotted on the control chart.

Process average $\bar{c}$ = (total number of defects)/(number of batches inspected)

Standard deviation $= \sigma = \sqrt{(\bar{c})}$

As previously:
Action limits occur at $\bar{c} \pm 3\sigma$
Warning limits occur at $\bar{c} \pm 2\sigma$

U-Chart. This is used with a variable sample size and computes the number of defects per item (it is more practical to compute defects per 100 or even per 1000 items). The operation of these charts is similar to the *C*-charts. That is, samples are taken at various time intervals instead of batch sizes; the number of non-conformities per item (or per 100 or 1000 items) are computed and the number of non-conformities are plotted onto the chart where:

$\bar{u}$=(total number of non-conformities)/(total number of items inspected)
$\bar{n}$=(total number of items inspected)/(number of batches inspected)

Standard deviation $(\sigma) = \sqrt{(\bar{u}/\bar{n})}$
Control limits are at $\pm 2\sigma$ and $\pm 3\sigma$ as previously
As with *p*-charts, the range of sample sizes has to be within ±25%.

Note

- It may seem strange with attribute charts that a lower limit is not only set but is actually used by inspectors; strange because the nearer a reading is to zero defects, then the nearer the process is approaching perfection. However, due to the inherent variability of any process, perfection is greeted with suspicion in quality control and any practitioner will want to check the reading being plotted.
- Some practitioners do not use the warning limits on attribute charts but prefer to use the '7-rule' instead. This states that a change has taken place in the process average if seven consecutive readings are

plotted either above or below the established process average or seven consecutive readings are seen to be either rising or falling with regard to the last point plotted.

12.29 Notes for charts for variables

Warning and action limits do not depend upon design limits; they are totally dependent upon the process variability. Therefore, designers must take account of the capability of a process when allocating toleranced work onto specific machines. However, there is a relationship which can be used to compare process variability with design limits. This is the capability index (C_p), introduced in section 12.24. It is also known as the relative precision index (RPI). There are three classifications (see BS 5700).

- Medium relative capability: where the variability just meets the design specification with a C_p between 1.0 and 1.3. The control chart average must be set at the centre of the limits and strict control exercised.
- High relative capability: where C_p is greater than 1.3 and the variability does not fill the design limits.
- Low relative capability where C_p is less than 1.0. Variability is so great that a high level of non-conformity can be expected.

12.30 Cost of quality

Quality is no different from any other function associated with production; there is a cost which must be taken into account and a high level of quality can absorb valuable resources. Like any other function, the costs must be analysed and budgeted. Where TQM is being practised it can be said that every function in the company is associated with quality but the actual costs can be considered in three broad categories.

Failure costs (internal)

The internal failure costs are any cost within the process of production up to the point at which the customer takes delivery. For example:

- scrap: items which must be discarded;
- re-work: the cost of any correction processes;
- re-inspection: the inspection of any work that has been corrected;
- down-grading: where items can be sold as seconds at a lower price provided they do not detract from the market for first-class items;
- others: must include any loss of production capacity caused by resetting of the process to regain conformance together with the cost of any investigation into non-conformance even to the point of a re-design.

Failure costs (external)

External failure costs include any costs occurring after delivery to the customer. For example:

- loss of reputation and possible future orders;
- recalling of goods for rectification or replacement;
- expenses for specialists to travel to the customer to rectify defective goods;
- product liability and the possible cost of litigation;
- maintenance of higher stock levels to cover the replacement of faulty goods.

Loss of reputation is the most damaging as the word soon gets around an industry and it takes considerable time and effort to regain customer confidence.

Prevention costs

Prevention costs include any costs brought about to prevent defective work being produced in the first place. This means the implementation of a 'right first time' philosophy so that no scrap or re-working is involved. This is also known as a policy of 'zero defectives' in the USA.

There must be financial advantages in the pursuance of such an ideal as 'right first time', and now, with the advent of 'just-in-time' production, the drive for minimum defects is imperative. JIT and low-quality production are not compatible, since the former is not possible without great attention being paid to the quality of production. Prevention costs can include:

- the management costs of setting up TQM;
- the provision of inspection equipment;
- training programmes for all personnel;
- vetting incoming supplies and vendor appraisals;
- method study to ensure that the most cost-effective, fool-proof methods are in operation;
- high-quality maintenance of process plant, measuring and test equipment.

The cost of quality, like any other organisational function, must be compatible with the selling price of the goods or services. In the quality control of any production system, care must be taken to ensure that there is not more 'harness than horse'.

Assignments

1. Describe the role of an *independent Calibration Service* in the delivery of quality. Explain how it could benefit your own company.
2. Explain how the profitability of a company could benefit by operating a *supplier quality assurance* scheme.
3. Describe an example from your own experience of one defect in each of the classes:

(a) critical;
(b) major;
(c) minor.
Give reasons why each should be placed a particular class.

4. A process machine is given a routine maintenance service during which the average setting is reduced and a decrease in variability is noted. Show on a control chart how these changes would be seen.
5. Explain the reasons for using standard specifications in the design process, stating any advantages derived from the use of these specifications.
6. Define process capability analysis and relate its use to the product designer.
7. Explain how the output of a machine or process is controlled by either 'first-off inspection' or 'patrol inspection'.
8. In a company of your choice, outline the steps being taken to ensure that the customer receives a 'high quality product or service'.
9. In a company of your choice which is seeking accreditation under BS5750, describe how that company is implementing the requirements of such accreditation.
10. Explain how a system of Quality Circles could be set up in your own place of work. Describe the membership of the Circle which you would recommend and give your reasons for that particular comination of persons.

Index